ce does

MARINE ZOOGEOGRAPHY

McGRAW-HILL SERIES IN POPULATION BIOLOGY
Consulting Editors
Paul R. Ehrlich, Stanford University
Richard W. Holm, Stanford University

Briggs: Marine Zoogeography
Edmunds and Letey: Environmental Administration
Ehrlich and Holm: The Process of Evolution
George and McKinley: Urban Ecology:
                                      In Search of an Asphalt Rose
Hamilton: Life's Color Code
Poole: An Introduction to Quantitative Ecology
Stahl: Problems in Vertebrate Evolution
Watt: Ecology and Resource Management
Watt: Principles of Environmental Science
Weller: The Course of Evolution

MARINE ZOOGEOGRAPHY

John C. Briggs

Director, Graduate Studies
University of South Florida

McGRAW-HILL BOOK COMPANY

New York / St. Louis / San Francisco
Düsseldorf / Johannesburg / Kuala Lumpur
London / Mexico / Montreal
New Delhi / Panama / Paris / São Paulo
Singapore / Sydney / Tokyo / Toronto

MARINE ZOOGEOGRAPHY

1234567890 MAMM 7987654

This book was set in Helvetica Light by Black Dot, Inc.
The editors were William J. Willey, Renée E. Beach, and Richard S. Laufer;
the designer was J. Paul Kirouac;
and the production supervisor was Bill Greenwood.
The drawings were done by John Cordes, J & R Technical Services, Inc.
The Maple Press Company was printer and binder.

Library of Congress Cataloging in Publication Data

Briggs, John C
    Marine zoogeography.

    (McGraw-Hill series in population biology)
    1.   Marine fauna—Geographical distribution.
I.   Title.
QL121.B77   591.9′2   73-14502
ISBN 0-07-007800-9

Contents

Preface

I was fortunate to be raised near the sea and to become familiar at a young age with the California beaches and rocky headlands. Later, I had the privilege of attending two coastal universities, Oregon State as an undergraduate and Stanford as a graduate student. At Stanford, I was exposed to an exceptional faculty and a stimulating group of fellow graduate students. My research interests were most strongly influenced by the courses taught and the personal examples set by two of my Stanford professors, George S. Myers and Rolf L. Bolin.

Following my student days, I was able to continue my research in marine biology—with an emphasis on fishes—first at Stanford on a postdoctoral appointment and then as a faculty member at the University of Florida, University of British Columbia, University of Texas, and now at the University of South Florida.

The research for this book started in 1962 when I was on the staff of the Marine Science Institute of the University of Texas at Port Aransas. My move to the University of South Florida, a new state university, in 1964 involved the assumption of considerable administrative responsibility. Consequently, there was less time available for research, and progress on the book became discouragingly slow. Now, after almost 10 years, I can with great relief add the preface and send it off to the publisher.

The greater part of this volume is devoted to the various marine life zones and to the delineation and characterization of the many zoogeographic regions and provinces that are found within them. This had not not previously been done in a detailed manner, so it was a necessary task. In the last chapter, it was possible to trace the history of a few of the faunas to some extent and to examine certain aspects of the relationship between zoogeography and evolution.

Zoogeographic conclusions, if they are to have lasting value, must be based on dependable systematic work. Consequently, I have attempted to be selective about the works I have utilized for analysis. The reader will find that I have placed considerable emphasis on the fishes. There are two reasons for this. First, I am primarily an ichthyologist and am therefore better able to evaluate systematic works on this group. Second, in many parts of the world the fishes happen to be better known than the major groups of invertebrates. Among the phyla of marine animals, the general distributional patterns seem to coincide remarkably well so that good data for one group are likely to have significance for the rest.

In the introduction to his valuable little volume, *The Natural History of European Seas*, published in 1859, Edward Forbes remarked, "In this age of volumes, a man needs to offer a good excuse before adding a new book, even though it be a small one, to the heap already accumulated." If Forbes were alive today, what would he have thought about the need of adding one more volume to an annual world production of something on the order of 100,000 titles?

I would like to think that I am offering this book in the same spirit that Jean Baptiste Lamarck did his *Zoological Philosophy* in 1809. In the introduction to that monumental work he said, "In publishing these observations, together with the conclusions that I have drawn from them, my purpose is to invite enlightened men who love the study of nature to follow them out, verify them, and draw from them on their side whatever conclusions they think justified."

JOHN C. BRIGGS

Acknowledgments

Since a fair amount of the material in this book has been dealt with in a series of my published papers or has been presented as talks at seminars and scientific meetings, I have had the benefit of criticisms and suggestions from a large number of colleagues. I hope I will be forgiven for not attempting to thank each person at this time. To do so would not only result in a long list of names but, since this process went on for about 10 years, I'm afraid I might leave someone out. Let me say only that I am grateful for all the help that I have received.

I am indebted particularly to Joseph L. Simon for taking the time to read the entire finished manuscript. Also, four people were kind enough to read complete chapters for me: Margaret M. Smith provided useful comments on Chapter 6 (Southern Hemisphere Warm-Temperate Regions), Thomas L. Hopkins and Ronald C. Baird on Chapter 10 (The Pelagic Realm), and Daniel M. Cohen on Chapter 11 (The Deep Benthic Realm). I was fortunate to have the services of two fine artists, Henry W. Compton and Nancy Smith. Mr. Compton executed the beautiful colored drawing of *Hemanthias leptus*. Finally, I am indebted to the National Science Foundation for their support of the research that led to this book in the form of two grants, GB 2866 and GB 4330.

The reader will find that some of the material in Chapter 5 (Relationships of the Tropical Shelf Regions) has been taken, almost verbatim, from two of my papers on the subject: Dispersal of tropical marine shore animals: coriolis parameters or competition? (1967, *Nature*, 216(5113):350) and Tropical shelf zoogeography (1970, *Proc. Calif. Acad. Sci.*, 38(7):131–138). Part I of Chapter 12, on the history of marine life, contains material from my article, A faunal history of the North Atlantic Ocean (1970, *Syst. Zool.*, 19(1):19–34); and part II of Chapter 12, on Worldwide patterns, includes material from my article, Zoogeography and evolution (1966, *Evolution*, 20(3):282–289). I am indebted to the editors of the indicated journals for their permission to utilize these publications.

JOHN C. BRIGGS

MARINE ZOOGEOGRAPHY

Introduction

The design of a book is the pattern of a reality controlled and shaped by the mind of the writer. This is completely understood about poetry or fiction, but it is too seldom realized about books of fact. And yet the impulse which drives a man to poetry will send another man into the tide pools and force him to try to report what he finds there.

John Steinbeck in Sea of Cortez, *1941*

As viewed from outer space, the planet we inhabit is a beautiful, blue and white sphere. The blue is reflected from the oceans which cover 71 percent of the surface, and the white comes from the clouds which float above both oceans and land. In our solar system, only earth has liquid water. The other planets are either too hot or too cold. Our total volume of water has been estimated at 1,500 million cu km, about 97 percent of it forming the oceans. Most of the rest exists in the form of ice. If all the water were liquid and the surface of the earth perfectly smooth, our planet would be covered by a layer of water about 3,000 meters deep (Penman 1970). In the words of Ray (1970), "Earth is different from the other planets. Earth is unique. Earth has water."

Our Water Planet

Life on earth is confined to the globe's fluid envelope, or covering. This envelope is divided in two parts—a gaseous portion called the atmosphere and a liquid portion called the hydrosphere. The atmosphere is so thin that very few living things have been able to achieve a purely aerial existence. Almost all atmospheric organisms are obliged to depend on the physical support of the substrate that forms the lower boundary of the atmosphere. So, from the standpoint of their overall distribution, we can see that such organisms occupy a thin, two-dimensional layer on the earth's surface.

In contrast, the liquid of the hydrosphere is viscous enough to provide physical support and contains sufficient nutrients in suspension so that almost all its volume is continually occupied by a vast array of living things. Here, we find that life exists on a truly three-dimensional scale with its vertical component extending from the surface to the greatest depths—almost 11,000 meters. If we call that portion of the earth inhabited by life the biosphere, it is clear that the oceans comprise by far the greatest part of it. Thorson (1971:20) estimated that the total oceanic living space is roughly 300 times larger than that available for life on land.

Although the number of species of animals described from the terrestrial environment greatly exceeds that reported from the sea or from fresh water, such superior terrestrial diversity is due mainly to the enormous number of species of arthropods, especially arachnids and insects. However, if one considers diversity at the level of the phylum or class, a much different picture results. No less than 34 of the 37 phyla of living animals are found in the sea, while only 17 occur in fresh water,

The Marine Fauna

and 15 occur on land; at the class level, 73 occur in the seas, 35 in fresh water, and 33 on land (Nicol 1971).

Not only did life apparently originate in the sea, but as far as fundamental structure is concerned, it exists in far greater variety there today. Most of the basic differentiation in the animal kingdom took place in the early Paleozoic era, and many of these marine groups have never developed the ability to invade fresh water or land. Of the large phyla that have entered the latter two environments, the phylogeny of most reflect a long history of life in the sea. For example, of the seven living classes of mollusks, all live in the sea, only two have also invaded fresh water, and only one of the latter (the gastropods) is also found on land.

Vertebrate and invertebrate zoology, and to a considerable extent ecology, can be taught best at the edge of the sea where one may gain a measure of appreciation for the vast complex of animal life in our world. Here, students can collect and study fresh or living examples of a wide variety of interesting creatures instead of concentrating upon pickled specimens of a few so-called "representative types." A number of years ago, Wheeler (1923) spoke in disparaging terms about the "present depauperate glacial fauna of the Laboratory," a condition that is unfortunately still with us.

Historical Marine Zoogeography

Most of our knowledge about the distribution of animals within the oceanic part of the biosphere is recent and, compared to what is known about the terrestrial environment, still very fragmentary. People have been interested in marine animals for a long time, but the enormous size of the world ocean, the three-dimensional pattern of life within it, and the difficulty of making adequate observations in a habitat relatively hostile to man have prevented the accumulation of needed data. For example, the limits of the terrestrial zoogeographic regions were essentially established by Sclater in 1858, but only now can we attempt to do the same thing for the continental shelves. Distributional patterns in the deep sea and the open sea are still relatively poorly known, and it will be many years before they are thoroughly worked out.

Marine zoogeography got its start with the work of a young man who was later to become a world-famous geologist. James D. Dana, who participated in the United States Exploring Expedition, 1838 to 1842, made observations on the distribution of corals and crustaceans that led him to some important conclusions. He was able to divide the surface waters of the world into several different zones based on temperature and used isocrymes (lines of mean minimum temperature) to separate them. His plan was published as a brief paper in the *American Journal of Science* in 1853.

In 1856, Edward Forbes published his "Map of the Distribution of Marine Life," together with a descriptive text, in Alexander K. Johnston's *The Physical Atlas of Natural Phenomena*. In this, the first comprehensive work on marine zoogeography, the world was divided into 25 provinces which were located within a series of 9 horizontal, "homoizoic belts." A series of five depth zones was also recognized. In the same year, Samuel P. Woodward, the famous malacologist, published part three of his *Manual of the Mollusca*, which dealt with the worldwide distribution of this group.

Three years later, in 1859, Forbes' small but skillfully written volume, *The Natural History of European Seas*, was published posthumously by Robert Godwin-Austen. This work constitutes an important contribution to both marine zoogeography and marine ecology. Also in 1859, Charles Darwin in his *On the Origin of Species* devoted two chapters to geographic distribution. He pointed out the interesting relationship between distributional and evolutionary patterns and gave some examples from the marine environment. In 1880, Albert Günther included in his book, *An Introduction to the Study of Fishes*, an analysis of shore fish distribution. In 1896, Arnold Ortmann published his *Grundzüge der Marinen Tiergeographie* based on the distribution of the decapod crustaceans. The following year, in 1897, Philip L. Sclater published a paper in the *Proceedings of the Zoological Society of London* on the general distribution of marine mammals. In 1901, David Starr Jordan, in an article in the journal *Science*, provided an improvement to Günther's scheme for the distribution of fishes.

The publication of the physical *Atlas of Zoogeography* (Bartholomew et al.) in 1911 was a significant event. It presented a compendium of knowledge about the distribution of 700 families of animals, both marine and terrestrial, on beautifully executed, color-plate maps. In 1935, Sven Ekman, a professor at the University of Upsala in Sweden, completed the huge task of analyzing all of the pertinent literature on marine animal distribution and published his results in a book entitled *Tiergeographie des Meeres*. This well-documented presentation immediately became the definitive work on the subject. In 1953, a second edition was published in English. Regrettably, the 1953 edition was not brought entirely up to date so that much of it was only a translation of the 1935 work.

In 1957, the ecology volume of the comprehensive *Treatise on Marine Ecology and Paleoecology* was published by the Geological Society of America. A chapter in this volume, written by Joel W. Hedgpeth, was devoted to marine biogeography. Although many other books, dealing with marine zoogeography to some extent, were published and hundreds of smaller contributions appeared, I believe those mentioned have advanced our knowledge the most.

Modern Marine Zoogeography

The need for an up-to-date treatment of marine zoogeography along the lines of Ekman's original 1935 book has existed for a long time. In the past 20 years, especially, an enormous amount of useful work has been accomplished. Ekman felt that the ultimate aim of zoogeography (and zoogeographers) was not only the understanding of the present regional system but the revealing of the history which led to its establishment, the history of the faunas.

So far, it can be said that two goals of zoogeographic research have been generally recognized: (1) the delineation and characterization of each distinct faunal area and (2) the attempt to trace the history of the faunas. A third goal needs to be added: the use of zoogeographic data to augment our knowledge about the course of evolution. Several people who have had the privilege of doing systematic work on groups of marine animals that are widespread have found a close relationship between distributional and evolutionary patterns. This means that the accumulation of detailed knowledge about the distribution of species and higher categories can often be very useful in reconstructing the path of evolutionary change within a given group.

Although many ecologists are prone to consider zoogeography a segment of their discipline, there are important differences between the two fields. Ecological research is devoted primarily to community or ecosystem structure including trophic relationships, energy flow, succession, and the characteristics of ecological niches. If distribution is studied at all, it is done so on a local scale with regard to the relationship of species to their immediate physical, chemical, or biological surroundings. Important as such investigations are, they are not zoogeography. Zoogeographers are interested in the various faunal areas of the world and in the distributional patterns of individual animal groups. Studies along these lines are rewarding, since they lead to a better understanding of the faunal areas, their history, and—perhaps most important of all—the phylogeny of certain widespread groups of animals.

Bartholomew, J. G., W. E. Clark, and P. H. Grimshaw. 1911. *Atlas of zooge-ography*. John Bartholomew and Co., Edinburgh, viii + 67 + xi pp., 36 plates.

Dana, James D. 1853. On an isothermal oceanic chart illustrating the geographical distribution of marine animals. *Amer. J. Sci.*, 16:314–327, 1 map.

Ekman, S. 1935. *Tiergeographie des Meeres*. Akademische Verlagsgesellschaft Leipzig, xii + 512 pp.

———. 1953. *Zoogeography of the sea*. Sidgwick & Jackson, London, xiv + 417 pp., 121 figs.

Forbes, E. 1856. Map of the distribution of marine life. *in* Alexander K. Johnston, *The physical atlas of natural phenomena* (new edition). W. and A. K. Johnston, Edinburgh and London, plate no. 31.

———. 1859 *The natural history of European seas* (edited and continued by Robert Godwin-Austen). John Van Voorst, London, viii + 306 pp., 1 map.

Günther, A. C. L. G. 1880. *An introduction to the study of fishes*. Adam and Charles Black, Edinburgh, xvi + 720 pp., 320 figs.

Hedgpeth, J. W. 1957. Marine biogeography. *in Treatise on marine ecology and paleoecology*, vol. 1. *Mem. Geol. Soc. Amer.* (67):359–382, 16 figs, 1 plate.

Jordan, D. S. 1901. The fish fauna of Japan, with observations on the geographical distribution of fishes. *Science*, new series 14(354):545–567.

Nicol, D. 1971. Species, class, and phylum diversity of animals. *Quart. J. Florida Acad. Sci.*, 34(3–4):191–194.

Ortmann, A. E. 1896. *Grundzüge der marinen Tiergeographie*. Gustav Fischer, Jena, iv + 96 pp., 1 map.

Penman, H. L. 1970. The water cycle. *Sci. Amer.*, 223(3):98–108, illus.

Ray, D. L. 1970. Water water everywhere—but what if it is bad? *Stanford Almanac*, 8(5):6–7, 1 fig.

Sclater, P. L. 1858. On the general geographical distribution of the members of the class Aves. *J. Proc. Linn. Soc. Zool.*, 2:130–145.

———. 1897. On the distribution of marine mammals. *Proc. Zool. Soc., London*, pp. 349–359.

Thorson, G. 1971. *Life in the sea* (translated from the Danish). World University Library, McGraw-Hill, New York, pp. 1–256.

Wheeler, W. M. 1923. The dry-rot of our academic biology. *Science*, 57:61–71.

Woodward, S. P. 1851–1856. *A manual of the mollusca*. John Weale, London, viii + 1–486 pp., 24 plates + 24 pp., 1 map.

Literature Cited

Life
on the
Continental
Shelves

The Tropical Ocean

The Indo-West Pacific Region

. . . many fish range from the Pacific into the Indian Ocean, and many shells are common to the eastern islands of the Pacific and the eastern shores of Africa, on almost exactly opposite meridians of longitude.

Charles Darwin *in* On the Origin of Species, *1859*

The tropical seas, as well as the better-known terrestrial environment of the tropics, support an astonishing variety of animal life. Why is this so? Ecologists have recognized for a long time that animal populations existing in the temperate parts of the world may be profoundly affected by seasonal variations in the weather. Of course, weather affects animals directly, but often more important, it also can control such vital necessities as food and shelter. Consequently, it has been suggested (Dobzhansky 1950) that natural selection in the temperate zones operates mainly in response to the exigencies of the physical environment, but ·in the tropics, where weather is relatively stable, natural selection appears to be closely affected by the competition among organisms; that is, it responds mainly to the biological rather than the physical environment.

Tropical Diversity

We should also recognize, as Sanders (1968) did in his thoughtful study of marine benthic diversity, that in order for biological competition to have evolutionary effects, it must be able to operate under conditions of long-term physical stability. Thus, increased diversity appears to represent the slow but inevitable response of living organisms, via the mechanism of natural selection, to increased climatic stability. Since the equatorial regions are the only parts of the world that have over the ages remained relatively unaffected by climatic change, we might expect to find a greater number of species per unit area than in regions with less stable histories.

Because evolution in the tropics has apparently been controlled mainly by biological competition, the species have tended to become highly specialized, and many have developed intricate interdependencies. This tendency is nicely illustrated by a study of the latitudinal differences in the feeding habits of shallow-water marine animals (Bakus 1969). As a result, there are in the tropics numerous niches being exploited that are simply unoccupied or inefficiently utilized in temperate latitudes.

There is, in addition to climatic stability, a second factor that appears to have a profound effect upon species diversity. This is the geographic size of the habitable area. Some biologists who have approached the problem of diversity from the standpoint of community structure (Sanders 1968; Johnson 1970) apparently did not consider the possibility of an area effect. Yet, it is true that in small isolated places, such as

The Effect of Area

oceanic islands, there is a close relationship between diversity and area (McArthur and Wilson 1963, 1967; Johnson, Mason and Raven 1968).

It seems that an island of a given size can only support a limited number of species, and when the saturation point is reached further colonization must be balanced by extinction. Although almost all investigations along this line have been done in the terrestrial habitat, there is no reason why the results should not be just as meaningful for the marine environment. In fact, Taylor (1971:502), in his study of the reef-associated molluscan assemblages in the western Indian Ocean, noted that a positive correlation seemed to exist between the number of shallow-water molluscan species and the area of shallow water around an island.

Although zoogeographers, e.g., Darlington (1959), had previously called attention to the existence of a large-scale relationship between species diversity and area, their contributions have gone unnoticed by many who have discussed the diversity problem. More recently, Preston (1962) cited many examples of area effect, and Simpson (1965:227), in discussing mammalian evolution on the southern continents, referred to an "equilibrium" in which the species diversity depends on the size of the continent. In a more detailed comparison, Keast (1969) found that the numbers of mammalian species per 100,000 miles of inhabitable area were quite comparable (Africa 9.39, Neotropica 11.3, Australia 11.03).

In the shallow-water habitat of the continental shelf, there also appears to exist a large-scale relationship between species diversity and area. Newell (1971) compared the coral reef habitat of the Indo-Pacific, estimated at 125,000 sq km, with the same habitat in the Caribbean, estimated at 25,000 sq km. He then estimated the number of species in each area for fishes and four groups of invertebrates (corals, shelled mollusks, cidarid echinoids, cypraeid gastropods) and concluded that the species diversity may be roughly proportional to the area of occupancy.

Despite the fact that documentation of a positive relationship between species diversity and area (whether small or large) is now quite convincing, we need to recognize that this concept does not seem to be compatible with a current theory about the evolution of diversity. Some investigators (Whittaker 1969, Whittaker and Woodwell 1972), having studied diversity in various communities, maintain that there is no general saturation level and that, with time and under conditions of environmental stability, diversity will go on increasing without a limit or ceiling. If this theory were true, it does not seem probable that there could also be a consistent relationship between species diversity and area.

The ecologist finds that, in his study of particular habitats, the number of species present will vary greatly according to such factors as substrate complexity, the stage of succession in the community being examined, and a variety of other local causes. However, important as these effects are in certain places, they tend to become relatively insignificant when diversity is measured in large areas such as most zoogeographic regions or provinces. The long-continued existence of biological competition as the principal selective factor, when it operates in large geographic areas, can probably account for the greater species diversity that exists in most parts of the tropics.

On a worldwide scale, there exist from the lower to the higher latitudes increasing gradients in climatic severity. Because the organisms comprising the total biotas of the major geographic areas demonstrate evolutionary responses to climatic factors, there exist from the lower to the higher latitudes decreasing gradients in species diversity. Moreover, this negative correlation of species diversity with latitude does not appear to progress smoothly but takes place in a series of steps (Fischer 1960).

Latitudinal Gradients

When one considers the succession of major zoogeographic boundaries that occurs with increasing latitude, it can be seen that as each boundary is passed a drop in species diversity takes place. Each such drop can be considered a shift in the direction of natural selection from adaptation to the biological environment toward adaptation to the physical environment. Polar species struggle for existence primarily against the elements; tropical species struggle mainly against one another. It may be added that intraspecific competition among humans seems to demonstrate the same kind of shift.

The various hypotheses that have been utilized to explain the existence of latitudinal gradients in species diversity have been reviewed by Pianka (1966). It now seems apparent that such gradients can be explained by a combination of two of these hypotheses plus a third factor that may have been overlooked because it was taken for granted: (1) the competition theory (Dobzhansky 1950), (2) the theory of climatic stability (Klopfer 1959) or the stability-time hypothesis (Sanders 1968), and (3) the availability of comparatively large geographic areas.

The Tropical Shelf Regions
The richest (most diverse) marine fauna is found in the shallow waters of the tropical oceans at depths generally less than 200 meters. Zoogeographically, four great regions may be identified: the Indo-West Pacific, the Eastern Pacific, the Western Atlantic, and the Eastern Atlantic. Each region may, in turn, be subdivided into provinces. To the north and south, the marine tropics are bounded by the 20° isotherm for the coldest month of the year. Longitudinally, the tropical regions are separated from one another by barriers that are very effective, since each region possesses, at the species level, a fauna that is highly endemic.

Indo-West Pacific
The shelf waters of the Indo-West Pacific Region are spread over an enormous geographic area extending horizontally more than halfway around the world and vertically through about 60° of latitude. Furthermore, its fauna is incredibly rich with a species diversity that far exceeds even that of the other tropical regions.

It has been estimated that the Indo-West Pacific possesses about 500 species of hermatypic (reef-building) corals, about 10 times the number found in the Western Atlantic, the next richest area (Vaughan and Wells 1943). There are over 1,000 species of bivalve mollusks, more than twice as many as the next richest area (Stehli, McAlester, and Helsley 1967); about four times as many cowries (family Cypraeidae) (Schilder 1965); and about five times as many conchs (Strombidae) (Abbott 1960). There are probably more than 3,000 species of shore fishes compared to less than 1,000 for each of the other tropical regions.

Not only does the Indo-West Pacific have the most species belonging to the major, widespread groups of tropical shore animals, but it also possesses many families that are not found elsewhere. For example, the giant clams (Tridacnidae) are confined to this region (Rosewater 1965) as well as such fish groups as the sea moths (Pegasidae), whitings (Silliginidae), sand fishes (Kraemeriidae), rabbit fishes (Siganidae), and plesiopids (Plesiopidae). There are about 50 known species of sea snakes (Hydrophidae), and all but one (*Pelamis platurus*, which reaches the Eastern Pacific) are restricted to the Indo-West Pacific. The peculiar marine mammal called the dugong (Dugongidae), the only living representative of its family, is also an endemic (Rice and Scheffer 1968). The Indo-West Pacific possesses many such endemic families, while the other tropical regions possess very few.

Why does the Indo-West Pacific have such a marked superiority in diversity? It is noted (p. 427) that the Pacific Ocean has probably had a more stable climatic history than the Atlantic, but this cannot account for all the differences. The tropical eastern Pacific has a rather sparse fauna compared to the western side of that ocean. Since the climatic history of the two sides of the Pacific has apparently been similar, providing the same opportunity for diversity to evolve, the one evident difference is the disparity in the size of the two areas. This appears to constitute a good demonstration of the importance of the effect of geographic area on species diversity.

One of the most interesting features of the Indo-West Pacific is that, despite its basic homogeneity caused by the occurrence of many wide-ranging species, there are great differences in diversity among the various parts of the region. Over the years, almost every author who has studied one of the widespread, tropical groups has noted that there is a concentration of species in the comparatively small triangle formed by the Philippines, the Malay Peninsula, and New Guinea.

Ekman (1953:16) called the above fertile triangle the "Indo-Malayan region" and considered it to be a faunistic center from which the other subdivisions of the Indo-West Pacific Region recruited their faunas. The presence of this concentration of species is well supported by recent studies of the fauna in general (Cloud 1959, Usinger 1963) and of certain animal groups such as the mollusks (Abbott 1960, Schilder 1965, Salvat 1967), crustaceans (Hall 1962) and fishes (Springer 1967, Whitehead 1967, Greenfield 1968). Ladd (1960), who studied recent and fossil mollusks, thought that the original faunistic center was located in the islands of the central Pacific and that prevailing winds and currents had carried species into the Indo-Malayan area. But this theory has not received wide acceptance.

As one leaves the Indo-Malayan center and considers the faunas of the peripheral areas, there is a notable decrease in diversity that is correlated with distance. For example, about 2,000 species of shore fishes may be found in the shallow waters of the Philippines (Herre 1953), but Harry (1953) succeeded in collecting only about 400 species at Raroia Atoll in the Tuamotus, and Gosline and Brock (1960) have recorded but 384 species at Hawaii. About 2,000 species of mollusks occur at the New Hebrides but farther out at the Tuamotus and Gambier Islands only about 500 species have been found (Salvat, 1967:15).

Except for extremely isolated areas, such as the Hawaiian archipelago and Easter Island, where many endemic species are found, the great majority of species in the peripheral localities appear to be common, widespread, Indo-West Pacific animals. Demond (1957), in her careful study of the reef-associated gastropods in Micronesia, obtained distributional data on 174 species and found that 136 of them were known to range broadly in the Indo-West Pacific and that only 37 were apparently confined to the western Pacific; only a single species was considered to be a Micronesian endemic. A total of 379 littoral mollusk species occurs at the Cocos-Keeling Islands in the eastern Indian Ocean, and 82 percent of them are widespread Indo-West Pacific forms (Maes 1967); no endemics were found.

Forest and Guinot (1962), who studied the brachyuran crabs of the Society and Tuamotu Islands, found that 112 of their 186 species (60 percent) ranged all the way to the western edge of the Indian Ocean. Abe (1939:524), in discussing the fishes of the Palau Islands, found that they were for the most part widely distributed throughout the Indo-West Pacific. A similar observation can be made about the fishes of the Tuamotus (Harry 1953), the Gilbert Islands (Randall 1955), and the Marshall and Marianas Islands (Schultz et al. 1953–1966).

The Problem of Subdivisions

Forbes (1856), in his significant work on the provinces of marine life—the first comprehensive marine zoogeography—recognized an "Indo-Pacific Province" extending from the east coast of Africa to the outermost limits of Polynesia. He described it as the "realm of reef-building corals, and of the wondrously beautiful assemblage of animals, vertebrate and invertebrate, that live among them, or prey upon them." In the same year, Woodward (1851–1856) published the third and final part of the first edition of his *Manual of the Mollusca* which dealt with geographic distribution in this phylum. He also recognized a single, extensive Indo-Pacific Province covering the same general area.

Ortmann (1896), on the basis of his work on the decapod crustaceans, recognized an "Indo-pacifische Litoralregion" occupying about the same longitudinal area described by the previous authors but extending latitudinally into areas now regarded as warm-temperate rather than tropical. In his two comprehensive works on marine zoogeography, Ekman (1935, 1953), in identifying an area roughly equivalent to that called the Indo-Pacific Province by Forbes and Woodward, termed it the "Indo-West-Pacific Region." Schilder (1956) delineated about the same area but called it the "Eotropisch Region." The name Indo-West Pacific, since it does serve to eliminate confusion with the tropical Eastern Pacific, a distinct region in its own right, appears to be the most appropriate.

Within each of the tropical regions, minor barriers to gene flow exist that, over extensive periods of time, have resulted in the formation of local endemic species. In places where the endemism has developed to the extent that the fauna has taken on a definite provincial character, we should consider some further zoogeographic subdivision. In such instances, it seems reasonable to identify provinces, but the difficult question is, how much does a local fauna have to differ from the parent in order to merit formal recognition as a province? In this work, an admittedly arbitrary decision has been made: if there is evidence that 10 percent or more of the species are endemic to a given area, it is designated as a separate province.

In regard to the Indo-West Pacific, it has been only rather recently that subdivisions have been commonly utilized. Ekman (1953:17) discussed several subregions in general terms but felt that the knowledge of that time did not permit the precise delimitation of such areas. Chabanaud (1949), on the basis of fish distribution, recognized no less than 7 regions divided into 17 subregions; Schultz (1957:415), who also worked on fishes, recognized 7 subfaunas; Powell (1957), who worked on mollusks, suggested 3 large provinces divided into 18 regions; Knox (1957), after studying polychaete distribution, advocated 8 different regions; Okada and Mawatari (1957), who worked with bryozoans, recognized 3 subregions and 8 provinces; Forest and Guinot (1962), on the basis of brachyuran crab distribution, delineated 17 different regions; Schilder (1965), who worked on mollusks (cowries), recognized 17 regions in 3 provinces; and Kramp (1968), in his work on the neritic hydromedusae, used 6 different subregions.

In general, it can be said that the foregoing schemes show little agreement with one another, yet when one examines systematic works that give detailed information about the distribution of species, a remarkable similarity of pattern among the various animal groups is revealed. This similarity also extends to the amount of endemism found in the better-known faunal subdivisions. That is, when, in a given area, a certain level of endemism is found for one group of shelf animals, the chances are that most of the other major groups (phyla) will demonstrate endemism of about the same order of magnitude. These factors indicate that, despite the current lack of agreement among biologists, the animals themselves show by their distributions that they are responding to common geographic and evolutionary pressures. Therefore, it is possible to devise a general zoogeographic arrangement that should be useful for most animal groups and, perhaps to some extent, for marine plants as well.

In comparison to most of the works listed above, the following arrangement of provinces may seem to be rather conservative. However, provinces are not recognized unless there appears to be good evidence, in the form of endemism, for doing so. Since the shelf fauna of the Indo-West Pacific is still not well known, this arrangement of provinces may need future modifications.

The Provinces
Western Indian
Ocean

The southern part of the western Indian Ocean, that area around Madagascar and the southeastern African coast, is under the constant influence of the South Equatorial Current. As the main current approaches Madagascar from the east, it divides, with one part flowing around the north tip of that island and thence southward through the Mozambique Channel, while the other part extends in a southwesterly direction toward the tip of South Africa (Fig. 1-1). The part that flows through the Mozambique Channel and southward along the shelf is called the Agulhas Current. Although it gradually becomes cooled as it reaches the higher latitudes, its effect permits the tropical fauna to spread far to the south.

The northern part of the Indian Ocean is subject to considerable change in the structure of its currents. From November to March, the northeast tradewinds blow and maintain a North Equatorial Current that affects the entire shelf north of the equator. Starting in April, there is a complete change in the wind system whereby the northeast tradewinds are replaced by the southwest monsoon (Pickard 1964). The wind change sets a Southwest Monsoon Current that flows across the northern Indian Ocean in an opposite direction to the former North Equatorial Current. The monsoon winds also cause the formation of a Somali Current that flows northward along the northeastern African coast and into the Arabian Sea instead of a coastal current in the opposite direction.

Since there are only a few modern works that have examined East African and Arabian Peninsula marine animal groups in the light of their relationship to other parts of the Indo-West Pacific Region, the overall amount of endemism is difficult to predict. There are in this area about 24 species and subspecies belonging to the molluscan family Strombidae (genera *Strombus* and *Lambis*), and 4 of them, or about 19 percent, appear to be endemics (Abbott 1960). There are about 59 species of cowries (family Cypraeidae) with 15, or about 25 percent, of them endemic (Schilder 1965). Of 312 species of echinoderms, 75 (about 24 percent) appear to be endemic (Clark and Rowe 1971).

The shallow-water fishes are somewhat better known than the in-

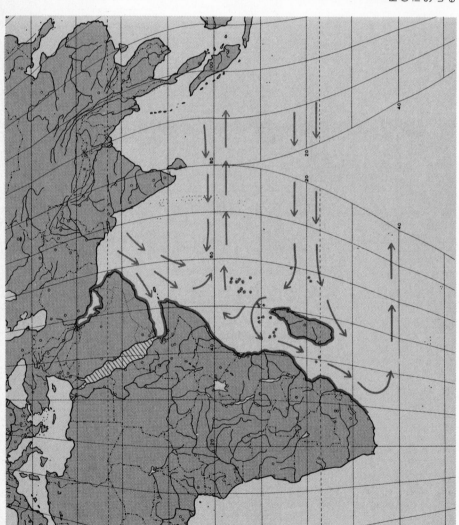

Figure 1-1   Western Indian
Ocean and Red Sea Prov-
inces. Currents in Arabian
Sea depict prevailing pattern
under influence of the northeast tradewinds.

vertebrates. Smith (1958, 1959a, 1959b) found 9 of 27 species in the family Blenniidae, 15 of 40 in the Salariidae, and 47 of 85 in the Gobiidae to be endemic. These works indicate an average endemism of about 42 percent. The parrotfishes (family Scaridae) have been studied by both Smith (1956, 1959c) and Schultz (1958), who have shown that about 42 species are present of which 15 are probably autochthonous, giving an endemic level of about 36 percent.

In view of the fact that there is an extended, tropical coastline, without any obvious physical barrier, connecting East Africa to the Malay Peninsula and that the shore fauna of the northern Indian Ocean is also poorly known, the foregoing estimates of endemism must be used with caution. The status of many of the East African endemics has been determined by means of figures and descriptions of related forms from various western Pacific localities, some as far away as the Philippines or Samoa. In some of these cases, it is probable that intermediate forms will be taken in the northern or eastern Indian Ocean. Thus, we could have a situation in which two species, presently considered to be distinct, may actually represent the ends of a cline. If many such pairs are found to be only single, widely distributed species, the actual rate of endemism may be considerably lower than is presently estimated.

Some of the work on the marine fishes of India and Ceylon gives the impression of a strong relationship to the fauna of the Malay Peninsula and the western Pacific. For example, of the 61 nonendemic species of Indian clupioid fishes, 44 appear to range to the Malay Peninsula or beyond, but only 17 reach East Africa (Misra 1947). More than 60 species of elasmobranch fishes (sharks and rays) range from the Indian Peninsula to the western Pacific, but only about half of them also extend westward to the East African coast (Misra and Menon 1958). This information seems to indicate the presence of a zoogeographic boundary in the northern Arabian Sea or the Persian Gulf instead of somewhere in the Bay of Bengal. Accordingly, until more local distribution data are available, the northeastern boundary of the Western Indian Ocean Province is tentatively considered to occur at the entrance to the Persian Gulf (the Gulf of Oman).

In regard to the southwestern boundary of the province, which is also the boundary between the tropical and warm-temperate faunas, several opinions have been expressed: Stephenson (1947:228–229) observed that the South African warm-temperate, intertidal fauna extended in its "purest form" as far north as Algoa Bay. Smith (1949:9) considered that the area about the mouth of the Kei River could be broadly taken as marking the division between the tropical and Cape of Good Hope fish faunas. Ekman (1953:11) mentioned a boundary location for the tropical fauna a little south of Durban. In his monograph on the polychaetes of southern Africa, Day (1967:12) recognized a boundary at the mouth of the Bashee River. Penrith (1970:140), who analyzed the distribution of the fish family Clinidae, presented tables that indicated a boundary effect at about the mouth of the Kei River. The latter location is also regarded as a boundary for shore fishes by Margaret M. Smith (personal communication). So, the evidence seems to favor the Kei River mouth, about 45 miles south of the Bashee River and 40 miles north of East London, as the southwestern boundary of the Western Indian Ocean Province.

Although the Comoro Islands, between northern Madagascar and the East African mainland, have received considerable attention because of the presence of the only known population of the modern coelacanth (*Latimeria chalumnae*), the fish fauna in general (Fourmanoir 1955) is simply a continuation of that of the Western Indian Ocean Province. The same is true for Madagascar itself (Fourmanoir 1957), the Seychelles (Smith and Smith 1963), Aldabra (Smith 1955–1958, Arnoult, Bauchot-Boutin, and Roux-Estéve 1958), and the famous home of the dodo, the Mascarene Islands (Baissac 1949–1957, Blanc and Postel 1958).

Madagascar and its associated island groups have recently been given an independent designation by Forest and Guinot (1962:42), who recognized a "Région malgache" on the basis of brachyuran crab distribution, and by Schilder (1965:174), who identified a "Lemurian region" from his work on cowries (Cypraeidae). However, the amount of endemism for the shelf fauna in general does not, at this stage in our knowledge, appear to be sufficient to warrant such recognition.

Red Sea

The Red Sea acts as an evaporation basin because the climate is dry and there are no large tributary streams. As a consequence, salinities up to 42.5 parts per thousand (^{o}/oo) occur at the northern end of the sea (Pickard 1964:176). The southern end is partially blocked by a shallow sill that lies at about 125 meters. A subsurface current above the sill carries warm, saline water into the Indian Ocean while a compensating inflow takes place at the surface. During the glacial periods of the Pleistocene, the glacio-eustatic level of the world ocean probably dropped to the extent that the Red Sea was completely isolated.

Although the fauna of the Red Sea has been investigated by a series of competent biologists beginning with Forsskål (1775), who was a member of the ill-fated Danish expedition of 1761 to 1767 (Hansen

1962), it is still not well known. Thus, the task of evaluating the relationship of the Red Sea fauna to that of the Western Indian Ocean is not easy. About 75 species of reef corals have been reported, of which 15 (25 percent) are endemic (Vaughan and Wells 1943:78). Ekman (1953:29) related, from literature sources, that 31 to 33 percent of the macruran and brachyuran decapod crustaceans were endemic. Adam (1960:3) pointed out that endemism in the cephalopods extended to about half the recorded species. Among the gastropods, 3 of the 9 species of Strombus (Abbott 1960) and 5 of the 33 species (15 percent) of the cowries (Schilder 1965) are endemic. Two of the 10 species of neritic hydromedusae are apparently endemic (Kramp 1968:167). The work by Clark and Rowe (1971) indicated that the echinoderm fauna consists of 181 species of which 28, or about 15 percent, are endemic.

The literature on the fishes of the Red Sea is voluminous. Many of the common Indo-West Pacific species were first described from this area in the fine, illustrated works by Rüppell (1826, 1835) and Klunzinger (1884). Modern ichthyological research has indicated that the Red Sea fauna is not entirely a continuation of that found in the Western Indian Ocean. Marshall (1952a, 1952b) and Steinitz and Ben-Tuvia (1955:11– 12) have noted a definite tendency toward endemism at the species and subspecies levels. Marshall (1952b) mentioned that 10 to 15 percent of the fishes taken in the Gulf of Aqaba were Red Sea endemics and suggested it was likely that at least 15 percent of the Sudanese fishes would prove to be in this category. Gohar (1954:25) estimated that about 15 percent of the Red Sea fish species were endemics. The two known species of clingfishes (Gobiesocidae) are both endemics (Briggs 1966a) and so is the single known species of sand diver (Trichonotidae) (Clark and von Schmidt 1966).

Although the Red Sea has not been recognized as a distinct zoogeographic area by many of those who have dealt with distributions in the Indo-West Pacific, there appears to be good reason for doing so. The extent of endemism in some of the recently investigated invertebrate groups is considerable, and that in the shore fishes may be even greater than has so far been reported.

Indo-Polynesian   As noted earlier (p. 14), the richest, most diverse marine faunal area in the world is the triangle formed by the Philippines, the Malay Peninsula, and New Guinea. Can this area by itself be considered a distinct province? The answer is "no," because the peripheral areas, although demonstrating notable reductions in diversity, do not as a rule possess species that are not also found in the central triangle. Considering that such distant places as the Tuamotu Archipelago, India and Ceylon, Taiwan (Formosa), and the Great Barrier Reef of Australia are populated primarily by wide-ranging species and possess few local endemics, all should be placed in the same province.

So, we find that the heart as well as the greater part of the Indo-West Pacific Region is occupied by one huge province that is larger than any of the regions in other parts of the world. The Indo-Polynesian Province extends all the way from the entrance to the Persian Gulf to the Tuamotu Archipelago, and from Sandy Cape on the coast of eastern Australia to the Amami Islands in southern Japan. All of Polynesia is included with the exception of the Hawaiian archipelago, the Marquesas, and Easter Island.

Because of its size and complexity, the Indo-Polynesian Province comes under the influence of several oceanic current systems. In the northern Indian Ocean, as was noted previously (p. 17), the surface currents may be reversed when one seasonal wind system gives way to the other. The shores of the northwestern Pacific from the Philippines to Japan are affected by the North Equatorial Current which, as it turns northward, becomes the Kuroshio Current. Polynesian island groups that are closer to the equator, such as the Marshalls and the Line Islands, are apt to be under the influence of the North Equatorial Countercurrent. Northern Australia and much of the Indo-Australian archipelago appear to be supplied mainly by the South Equatorial Current. When the main stream of the latter current turns southward to parallel the northeastern coast of Australia, it is called the East Australian Current. Finally, there is a South Equatorial Countercurrent which may have some effect on island groups, such as Samoa and the Marquesas (Fig. 1-5).

The fauna of the Indo-Polynesian Province is so diverse, and spread over such a large area, and many of the animal groups are still so poorly known, that it almost defies analysis. At this stage in our knowledge, we can perhaps obtain only a very general indication of endemism by examining the results of some recent systematic work on certain groups of mollusks, echinoderms, and fishes.

The work of Abbott (1960) on the gastropod genus *Strombus* showed that 43 species and subspecies were present in the province and that 22 of them (slightly over 50 percent) were apparently endemic. Schilder (1965) listed 127 species and subspecies of cowries (Cypraeidae) of which 47 (37 percent) appeared to be endemic. Of the 6 species of giant clams (Tridacnidae), 3 are probably endemic (Rosewater 1965). A

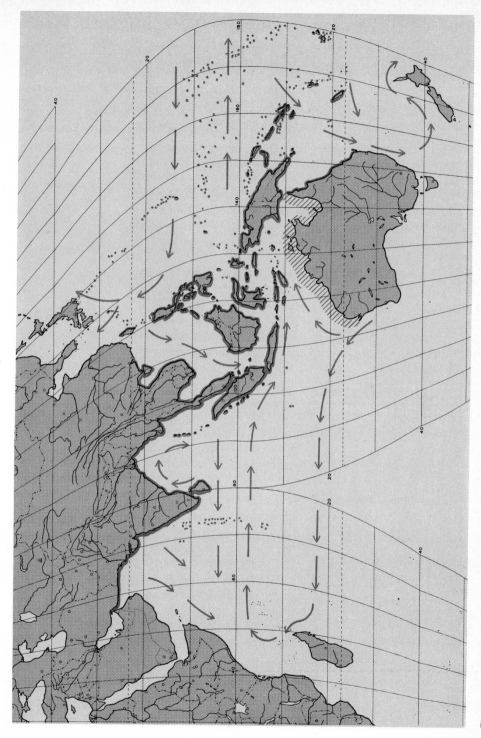

Figure 1-2 Indo-Polynesian and Northwestern Australia Provinces. Former extends eastward to Tuamotu Archipelago (beyond map).

monograph on the recent volutes (Volutidae) indicated the presence of 39 shelf species with 30 of them (85 percent) being endemic (Weaver and du Pont 1970). Among the echinoderms, the echinoids and the crinoids, the two classes that have been most recently investigated, demonstrate endemism levels of about 50 and 90 percent, respectively (Clark and Rowe 1971).

A revision of the fish genus *Entomacrodus* (Springer 1967) showed that 11 species are found in the province with 5 of them (about 45 percent) being endemic. In the widespread fish genus *Myripristis* (Greenfield 1968) only 2 of the 13 species appear to be endemic. My own work on the clingfishes (Gobiesocidae) indicates that, at the time of this writing, 6 of the 10 known species are endemic.

It is interesting to see that the vicinity of Sandy Cape (Fraser Island), on the coast of Queensland, Australia, was recognized as a major zoogeographic boundary over 100 years ago by Woodward (1851–1856) in his *Manual of the Mollusca*. Whitley (1932) was of the same opinion, since he considered his "Banksian" province to terminate at about this point. H. L. Clark (1946), in his book on the echinoderm fauna of Australia, indicated that he agreed with an earlier designation by Hedley (1926) of Wide Bay as the boundary line. However, since Wide Bay lies at the south end of Great Sandy Island and since the north end of this island forms Sandy Cape, the two localities are very close. In their intensive survey of the intertidal fauna of the Queensland coast, Endean, Kenny, and Stephenson (1956) called attention to a very sharp zoogeographic boundary with very little overlap at 25°S (close to Sandy Cape).

Boundaries

A brief article (Briggs, 1966b) has been published on the composition and geographic extent of the warm-temperate marine fauna of Japan, Taiwan, and the eastern China coast. It was suggested that the southern boundaries of this fauna, and thus the northern boundaries for the tropics, were probably located just north of the Amami Islands in the Ryukyu chain, on Taiwan (the western coast being warm-temperate and the eastern coast tropical), and at about Hong Kong on the mainland coast. Evidence for the placement of these boundaries was based mainly on works dealing with the local distribution of fishes (J. T. F. Chen 1951–1953, T. Chen 1960, Chu 1957, Kamohara 1957, Kamohara and Yamakawa 1965, Fowler 1930–1962).

The West Australian Current, the eastern limb of the main gyre in the southern Indian Ocean, brings cool water northward along the west

Northwestern Australia

coast of Australia to about 25°S where it appears to leave the coast, permitting a replacement by tropical waters from the north (Fig. 1-2).

Many Australian zoologists have been interested in the distribution of the littoral marine fauna in their part of the Pacific. According to Whitley (1932:166) and H. L. Clark (1946:465), the conchologist Charles Hedley (1904, 1926) was the first to recognize that the continent may be divided into several distinct zoogeographic areas. He described four provinces with boundaries that are very close to those accepted by most modern workers.

In 1932, Whitley provided a map of the marine zoogeographic regions of Australia based on information given in the works of Hedley (op. cit.) and others interested in invertebrate distribution as well as on his own knowledge of the fish fauna. In regard to the tropical waters, he delineated a northwestern "Dampierian" Province, extending from Cape York to a point just south of Geraldton, and a northeastern area that was divided into a "Solanderian" Province for the Great Barrier Reef, and a "Banksian" Province for the coastal region from Cape York to Sandy Cape.

More recently, Whitley (1948) published a distributional list of the fishes of Western Australia. From this, it can be seen that the dividing line between the tropical and warm-temperate faunas on this side of the continent should be located somewhat farther to the north, perhaps in the vicinity of Shark Bay, rather than south of Geraldton. H. L. Clark (1946), in his extensive survey of the Australian echinoderm fauna, did "arbitrarily" set the boundary at Geraldton. The map published by Bennett and Pope (1953:139) indicates a gradual change in the intertidal fauna and flora between Geraldton and Fremantle. However, Munro (1949), in his distributional study of the fish genera *Mylio* and *Rhabdosargus*, found a species break at Shark Bay. Also, the work on the gastropod family Cypraeidae in Western Australia by Wilson and McComb (1967) lends support to the idea of a boundary at Shark Bay.

Since the placement of an eastern boundary at Cape York on the Torres Strait has been recognized for many years, and since many of the Indo-Polynesian species do not extend westward past this point, its position seems to be well established. The Torres Strait may have been above water for much of the Tertiary and Quaternary and thus served as a land bridge between Australia and New Guinea. The present distributions of many of the shelf species apparently reflect this historic block to migration.

For the province, the amount of endemism in echinoderms is apparently quite high. According to Endean's (1957:265) interpretation of H. L.

Clark's (1946) data, about 40 percent of the well-authenticated species are indigenous. The work by Laseron (1956) on the small marine gastropods also suggests a relatively high percentage of endemism. Several of the sponges may also be endemic (Bergquist 1967). On the other hand, the cowries are not especially distinct; according to Schilder's (1965) work, 56 species are present but only 8, or about 11 percent, are endemics. The shore fishes are still so poorly known that it is not worthwhile attempting an estimate of their endemism; however, according to Whitley's (1948) list, an impressive number of species have been taken nowhere else.

For the reasons given above, it appears desirable to recognize a Northwestern Australia Province extending from Cape York, Queensland to Shark Bay, Western Australia.

Four oceanic localities in the Tasman Sea are of considerable zoogeographic interest. The standard surface temperature charts indicate that Lord Howe Island, Norfolk Island, Middleton Reef, and Elizabeth Reef have winter temperatures too low to support tropical marine faunas. Yet, judging from the species that have been reported from these areas, the faunas are primarily tropical rather than warm-temperate. In fact, Pope (1959:141–159), in discussing Lord Howe Island, the most southerly of the localities, called attention to its well-developed coral reefs.

Lord Howe–Norfolk

The case of Lord Howe Island is particularly interesting. It is located at 31°33′S, 159°5′E, some 300 miles off the east coast of Australia. Early Australian ichthyologists soon became interested in the island. The first comprehensive report on its fauna, both terrestrial and marine, and its geology was published as a memoir by the Australian Museum in 1889. In this work, the general zoology was reported by Etheridge (1889) and the reptiles and fishes by Ogilby (1889).

Both the general observations of Pope (op. cit.), who described the island as the Pacific counterpart of Bermuda, and the tropical nature of the fish fauna strongly indicate that Lord Howe Island should be included within the Indo-West Pacific Region. A salubrious temperature is provided by the East Australian Current, a southwesterly running branch of the South Equatorial Current.

The distributional relationships of the four localities still need to be clarified. Whitley (1932) included Lord Howe and Norfolk in a separate "Phillipian" province. Iredale and Allan (1940) maintained, from their studies of the marine and terrestrial mollusks, that Lord Howe showed

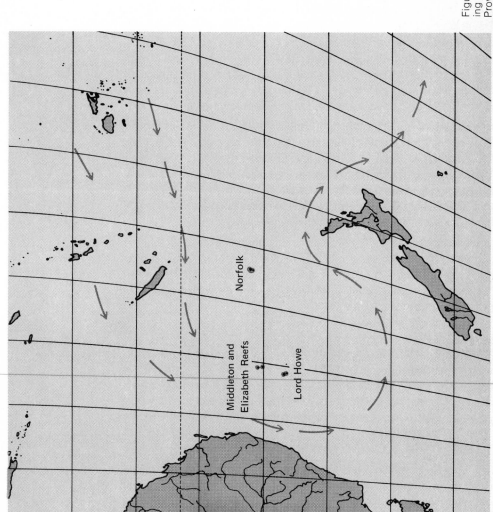

Figure 1-3 Islands comprising the Lord Howe–Norfolk Province.

Norfolk

Middleton and
Elizabeth Reefs

Lord Howe

little relationship to Norfolk (or the Kermadecs) and that it was an "outlier" of New Caledonia. These authors even provided a map (op. cit., 445) showing a hypothetical long, narrow land bridge from New Caledonia to Lord Howe Island and a similar connection from the Fijis to Norfolk Island.

Dell (1957) also discussed molluscan relationships and concluded that the Kermadecs[1] and probably Lord Howe and Norfolk Islands were separate provinces. H. L. Clark (1946) pointed out that 10 of the 59 Lord Howe echinoderm species were endemic, but he did not examine collections from Norfolk or the Middleton and Elizabeth Reefs.

[1]A warm-temperate island group.

The best ichthyogeographic data were provided by Waite (1916), who published a list of Norfolk fishes. Of a total of 64 shore species, 50 were found to be shared with Lord Howe, 30 with eastern Australia, 11 with New Zealand, and 13 with the Kermadecs. The degree of endemism for Norfolk itself appeared to be only about 4 percent, but the Lord Howe–Norfolk endemics comprised about 22 percent of the total.

Norfolk Island, at 29°2'S, 167°57'E about 400 miles northwest of New Zealand, lies in the path of the same tropical current that affects Lord Howe Island. Also the Middleton and Elizabeth Reefs, located close together and not far from Lord Howe (a little over 100 miles to the north), apparently are under the influence of the same current. Whitley (1937) concluded that the fish fauna of the two reefs was closely allied to that of Lord Howe and Norfolk.

Considering the apparent rather close faunal relationships among the four areas in question along with the fact that they are all influenced by the same current system, it would seem logical to treat them as members of the same province. For these reasons, a Lord Howe–Norfolk Province is recognized.

Hawaiian

The Hawaiian Islands consist of a chain of 20 islands forming an archipelago extending from 18°55' to 23°N and 154°40' to 162°W. At least some of the islands are relatively old, since it has been determined that the initial building above sea level of most of the Hawaiian volcanoes took place in the Pliocene about 12 million years ago (Sterns 1966:xi). The archipelago appears to be mainly under the influence of the North Equatorial Current which flows from east to west.

Although the Hawaiian Islands were first discovered by Captain James Cook in 1778, the first notable collection of marine animals was that obtained by the French corvette *Uranie* in 1819. This material, including fishes, mollusks, and crustaceans, was described by Quoy and Gai-

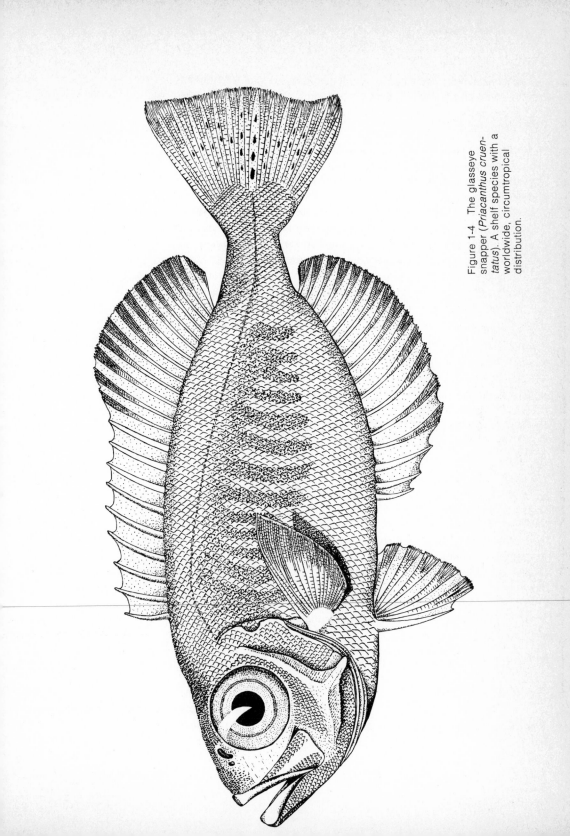

Figure 1-4 The glasseye snapper (*Priacanthus cruentatus*). A shelf species with a worldwide, circumtropical distribution.

mard (1824) in their beautifully illustrated report on the zoology of the voyage. Many additional small collections were made, and the animals were described in various publications, during the remainder of the nineteenth century.

In 1901 and 1902, the United States Fish Commission undertook a comprehensive inventory of the aquatic resources of the Hawaiian Islands. The work was carried out under the direction of David Starr Jordan, then president of Stanford University, and Barton Warren Evermann, ichthyologist of the Fish Commission. The shallow waters were collected from shore stations and the deeper waters by means of trawls and dredges from the Fish Commission steamer *Albatross*. This work resulted in extensive collections that made possible significant reports on the Hawaiian fishes, crabs, isopods, hydroids, nemerteans, starfishes, medusae, and polychaetes. These publications appeared in the *Bulletin of the U. S. Fish Commission* from 1903 to 1906.

The Hawaiian marine fauna, being very rich compared to temperate localities, has continued to be of interest to the biologist. In recent years, much additional work has been done, especially by investigators at the University of Hawaii and at the Bernice P. Bishop Museum. Now, it can be said that the fauna is quite well known, probably better than that of any other area in the tropics. It is interesting to see that the amount of endemism is quite high. About 34 percent of the shore fishes (Gosline and Brock 1960), 30 percent of the shallow-water echinoderms (Asteroidea and Ophiuroidea) (Ely 1942), 45 percent of the crustacean family Crangonidae (Banner 1953), and 20 percent of the mollusks (Kay 1967) are endemic species. More recently, a publication on the dorid nudibranchia (Kay and Young 1969) indicated a total of 49 species with only 6 being listed as possible endemics.

The relatively high degree of overall endemism may be attributed to a long, stable climatic history and a considerable geographic isolation. Except for Johnston Island, which lies about 450 miles to the south, the Hawaiian chain is separated from the rest of Polynesia by a deep water gap of some 900 miles. Gosline (1955), who investigated the fish fauna of Johnston Island, found that its faunal relationships were closer to Hawaii than to any of the other island groups and suggested that Johnston Island may have served as an important steppingstone in the migratory movement of shore fishes to Hawaii.

So far, the numerous scientific visitors to Easter Island have apparently been so struck by its anthropological mysteries that very little attention

Easter Island

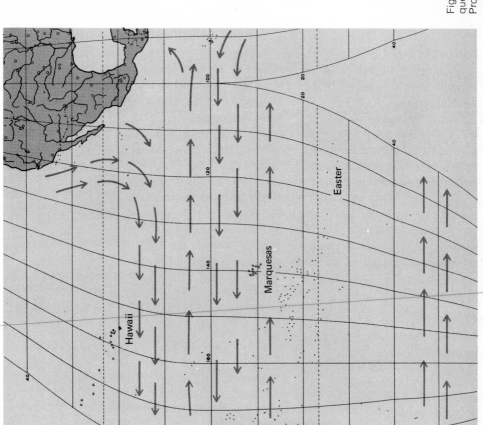

Figure 1-5 Hawaiian, Marquesas, and Easter Island Provinces.

has been paid to its marine fauna. This is unfortunate, for this island, together with Sala y Gómez some 300 miles farther east, is the most geographically isolated portion of the entire Indo-West Pacific Region. It lies at 27°6'S and 109°17'W, over 900 miles east of Ducie Island, the nearest Polynesian locality, and about 1,400 miles west of Juan Fernández Island.

Although Easter Island may have been first discovered by the English buccaneer Edward Davis in the 1680s, the Dutch admiral Roggeveen rediscovered it on Easter Day in 1772 and gave it its name. The first important collection of marine animals was not made until the visit of the U. S. Fish Commission steamer *Albatross* in 1904. This vessel, under the direction of Alexander Agassiz, collected a relatively large number of fishes (Kendall and Radcliffe 1912) and apparently lesser numbers of invertebrates. This material, together with collections made in many other parts of the tropical Eastern Pacific, was described in a series of papers that appeared in the *Memoirs* and the *Bulletin of the Museum of Comparative Zoology*, Harvard University.

Despite the good start made by the *Albatross*, few additional collections of scientific value, except for fishes, have apparently been made. The Canadian Medical Expedition of 1964 to 1965 (Reid 1965) reportedly made good biological collections, but so far no publications have appeared. Since 1912, several papers on Easter Island fishes have been published.

Since the shelf fauna of Easter Island is still so poorly known, it is almost impossible to estimate the overall extent of endemism. In the fishes, there are obviously many endemics, possibly 30 to 40 percent of the total fish fauna. When the entire marine fauna of Easter Island and Sala y Gómez becomes reasonably well known, it will be most interesting to compare it with that of Hawaii. Will the endemism be greater or less? How many of the nonendemics are shared? Are there parallel evolutionary trends? Someday, we may have the answers to such questions.

The Marquesas, comprising a group of 13 islands between 8° to 11°S Marquesas and about 140°W, lie about 250 miles north of the nearest islands in the Tuamotu Archipelago. Although the shelf fauna of the Marquesas is very poorly known, there are good indications of a significant degree of endemism.

Rehder (1968), who reported on the molluscan fauna, pointed out that, although many of the species were widely distributed in the Indo-West Pacific, about 20 percent were apparently endemic. In his revision of the shore fish genus *Entomacrodus* (family Blenniidae), Springer (1967) noted that all three of the Marquesan species were endemics. Forest and Guinot (1962), who worked on the decapod crustaceans of the Society and Tuamotu Islands, recognized a separate "Marquesas region."

Although the 250-mile separation from the Tuamotus is not a great distance for many shore animals, the prevailing surface currents may contribute an additional isolating factor. Apparently, the South Equatorial Countercurrent extends between the Marquesas and Tuamotus (Pickard 1964). This current may make it difficult for marine shore animals to migrate between the two island groups.

Other Isolated Areas

The most isolated group of islands in the tropical part of the Indian Ocean is called Cocos or Cocos-Keeling. Its location at 12°5′S and 96°53′E is about 580 miles southwest of Java. Marshall (1950) recorded 189 species of fishes but considered only 1 species could possibly be an endemic. Maes (1967) identified 379 species of mollusks and found no endemism. Why so little endemism in an area that is so well isolated? Wallace (1892:285), upon observing that the terrestrial flora and fauna of Cocos-Keeling were comprised entirely of common species found elsewhere, spoke of the comparatively recent origin of the islands. This seems to be the best explanation of their low endemism.

In contrast to Cocos-Keeling, Christmas Island, also lying in the eastern Indian Ocean, at 10°30′S and 105°34′E about 225 miles south of the west end of Java, is quite old, and in general its terrestrial fauna shows a high degree of endemism (Andrews 1900). However, the fishes (Palmer 1950) seem to be common Indo-Polynesian species. The fauna of the Chagos Archipelago, a group lying in the north central Indian Ocean about 340 miles south of Maldives, would be well worth investigating, for apparently a good collection of marine animals has never been made.

In the Western Pacific, it is possible that localities such as Wake and Marcus Islands might demonstrate a significant degree of endemism. However, only a few marine animals have ever been identified from either place. Although the work by Bergquist (1965) on the sponges of the Palau archipelago indicates a considerable amount of endemism (14 of the 50 species reported), the sponges of the other parts of the Indo-West Pacific are so poorly known, as the author pointed out, that the actual amount of endemism may be quite low. Moreover, there is apparently very little endemism among the shore fishes of the Palaus (Abe 1939).

Although many isolated islands are found in the tropical Indo-West Pacific, some are not zoogeographically distinct, and the marine animals of others are not well enough known to provide dependable data on faunal relationships.

Literature Cited

Abbott, R. T. 1960. The genus *Strombus* in the Indo-Pacific. *Indo-Pacific Mollusca*, 1(2):33–146, 117 plates.

Abe, T. 1939. A list of the fishes of the Palao Islands. *Palao Trop. Biol. Sta. Studies*, 4:523–583.

Adam, W. 1960. Cephalopoda from the Gulf of Aqaba. *Bull. Sea Fish. Res. Sta., Haifa*, (26):1–26, 10 figs., 1 plate.

Arnoult, J., M. L. Bauchot-Boutin, and R. Roux-Estéve. 1958. Les poissons de l'île Aldabra. Resultats scientifique des Compagnes de la "Calypso." *Ann. Inst. Oceanog.*, 34(3):47–90.

Andrews, C. W. 1900. *Monograph of Christmas Island (Indian Ocean)*. British Museum of Natural History, London, xiv + 337 pp., 21 plates.

Baissac, J. de B. 1949–1957. Contribution a l'étude des poissons de l'île Maurice, I–VII. *Trans. Roy. Soc. Mauritius*, (15):19–36; *Proc. Roy. Soc. Arts Sci. Mauritius*, 1(1):25–31; 1(2):123–152, 6 figs.; 1(3):185–240; 1(4):319–365; 2(1):1–37.

Bakus, G. J. 1969. Energetics and feeding in shallow marine waters. *Int. Rev. Gen. Exptl. Zool.*, 4:275–369, 9 figs.

Banner, A. H. 1953. The Crangonidae, or snapping shrimp, of Hawaii. *Pacific Sci.*, 7(1):3–147, 50 figs., 1 plate.

Bennett, I., and E. C. Pope. 1953. Intertidal zonation of the exposed rocky shores of Victoria, together with a rearrangement of the biogeographical provinces of temperate Australian shores. *Australian J. Marine Freshwater Res.*, 4(1):105–159, 5 figs., 6 plates.

Bergquist, P. R. 1965. The sponges of Micronesia. Part I. The Palau Archipelago. *Pacific Sci.*, 19(2):123–204.

———. 1967. Australian intertidal sponges from the Darwin area. *Micronesica*, 3:175–202, 4 figs., 6 plates.

Blanc, M., and E. Postel. 1958. Sur une petite collection de poissons de la Reunion. *Mem. Inst. Sci. Madagascar, Ser. F. Oceanog.*, 2:367–376.

Briggs, J. C. 1966a. A new clingfish of the genus *Lepadichthys* from the Red Sea. *Bull. Sea Fish. Res. Sta. Haifa, Israel*, (42):37–40, 2 figs.

———. 1966b. The warm-temperate marine fauna of Japan, Taiwan, and the east China coast. *Proc. 11th Pacific Sci. Congr.*, 7(7):7.

Chabanaud, P. 1949. Essai d'une division biogéographique du domaine oceanique. *C. R. 13th Int. Zool. Congr., Paris*, pp. 535–538.

Chen, J. T. F. 1951–1953. Check-list of the species of fishes known from Taiwan (Formosa). *Quart. J. Taiwan Museum*, 4(3–4):181–210; 5(4):305–341; 6(2):102–140.

Chen, T. 1960. Contributions to the fishes from Quemoy (Kinmen). *Quart. J. Taiwan Museum*, 13(3–4):191–213.

Chu, K. 1957. A list of fishes from Pescadore Islands. *Rep. Inst. Fish. Biol., Taiwan*, 1(2):14–22.

Clark, A. M., and F. W. E. Rowe. 1971. *Monograph of shallow-water Indo-West Pacific echinoderms*. British Museum, London, pp. 1–238, 100 figs., 31 plates.

Clark, E., and K. von Schmidt. 1966. A new species of *Trichonotus* (Pisces, Trichonotidae) from the Red Sea. *Bull. Sea Fish. Res. Sta. Haifa, Israel*, (42):29–36, 3 figs.

Clark, H. L. 1946. *The echinoderm fauna of Australia*. Carnegie Institute, Washington, pub. 566:iv + 567 pp.

Cloud, P. E., Jr. 1959. Geology of Saipan, Mariana Islands. Part 4. Submarine topography and shoalwater ecology. *U. S. Geol. Surv. Prof. Papers*, (280-K):vi + 361–445, 8 figs., 21 plates.

Darlington, P. J., Jr. 1959. Area, climate, and evolution. *Evolution*, 13(4):488–510, 8 figs.

Day, J. H. 1967. *A monograph on the Polychaeta of Southern Africa*. British Museum of Natural History, London, pp. viii + 878.

Dell, R. K. 1957. The marine mollusca of the Kermadec Islands in relation to the molluscan faunas in the south west Pacific. *Proc. 8th Pacific Sci. Congr.*, 3:499–503, map.

Demond, J. 1957. Micronesian reef-associated gastropods. *Pacific Sci.*, 11(3):275–341, 41 figs.

Dobzhansky, T. 1950. Evolution in the tropics. *Amer. Sci.*, 38:209–221.

Ekman, S. 1935. *Tiergeographie des Meeres*. Akademische Verlagsgesellschaft Leipzig, xii + 512 pp.

———. 1953. *Zoogeography of the sea*. Sidgwick & Jackson, London, xiv + 417 pp., 121 figs.

Ely, C. A. 1942. Shallow-water Asteroidea and Ophiuroidea of Hawaii. *Bull. Bernice P. Bishop Museum*, (176):1–63, 18 figs., 13 plates.

Endean, R. 1957. The biogeography of Queensland's shallow-water echinoderm fauna (excluding Crinoidea), with a rearrangement of the faunistic provinces of tropical Australia. *Australian J. Marine Freshwater Res.*, 8(3):233–273, 5 figs.

———, R. Kenny, and W. Stephenson. 1956. The ecology and distribution of intertidal organisms on the rocky shores of the Queensland mainland. *Australian J. Marine Freshwater Res.*, 7(1):88–146, 13 figs., 7 plates.

Etheridge, R., Jr. 1889. The general zoology of Lord Howe Island *in Lord Howe Island, its zoology, geology, and physical characters. Mem. Australian Museum*, (2):1–42.

Fischer, A. G. 1960. Latitudinal variations in organic diversity. *Evolution*, 14(1):64–81, 19 figs.

Forbes, E. 1856. Map of the distribution of marine life. *in* Alexander K. Johnston, *The physical atlas of natural phenomena* (new edition). W. and A. K. Johnston, Edinburgh and London, plate no. 31.

Forest, J., and D. Guinot. 1962. Remarques biogeographiques sur les crabes des archipels de la Société et des Tuamotu. *Cahiers du Pacifique*, (4):41–75.

Forsskål, P. 1775. *Descriptiones animalium.* Mölleri, Hauniae, 19 + xxxiv + 164 pp., 1 map.

Fourmanoir, P. 1955. Ichthyologie et pêche aux Comores. *Mem. Inst. Sci. Madagascar. Ser. A*, 9:187–239, 19 figs.

———. 1957. Poissons teleosteens des eaux Malgaches du canal de Mozambique. *Mem. Inst. Sci. Madagascar, Ser. F. Oceanog.*, 1:1–316, 195 figs., 17 plates.

Fowler, H. W. 1930–1962. A synopsis of the fishes of China. A series of more than 60 separate papers appeared under this title. They were published over a period of 32 years in *Hong Kong Nat., J. Hong Kong Fisheries Res. Sta.*, and *Quart. J. Taiwan Museum.*

Gohar, H. A. F. 1954. The place of the Red Sea between the Indian Ocean and the Mediterranean. *Pub. Hydrobiol. Res. Inst., Univ. Istanbul, Ser. B*, 2(2–3):1–38, map.

Gosline, W. A. 1955. The inshore fish fauna of Johnston Island, a Central Pacific atoll. *Pacific Sci.*, 9:442–480, 4 figs.

———, and V. E. Brock. 1960. *Handbook of Hawaiian fishes.* University of Hawaii Press, Honolulu, ix + 1–372 pp., 277 figs., 2 plates.

Greenfield, D. W. 1968. The zoogeography of *Myripristis* (Pisces: Holocentridae). *Syst. Zool.*, 17(1):76–87.

Hall, D. N. F. 1962. Observations on the taxonomy and biology of some Indo-West-Pacific Penaeidae (Crustacea, Decapoda). *Colonial Off., Fishery Pub.*, (17):1–229, 125 figs.

Hansen, T. 1962. Arabia felix. The Danish expedition of 1761–1767. Harper & Row, New York, pp. 1–381, illus.

Harry, R. R. 1953. *Ichthyological field data of Raroia Atoll, Tuamotu Archipelago.* Atoll Research Bulletin No. 18, Pacific Science Board, Washington, D. C., 190 pp., 7 figs.

Hedley, C. 1904. (title not available). Proc. Linn. Soc. N. S. W., 28:880. (not seen).

———. 1926. Zoogeography *in Australian Encyclopedia*, A. W. Jose and H. J. Carter (editors), 2:743–744, Angus and Robertson, Sydney, Australia.

Herre, A. W. C. T. 1953. Check list of Philippine fishes. *U. S. Fish Wildlife Serv., Res. Rep.*, (20):1–977.

Iredale, T., and J. Allan. 1940. A review of the relationships of the Mollusca of Lord Howe Island. *Australian Zool.*, 9:444–451.

Johnson, M. P., L. G. Mason, and P. H. Raven. 1968. Ecological parameters and plant species diversity. *Amer. Nat.*, 102(926):297–306, 1 fig.

Johnson, R. G. 1970. Variations in diversity within benthic marine communities. *Amer. Nat.*, 104(937):285–300, 2 figs.

Kamohara, T. 1957. List of the fishes from Amami-Oshima and adjacent regions, Kagoshima Prefecture, Japan. *Rep. Usa Marine Biol. Sta.*, 4(1):1–65, 38 figs.

——— and T. Yamakawa. 1965. Fishes from Amami-Oshima and adjacent regions. *Rep. Usa Marine Biol. Sta.*, 12(2):1–27, 10 figs.

Kay, E. A. 1967. The composition and relationships of marine molluscan fauna of the Hawaiian Islands. *Venus, Japan. J. Malacol.*, 25(3,4):94–104, 3 figs.

—— and D. K. Young. 1969. The Doridacea (Opisthobranchia; Mollusca) of the Hawaiian Islands. *Pacific Sci.*, 23(2):172–231, 82 figs.

Keast, A. 1969. Evolution of mammals on southern continents. VII. Comparisons of the contemporary mammalian faunas of the southern continents. *Quart. Rev. Biol.*, 44(2):121–167, 2 figs.

Kendall, W. C., and L. Radcliffe. 1912. The shore fishes. Scientific Results Expedition to Eastern Tropical Pacific No. 25. *Mem. Museum Comp. Zool.*, 35(3):77–172, 8 plates.

Klopfer, P. H. 1959. Environmental determinants of faunal diversity. *Amer. Nat.*, 93:337–342.

Klunzinger, C. B. 1884. *Die Fische des Roten Meeres. Eine kritische Revision mit Bestimmungs-Tabellen.* Teil I. *Acanthopteri veri Owen.* Stuttgart, 133 pp., illus. (not seen).

Knox, G. A. 1957. The distribution of polychaetes within the Indo-Pacific. *Proc. 8th Pacific Sci. Congr.*, 3:403–411.

Kramp, P. L. 1968. The Hydromedusae of the Pacific and Indian Oceans. *Dana-Rept.* (72):1–200, 367 figs.

Ladd, H. S. 1960. Origin of the Pacific Island molluscan fauna. *Amer. J. Sci.*, 258-A:137–150.

Laseron, C. F. 1956. The families Rissoinidae and Rissoidae (Mollusca) from the Solanderian and Dampierian provinces. *Australian J. Marine Freshwater Res.*, 7(3):384–485, 227 figs.

MacArthur. R. H., and E. O. Wilson. 1963. An equilibrium theory of insular zoogeography. *Evolution*, 17(4):373–387, 5 figs.

—— and ——. 1967. *The theory of island biogeography.* Monograph in Population Biol. I. Princeton University Press, Princeton, N. J., xi + 203 pp., 60 figs.

Maes, V. O. 1967. The littoral marine mollusks of Cocos-Keeling Islands (Indian Ocean). *Proc. Acad. Nat. Sci. Phila.*, 119(4):93–217, 3 figs., 26 plates.

Marshall, N. B. 1950. Fishes from the Cocos-Keeling Islands. *Bull. Raffles Museum Singapore*, (22):166–205, 2 plates.

——. 1952a. Recent biological investigations in the Red Sea. *Endeavour*, 11(43):reprint without pagination.

——. 1952b. Fishes: "Manihine" expedition to the Gulf of Aqaba, 1948–1949, IX. *Bull. Brit. Museum (Nat. Hist.), Zool.*, 1(8):221–252, 3 figs.

Misra, K. S. 1947. A check list of the fishes of India, Burma, and Ceylon. 2. Clupeiformes, Bathyclupeiformes, Galaxiiformes, Scopeliformes and Ateleopiformes. *Records Indian Museum*, 45(4):377–431.

——, and M. A. S. Menon. 1958. On the distribution of the elasmobranchs and chimaeras of the Indian region in relation to the mean annual isotherms. *Records Indian Museum*, 53(1–2):73–86.

Munro, I. 1949. Revision of Australian silver breams. *Mylio* and *Rhabdosargus*. *Mem. Queensland Museum*, 12(4):182–223, 5 figs., 4 plates.

Newell, N. D. 1971. An outline history of tropical organic reefs. *Amer. Museum Novitates*, (2465):1–37, 13 figs.

Ogilby, J. D. 1889. The reptiles and fishes of Lord Howe Island. *in Lord Howe Island, its zoology, geology, and physical characters.* R. Etheridge, Jr. (editor) *Mem. Australian Museum*, (2):51–74.

Okada, Y., and S. Mawatari. 1957. Distributional provinces of marine bryozoa in the Indo-Pacific region. *Proc. 8th Pacific Sci. Congr.*, 3:391–402, maps.

Ortmann, A. E. 1896. *Grundzüge der marinen Tiergeographie.* Gustav Fischer, Jena, iv + 96 pp., 1 map.

Palmer, G. 1950. Additions to the fauna of Christmas Island, Indian Ocean. *Bull. Raffles Museum Singapore*, (23):200–205.

Penrith, M.-L. 1970. The distribution of the fishes of the family Clinidae in southern Africa. *Ann. S. African Museum*, 55(3):135–150, 5 figs.

Pianka, E. R. 1966. Latitudinal gradients in species diversity: a review of concepts. *Amer. Nat.*, 100(910):33–46.

Pickard, G. L. 1964. *Descriptive physical oceanography.* Pergamon Press, New York, viii + 199 pp., 31 figs.

Pope, E. 1959. Lord Howe Island. *in* Keith Gillett and Frank McNeill, *The Great Barrier Reef and adjacent isles.* Coral Press, Sydney, pp. 141–159, 16 plates.

Powell, A. W. B. 1957. Marine provinces of the Indo-West Pacific. *Proc. 8th Pacific Sci. Congr.*, 3:359–362, 1 map.

Preston, F. W. 1962. The canonical distribution of commoness and rarity. *Ecology*, 43(2):185–215, 31 figs.; 43(3):410–432, 6 figs.

Quoy, J. R. C., and P. Gaimard. 1824. Voyage atour du monde . . . sur les corvetts *l'Uranie* et *La Physicienne* pendant les anneés 1817, 1818, 1819, et 1820. Zoologie. Pillet Aîné, Paris. pp. 1–712; atlas, 96 plates.

Randall, J. E. 1955. *Fishes of the Gilbert Islands.* Atoll Research Bulletin No. 47, Pacific Science Board, Washington, D. C., 243 pp.

Rehder, H. A. 1968. The marine molluscan fauna of the Marquesas Islands. *Bull. Amer. Malacol. Union*, (35):29–32.

Reid, H. E. 1965. *A world away; a Canadian adventure on Easter Island.* Ryerson Press, Toronto, pp. 1–165, illus.

Rice, D. W., and Scheffer, V. B. 1968. A list of the marine mammals of the world. *U. S. Fish Wildlife Serv., Spec. Sci. Rep. Fisheries*, (579):1–16.

Rosewater, J. 1965. The family Tridacnidae in the Indo-Pacific. *Indo-Pacific Mollusca*, 1(6):347–488, illus.

Rüppell, E. 1826. *Atlas zu der Reise im nördlichen Afrika. Zoologie.* Brönner, Frankfurt am Main. vi + 78 pp., 30 plates; 55 pp., 36 plates; 24 pp., 6 plates; 144 pp., 35 plates; 50 pp., 12 plates.

———. 1835. *Neue Wirbelthiere zu der Fauna von Abyssinien gehörig. Fische du rothen Meeres.* Schmerber, Frankfurt am Main, ii + 148 pp., 33 plates.

Salvat, B. 1967. Importance de la faune malacologique dans les atolls Polynésiens. *Cahiers du Pacifique*, (11):7–49, 7 figs., 6 plates.

Sanders, H. L. 1968. Marine benthic diversity: a comparative study. *Amer. Nat.*, 102(925):243–282, 18 figs.

Schilder, F. A. 1956. *Lehrbuch der Allgemeinen Zoogeographie.* Gustav Fischer, Jena, viii + 150 pp., 134 figs.

———. 1965. The geographical distribution of cowries (Mollusca: Gastropoda). *The Veliger*, 3(7):171–183, 2 figs.

Schultz, L. P. 1957. A new approach to the distribution of fishes in the Indo-West Pacific area. *Proc. 8th Pacific Sci. Congr.*, 3:413–416.

———. 1958. Review of the parrotfishes family Scaridae. *Bull. U. S. Nat. Museum*, (214):v + 1–143, 31 figs., 27 plates.

———, and collaborators. 1953–1966. Fishes of the Marshall and Marianas Islands. *Bull. U. S. Nat. Museum*, 1(202):xxxii + 685 pp., figs. 1–90, plates 1–74; 2:ix + 438 pp., figs. 91–132, plates 75–123; 3:vii + 176 pp., figs. 133–156, plates 124–148.

Simpson, G. G. 1965. *The geography of evolution.* Chilton Books, Philadelphia, xx + 249 pp., 45 figs.

Smith, J. L. B. 1949. *The sea fishes of southern Africa.* Central News Agency, Cape Town, xvi + 550 pp., 1,100 figs., 103 plates.

———. 1955–1958. The fishes of Aldabra. *Ann. Mag. Nat. Hist.*, Part I, ser. 12, 8:304–312, 1 plate; Part II, 8:689–697, 1 plate; Part III, 8:886–896, 1 plate; Part IV, 8:928–937, 1 fig., 1 plate; Part V, 9:721–729, 3 figs.; Part VI, 9:817–829, 5 figs.; Part VII, 9:888–892, 1 fig.; Part VIII, 10:395–400, 1 fig., 2 plates; Part IX, 10:833–842, 1 fig., 2 plates; Part X, ser. 13, 1:57–63, 1 plate.

———. 1956. The parrot fishes of the family Callyodontidae of the Western Indian Ocean. *Ichthyol. Bull. Rhodes Univ.*, (1):1–23, 5 plates.

———. 1958. The gunnellichthid fishes with descriptions of two new species from East Africa. *Ichthyol. Bull. Rhodes Univ.*, (9):123–129, 2 figs.

———. 1959a. Fishes of the families Blenniidae and Salariidae of the Western Indian Ocean. *Ichthyol. Bull. Rhodes Univ.*, (14):229–252, 16 figs., 5 plates.

———. 1959b. Gobioid fishes of the families Gobiidae, Periophthalmidae, Trypauchenidae, Taeniodidae, and Kraemeriidae of the Western Indian Ocean. *Ichthyol. Bull. Rhodes Univ.*, (13):185–225, 42 figs., 5 plates.

———. 1959c. The identity of *Scarus gibbus* Rüppell, 1828, and of other parrotfishes of the family Callyodontidae from the Red Sea and the Western Indian Ocean. *Ichthyol. Bull. Rhodes Univ.*, (16):265–282, 10 figs., 5 plates.

———, and M. M. Smith. 1963. *The fishes of the Seychelles.* Rhodes University, Grahamstown, South Africa, pp. 1–215, 98 plates.

Springer, V. G. 1967. Revision of the circumtropical shorefish genus *Entomacrodus* (Blenniidae: Salariinae). *Proc. U. S. Nat. Museum*, 122(3582):1–150, 11 figs., 30 plates.

Stearns, H. T. 1966. *Geology of the State of Hawaii.* Pacific Books, Palo Alto, Calif., xxii + 266 pp., 151 figs., 12 plates.

Stehli, F. G., A. L. McAlester, and C. E. Helsley. 1967. Taxonomic diversity of recent bivalves and some implications for geology. *Bull. Geol. Soc. Amer.*, 78(4):455–465, 10 figs.

Steinitz, H., and A. Ben-Tuvia. 1955. Fishes from Eylath (Gulf of Aqaba), Red Sea. *Israel Sea Fish. Res. Sta., Bull.*, 11, pp. 1–15.

Stephenson, T. A. 1947. The constitution of the intertidal fauna and flora of South Africa, part III. *Ann. Natal Museum*, 11(2):207–324, 11 figs., 2 plates.

Taylor, J. D. 1971. Reef associated molluscan assemblages in the western Indian Ocean. *Symp. Zool. Soc. London*, (28):501–534, 9 figs.

Usinger, R. L. 1963. Animal distribution patterns in the tropical Pacific. *in* J. L. Gressitt (editor), *Pacific basin biogeography, a symposium.* Bishop Museum Press, Honolulu, pp. 255–261.

Vaughan, T. W., and J. W. Wells. 1943. Revision of the suborders, families, and genera of the Scleractinia. *Geol. Soc. Amer., Spec. Papers*, (44):xv + 1–363, 39 figs., 51 plates.

Waite, E. R. 1916. A list of the fishes of Norfolk Island, and indication of their range to Lord Howe Island, Kermadec Island, Australia, and New Zealand. *Trans. Roy. Soc. S. Australia*, 40:452–458, 3 plates.

Wallace, A. R. 1892. *Island life.* 2d edition. Macmillan and Co., London, xx + 1–563 pp., 26 figs.

Weaver, C. S., and J. E. duPont. 1970. Living volutes, a monograph of the recent Volutidae of the world. *Delaware Museum Nat. Hist., Monograph Ser.*, (1):xv + 1–375, 43 figs., 79 plates.

Whitley, G. P. 1932. Marine zoogeographical regions of Australasia. *Australian Nat.*, 8(8):166–167, map.

———. 1937. The Middleton and Elizabeth Reefs, South Pacific Ocean. *Australian J. Zool.*, 8(4):199–273, 1 fig., 5 plates.

———. 1948. A list of the fishes of Western Australia. *Fish. Bull. No. 2, West Australian Fisheries Dept.*, 35 pp., 1 map.

Whitehead, P. J. P. 1967. The clupeoid fishes of Malaya. *J. Marine Biol. Ass. India*, 9(2):223–280, 58 figs.

Whittaker, R. H. 1969. Evolution of diversity in plant communities. *in* Diversity and stability in ecological systems. *Brookhaven Symp. Biol.*, (22):178–196, 6 figs.

———, and G. M. Woodwell. 1972. Evolution of natural communities. *in* J. A. Wiens (editor) *Ecosystem structure and function.* Oregon State University Press, pp. 137–156.

Wilson, B. R., and J. A. McComb. 1967. The genus *Cypraea* (subgenus *Zoila* Jousseaume). *Indo-Pacific Mollusca*, 1(8):457–484, illus.

Woodward, S. P. 1851–1856. *A manual of the mollusca.* John Weale, London, viii + 1–486 pp., 24 plates + 24 pp., 1 map.

Chapter
2

The
Eastern
Pacific Region

The investigation and determination of the provinces of marine life, have as yet been but little pursued, and there is no finer field for discovery in natural history, than that presented by the bed of the ocean, when examined with a view to the defining of its natural subdivisions.

Professor Edward Forbes in The Natural History of European Seas, *1859.*

The Eastern Pacific Zoogeographic Region includes the mainland coast from Magdalena Bay and the lower end of the Gulf of California to the southern Gulf of Guayaquil. The entire area covers about 28° of latitude and extends westward to include five offshore island localities. The most remote of these is Clipperton, an atoll lying off the coast of Costa Rica about 650 miles from the mainland. The general topography of the region offers a marked contrast to that of the Indo-West Pacific. Both the inshore and offshore islands and archipelagos are small, few in number, and widely scattered.

To the south, the cool Peru (Humboldt) Current comes far up the coast of Peru to prevent the tropical waters from ordinarily extending more than about 3° below the equator. However, to the north the corresponding California Current makes its main swing to the west off the southern part of Baja California. Apparently, the discrepancy in the behavior of the two main, cool, equatorial-bound currents of the Eastern Pacific is due to the difference in coastline topography. The South American coast and continental shelf line extends almost due north from the Antarctic while that of North America—at least from northern Mexico southward—runs to the southeast. This leaves most of the Mexican and all of the Central American coast untouched by cool water from the north (Fig. 2-2). The result is that almost all of the tropical waters of the Eastern Pacific shores are located to the north of the equator.

It is interesting to see that Forbes (1856), in his pioneering work on the general distribution of marine animals, described a "Panamanian Province" extending from about Magdalena Bay and the mouth of the Gulf of California to the Peruvian coast just south of the Gulf of Guayaquil, thus indicating a tropical area that is almost the same as that now called the Eastern Pacific Region. Ekman (1953:38) considered that the whole of the Gulf of California was included in the tropics but only the southern tip of the outer coast of Baja California (with a subtropical transition fauna to the north). To the south, he suggested a boundary at Point Aguja, Peru or farther north at Paita or possibly the Gulf of Guayaquil. Schilder (1956) recognized a "West-amerikanish Subregion" extending from about the United States–Mexican border to the southern Gulf of Guayaquil. Hedgpeth (1957) indicated a tropical fauna for most of the Gulf of California and, on the outer coast, from the lower part of Baja California to about Paita, Peru.

Some of the work on certain animal groups has helped to outline the geographic extent of the Eastern Pacific tropics. In regard to mollusks, Woodward (1856) recognized a "Panamic Province" between the Gulf of California and Paita; Keen (1958) referred to a tropical fauna as extending from Magdalena Bay to Guayaquil, Ecuador (but a fringe going on down to Paita); and Berry (1960) considered that such a fauna occupied the area from Magdalena Bay and the northern extremity of the Gulf of California to about as far as Paita. A region virtually identical to the latter was outlined by Ortmann (1896) in his study of the decapod crustaceans and called the "Westamerikanische Litoralregion." For nemerteans, Coe (1940) recognized a tropical zone from the Gulf of California and coast south to Ecuador. For bryozoans, Osburn (1950–1953) considered the tropical area to extend from Cedros Island and Point Eugenia, Baja California, all the way to San Juan Bay, Peru, at 15°20′S. For balanomorph barnacles, Zullo (1966) outlined a tropical fauna extending from Cape San Lucas to Guayaquil. The tropical fish fauna has been considered to reach from Magdalena Bay to Cape Blanco, Peru (Rosenblatt 1967).

Aside from the Galápagos Islands, which were soon recognized as a distinct faunal area (Woodward 1856, Günther 1880), most of the early biologists did not attempt to subdivide the Eastern Pacific tropics. Jordan (1901:566), on the basis of shore fish distribution, listed Sinaloan, Panaman, Revillagigedan, and Galápagan as "minor faunal areas," but since these areas were not defined, it is not known precisely what mainland stretches were meant by the first two terms. Bartsch (1912), after working on the distribution of the pyramidellid mollusks, recognized for the mainland shoreline a "Mazatlánic" fauna extending from Cape San Lucas to Acajutla, Guatemala, and a "Panamic" fauna from the latter to Point Aguja, Peru.

The Provinces
Mexican

Since the northern boundary of the province on the outer Baja California coast is also a place of contact with the warm-temperate waters of the California Region, there has been considerable speculation about its placement. Most of the works cited above indicated a boundary at either Magdalena Bay (24°40′N) or Cape San Lucas (23°N). In addition, the publications on mollusk distribution by Schenk and Keen (1936) and Newell (1948) showed a conspicuous species break at about 23°N. McKay (1943:114) found the crab genus *Cancer* to have its southern limit at Magdalena Bay, and Garth (1955) utilized the distributional patterns of all the brachyuran crabs to demonstrate the existence of a warm-temperate fauna that extended south to Point Entrada just outside Magdalena Bay.

Boundary recognitions based primarily on shore fish distribution have

Figure 2-1   The striped pargo (*Hoplopagrus guentheri*). A tropical shelf species confined to the Eastern Pacific Region. After Kumada (1937).

been made by C. L. Hubbs (1952:328) near Cape San Lucas, by C. Hubbs (1952:154) between Albreojos Point and Port San Bartolomé, by Briggs (1955:154) on Cedros Island or even farther south, by Springer (1958:485) at Sebastian Vizcaino Bay, and by Stephens (1963:115) above Cape San Lucas. C. L. Hubbs (1960), in his analysis of the distribution of all the marine vertebrates of the outer Baja California coast, spoke of an abrupt decline in the tropical elements between Cape San Lucas and Magdalena Bay.

Considering all the opinions listed above, and especially the distributional evidence that accompanied many of them, it appears that Magdalena Bay is the place where the rate of species change is the highest. Therefore, this is where the boundary should be placed. However, it should be emphasized, as C. L. Hubbs (1960) pointed out, that the west coast of Baja California is a complex faunal area. Many tropical species extend northward for a considerable distance, finding suitable conditions in the shallow, protected bays and inlets, while many northern cool-water forms are found within the influence of upwelling areas on the outer coast.

Since many of the earlier workers considered the fauna of the Gulf of California to be tropical, there has been less speculation about the location of zoogeographic boundaries within the Gulf. Here, information on the distribution of certain groups of shore fishes appears to be more precise than that for the invertebrates. On the mainland side, C. Hubbs (1952:154) mentioned a boundary north of Mazatlán, Briggs (1955:154) preferred one at Guaymas, and Springer (1958:484) expressed some agreement with both ideas, since some of his species found their range limits at Mazatlán and others at Guaymas. But Stephens (1963:113) found that the Isla San Ignacio de Farallon, off Topolobampo and about midway between Mazatlán and Guaymas, was inhabited by chaenopsids (Chaenopsidae) from the Mexican Province rather than from the Gulf, so the boundary must occur at least this far north. Accordingly, a tentative placement is made at Topolobampo.

To the west on the peninsular side of the Gulf, the fish fauna appears to undergo a rather sharp change at La Paz. This was first noted by C. Hubbs (1952:154) and then observed by others (Walker 1960:129, Stephens 1963:113). In regard to brachyuran crabs, Garth (1960:120) spoke of a boundary at Puerto Escondido at about 26°N. For bryozoans, Soule (1960:102) recognized a tropical assemblage extending north on the peninsula to about the same place. This leaves a discrepancy of about 2° of latitude between the evidence for fishes and that for the two

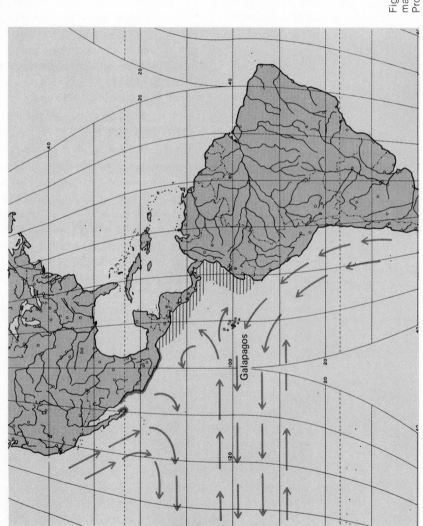

Figure 2-2 Mexican, Pana-
manian, and Galápagos
Provinces.

invertebrate groups. Since the evidence on fishes is much better documented, the La Paz location is preferred.

The location of the northern boundaries to the Mexican Province at Magdalena Bay, La Paz, and Topolobampo means that the tip of the Baja California peninsula exists as a spatially isolated portion of the province. Dispersal of species from the mainland on the east is limited by a deep-water gap of about 100 miles, and tropical animals cannot come down along the edges of the peninsula from the north because of the colder water temperatures in those areas. The fish fauna at the end of the peninsula demonstrates the effects of isolation. The number of species is reduced, there are several endemics, and there has been an invasion by some transpacific species.

It may be said that, although the cape area of Baja California is geographically the end of a peninsula, it is a zoogeographic island. However, despite its insular characteristics, the degree of endemism is quite low so that it needs to be regarded as a part of the Mexican Province. Walker (1960:132) called attention to the almost complete lack of northern elements, some endemism, and the presence of a large group of species that were known to occur at Mazatlán and southward.

Since the existence of the Mexican Province has been only recently recognized, there has been little discussion of where the southern boundary might be placed. Here again, most of the evidence is ichthyological. C. Hubbs (1952:153) was in favor of a line at Honda Bay, Panama; Briggs (1955:154) indicated a boundary at Tangola-Tangola Bay in the northern Gulf of Tehuantepec; Springer (1958:483) agreed with the latter placement; and Stephens (1963:115) put the boundary a little farther south at Salina Cruz. So far, Tangola-Tangola Bay seems the popular choice.

The Mexican Province does appear to possess an appreciable level of endemism, at least in some fish groups that have been revised with the help of a considerable amount of material from this coastline. A total of nine species of clingfishes are now known from the area (Briggs 1955, 1960), and six of them are virtually confined to it. An analysis of Springer's (1958) revision of the clinid fish genus *Malacoctenus* shows that, of the six Mexican species, four are endemic. The lack of other, modern systematic works on the shelf fauna of this area makes it impossible to predict the extent of endemism for the fauna in general.

Panamanian   The Panamanian Province extends from Tangola-Tangola Bay in the northern part of the Gulf of Tehuantepec to a southern boundary on the northwestern coast of South America. In the latter area there is, due to the local topography and the effect of the Peru Current in leaving the vicinity of the mainland coast, a rather abrupt change from a tropical to a warm-temperate fauna. Even so, the exact location of a southern boundary for the tropical fauna has not been agreed upon.

In the introduction to this chapter, a number of works were cited which indicated that the tropical forms extended down to the Gulf of Guayaquil, on the border between Ecuador and Peru, or Paita, Peru, or even farther south to Point Aguja, Peru. So far, it seems that more attention has been paid to the distribution of mollusks in this vicinity than to the other animal groups. Keen (1958:3) observed that, "To the south the Panamic fauna mostly ends along the Ecuador coast, at Guayaquil, but a fringe of durable forms extends southward to Paita, Peru—the absolute limit, beyond which the fauna is Peruvian in composition."

Olsson (1961:24), on the basis of his work on the Pelecypoda, placed the southern boundary for the tropical fauna at Cape Blanco which is at the lower end of the Gulf of Guayaquil. However, he did not find the warm-temperate (Peruvian) fauna to be well developed until a point farther south—Cape Aguja. The zone between the two capes he considered to be a transitional area naming it the "Paita Buffer Zone." The information on mollusk distribution appears to indicate that the most likely place for the boundary is at the southern end of the Gulf of Guayaquil.

Springer (1958:483) proposed that the area between Tangola-Tangola Bay, Mexico and the Gulf of Fonseca, Nicaragua be called the "Pacific Central American Faunal Gap" because no species of typical rocky shore fishes have been reported from there. While this designation may be appropriate for species confined to rocky shorelines, it cannot of course be applied to the fauna in general. There are additional indications in both the ichthyological and invertebrate literature that the distribution of species within the Panamanian Province is by no means uniform.

Further systematic work may show that a Central American Province, extending from the Gulf of Tehuantepec south as far as the Nicoya Peninsula in Costa Rica or even to Honda Bay in northern Panama, should be recognized. Newell's (1948) analysis of molluscan distribution shows a prominent "rate of change" peak near Honda Bay, and Haig's (1960) work on the porcellanid crabs indicates a number of range terminations in the vicinity of the Nicoya Peninsula. In fishes, there are apparent breaks in both places plus an impressive amount of endemism in the Gulf of Guayaquil. Finally, Keen (1958:3) points out that the Gulf of Panama is rich in endemic or short-ranging molluscan species.

On the basis of our present knowledge, it would be difficult to give a dependable estimate of provincial endemism. Work on some of the shore fish groups indicates that it may be quite high. In the clingfishes, 12 of the 14 species are apparently endemic (Briggs 1955, 1969a, 1969b); in the clinid genus *Malacoctenus*, 4 of the 5 species are endemics (Springer 1958) as are 5 of the 10 chaenopsid blennys (Stephens 1963).

Galápagos   The Galápagos Islands (Archipiélago de Colón), a group consisting of 13 large and many smaller islands, is located at 1°N to 1°30′S and about 89 to 92°W, which is approximately 600 miles off the coast of Ecuador. At present, about all that can be said on the age of the islands is that the oldest fossiliferous beds are probably Pliocene (Williams 1966:67).

The Galápagos Islands were first discovered in 1535, received many visitors during the seventeenth and eighteenth centuries, and were colonized by 1830. A complete history of the exploration of the Galápagos has been written by Slevin (1959). By the time the H. M. S. *Beagle*, with Charles Darwin aboard, arrived in 1835, there was a settlement on Charles Island (Santa María), with some 300 inhabitants. It was, however, Darwin's (1839) observations plus those published in the five parts of the *Zoology of the Beagle* (1839–1843, edited by Darwin) that successfully brought this island group to the attention of the scientific world. Darwin also made collections of the plants, and these were described in a historical series of papers by Sir Joseph Hooker (Wiggins 1966).

Many scientific expeditions have come to the Galápagos since 1835. From the standpoint of the marine fauna, some of the most important were the Hopkins-Stanford Expedition which worked in the islands from December 1898 to June 1899; the visit by the U. S. Fish Commission steamer *Albatross* in 1904; the California Academy of Sciences Expedition aboard the *Academy* from 1905 to 1906; the Crane Pacific Expedition in 1928; the Templeton Crocker Expedition aboard the *Zaca* in 1932; and several Allan Hancock Foundation trips that began in 1932. In more recent years, many additional visits have been made, and extensive collections of Galápagos specimens now reside in several institutions. Much of this material has never been reported on.

In his discussion of the factors influencing the zoogeographic affinities of the inshore marine fauna, Abbott (1966) called attention to the ocean currents that are responsible for transporting materials toward the Galápagos area. These are the Peru Oceanic Current, Peru Coastal Current, North Equatorial Countercurrent, and currents from the Gulf of Panama. Of these, the last, which are well developed in February to April, appear to be the most important, since the shore fauna is strongly related to that of the Panamanian Province.

The marine shore fauna shows evidence of having been isolated for a long time. Salvat (1967), utilizing data published by Hertlein and Strong (1955) and Emerson and Old (1965), noted that the mollusks were represented by 138 Panamanian species, 8 species from the Indo-West Pacific, and 29 (16 percent) endemics. Garth (1946) studied the brachyuran crabs and found a total of 120 species with 18 of them, or 15 percent, being endemic. He also noted that 5 of the nonendemics were Indo-West Pacific species, while 5 others were warm-temperate, being shared with the coast of Peru and/or Chile.

Durham (1966:125) recorded 32 species of stony corals and noted that 13 of them were apparently endemic; 7 of the total came from the Indo-West Pacific and the rest of the nonendemics were Panamanian species. Silva (1966:150) listed 311 species of marine algae and estimated that about 36 percent were endemic. There is a considerable body of literature on the shore fishes, but the report by Walker (1966) is the most recent. His data indicated a total of 223 species of which about 23 percent were considered to be endemics, 54 percent shared with other localities in the Eastern Pacific but did not occur elsewhere, and 12 percent derived from the Indo-West Pacific.

The rather high level of endemism indicates that, without a doubt, the Galápagos need to be placed in a separate province. Despite the high endemism, the great majority of species present are tropical forms that probably came from the Gulf of Panama or other parts of the Panamanian Province. Lesser numbers of species established themselves after having made their way across the Pacific or from the warm-temperate waters of the South American mainland.

Many of the Panamanian species that extend to the Galápagos also have cognate or analogous species in the tropical Western Atlantic. In addition, the Galápagos fauna appears to demonstrate a special, amphi-American relationship of its own. For example, the sparid (family Sparidae) fish *Archosargus pourtalesi*, a Galápagos endemic, has its closest relatives in the Western Atlantic; the clingfish (Gobiesocidae) *Arcos poecilophthalmus*, another endemic, is closest to the Atlantic *A.*

macrophthalmus; and the clinid (Clinidae) *Labrisomus dendriticus* is most closely related to the Atlantic *L. filamentosus* (Springer 1959:289). In a like manner, Garth's (1946) analysis of the brachyuran crabs shows that two of the Galápagos endemics have analogous species in the Atlantic and that two other species are shared with the Western Atlantic (and do not occur elsewhere in the Eastern Pacific). It appears that the mainland forms, from which the foregoing island species were derived, once inhabited the Eastern Pacific but have become extinct in that area.

Other Islands

Cocos

Cocos Island is located in the Eastern Pacific at 5°33'N and 86°59'W, which is about 300 miles west of Costa Rica and about 350 miles northeast of the Galápagos Islands. It is a small, steep island with the land area comprising only about 47 sq km. A good deal more has been written about its fame as a locale of buried treasure than about its natural history. Fournier (1966) related that over 400 expeditions have visited the island with the hope of finding treasure, but so far, none have succeeded. Since almost all the scientific expeditions to the Galápagos also visited Cocos, the history of biological exploration for the two areas is virtually the same.

It has been observed that Cocos Island lies within the path of the North Equatorial Countercurrent but that, occasionally, a southward shift takes place so that the island is exposed to currents from the mainland (Fournier 1966). Vinton (1951:373) suggested that the island probably originated during the Pleistocene, and Fournier (1966:185) observed that the low percentage of endemism in its terrestrial plants, as compared with those of the Galápagos, was probably due to the more recent origin of the former.

Hertlein (1963) published a useful compendium on the biogeography of Cocos Island, including a complete bibliography. From this, it can be seen that endemism in some of the major invertebrate groups is rather low: only 1 of 45 echinoderm species is indigenous, none of the 20 bryozoans, none of 9 annelids, and 5 of 57 crustaceans. Durham (1966:125) listed 18 species of stony corals, identifying 10 of them as Indo-West Pacific, 7 as Eastern Pacific, and 1 endemic. Salvat (1967:20) summarized the relationship of the mollusks as 88 percent Panamic, 7 percent Indo-West Pacific, and 5 percent endemic.

I have examined several collections of shore fishes from Cocos Island and have, so far, recorded a total of 82 species; of these, 46 are shared with the Galápagos and/or other Eastern Pacific localities, 21 are invaders from the Indo-West Pacific, and 6 (about 7 percent) are endemics. The presence of the 21 Indo-West Pacific species, compris-

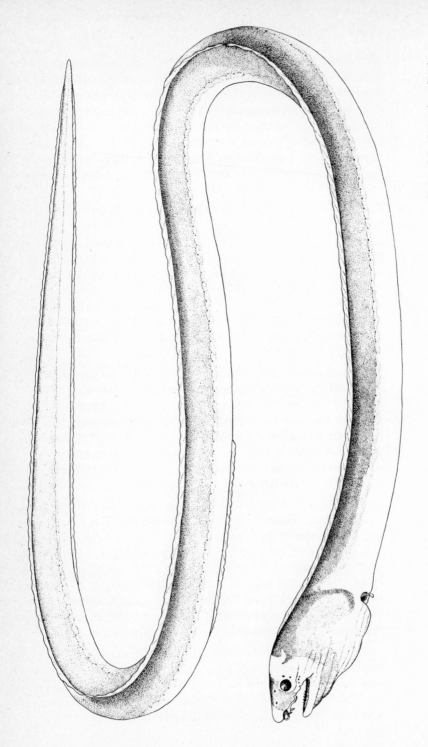

Figure 2-3 An undescribed shelf species of snake eel (*Bascanichthys sp.*). It is apparently confined to Cocos Island in the tropical Eastern Pacific.

ing about 26 percent of the total, indicates a surprisingly strong transpacific influence. This is close to the estimate of 23 percent Indo-West Pacific species that was made by Walker (1966:173).

Since the level of endemism in the various invertebrate groups is low and that in fishes is only about 7 percent, and since the great majority of the species occur along the mainland shores of the Eastern Pacific, Cocos Island is considered to be a part of the Panamanian Province.

Clipperton   Clipperton Island is a very low structure, an atoll about 2 miles in diameter, surrounded by a fringing reef. It lies at 10°18′N and 109°13′W, about 650 miles southwest of Acapulco, Mexico. It was apparently discovered by John Clipperton, an English pirate, in 1705, and the island was named after him by Captain James Cook. An account of the history of the island with a chronological catalog of known visits was published by Sachet (1960). In 1962, the same author published a monograph on the geology and biology of the island.

Since Clipperton Island is the only true atoll in the Eastern Pacific and since it is the most isolated of all the oceanic islands in this part of the world, the relationships of its fauna are of unusual interest. Although the sparse terrestrial flora and fauna are well known, the marine fauna is not. Only the mollusks and crabs have been worked on to any considerable extent.

The molluscan fauna now appears to comprise about 80 species with about 60 percent of them from the Indo-West Pacific, 35 percent from the Panamanian Province, and 5 percent endemic (Hertlein and Emerson 1953, 1957; Hertlein and Allison 1960, 1966). Chase (1962) who studied the nonbrachyuran, decapod crustaceans, recorded a total of 24 species; 14 of them were from the Indo-West Pacific, and the remaining 10 were Panamanian. Garth (1965), who worked on the brachyuran crabs, found 34 species, 16 of them transpacific and 18 American. None of the crab species were considered to be indigenous. A few additional invertebrate species (echinoderms and corals) were reported by Hertlein and Emerson (1957).

The fishes remain virtually unknown although extensive collections have been made (Baldwin 1963). Walker (1966) estimated that 26 percent of the species were Indo-West Pacific in origin. Durham (1963:362) remarked that the currently known marine fauna of the island is characterized by an almost equal representation of western American and Indo-West Pacific elements. Dawson (1959) stated that the marine algae species were "almost entirely" those that are common and widely distributed throughout the Indo-West Pacific Region. It is probable that when the fishes have been worked on, the majority of the marine animal species will prove to be American. Even so, the Indo-West Pacific influence is remarkably strong. So far, the amount of endemism appears to be small so that the island is considered to be a part of the Panamanian Province.

Although Clipperton apparently lies well to the north of the North Equatorial Countercurrent, this current is undoubtedly responsible for the original transport of the many Indo-West Pacific species. Since the island is an atoll with good coral reef development, it probably offers a more suitable habitat for Polynesian waifs than do the other Eastern Pacific islands.

Revillagigedos

The Revillagigedo archipelago consists of three volcanic islands and several smaller adjoining rocks located at 18°20′ to 19°20′N and 110°45′ to 114°50′W. This is about 250 miles south of the tip of Baja California, the nearest mainland locality.

The first collection of marine animals from the Revillagigedos was apparently that taken in 1880 by Lieutenant Henry E. Nichols aboard the U. S. Coast and Geodetic Survey steamer *Hassler*. A larger collection was taken by the U. S. Fish Commission steamer *Albatross* in 1889. Mr. R. C. McGregor, assistant naturalist aboard the schooner *H. C. Wahlberg*, made a good collection in 1897. More recent expeditions have called at these islands and collected biological specimens, but with the exception of the 1954 to 1958 cruises of the Canadian yacht *Marijean*, this material has not been described in the literature.

As things now stand, the fishes appear to be the best-known marine group. Snodgrass and Heller (1905) published a list of the species based on the specimens collected by the *Albatross* and by R. C. McGregor plus those that had been reported by earlier authors (Gilbert 1890, 1897; Jordan and McGregor 1899). Ricker (1959) reported on the material taken by the *Marijean*. It appears that the shore fish fauna consists of about 100 species with possibly 9 or 10 of them endemic, 23 from the Indo-West Pacific, and 58 from the Eastern Pacific.

The marine mollusks of the archipelago were reported on by Strong and Hanna (1930). They recognized a total of 61 species: 2 from the Indo-West Pacific, 59 from the Eastern Pacific, and no endemics. In view of the lack of endemism in the mollusks and the rather low amount (9 or 10 percent) in the fishes, the Revillagigedos are considered to be part of the Panamanian Province.

Malpelo

Melpelo Island is a small, barren mass of volcanic rock lying at 4°03′N and 81°36′W, about 250 miles west of the mouth of the San Juan River in

Colombia. Two gastropods and a barnacle (Hertlein 1932) and 14 fish species (Fowler, 1938) have been reported. There appears to be no endemism.

Although the Eastern Pacific Region includes a number of well-isolated oceanic islands, we may conclude that so far only one group, the Galápagos, possesses a shallow-water fauna distinctive enough to enable it to be recognized as a separate province.

Literature Cited

Abbott, D. P. 1966. Factors influencing the zoogeographic affinities of the Galápagos inshore marine fauna. *in* R. I. Bowman (editor), *The Galápagos.* University of California Press. Berkeley, pp. 108–122, 8 figs.

Baldwin, W. J. 1963. A new chaetodont fish, *Holacanthus limbaughi* from the Eastern Pacific. *Contrib. Sci., Los Angeles Co. Museum*, (74):1–8, 1 fig.

Bartsch, P. 1912. A zoogeographic study based on the pyramidellid mollusks of the west coast of America. *Proc. U. S. Nat. Museum*, 42(1906):297–349, plate 40 (map).

Berry, S. S. 1960. The nature and relationship of the Panamic fauna as manifested by the Mollusca. *Bull. Amer. Malacol. Union*, (26):44–45.

Briggs, J. C. 1955. A monograph of the clingfishes (Order Xenopterygii). *Stanford Ichthyol. Bull.*, 6:1–224, 114 figs.

———. 1960. A new clingfish of the genus *Gobiesox* from the Tres Marias Islands. *Copeia*, (3):215–217, 1 fig.

———. 1969a. A new clingfish (Gobiesocidae) of the genus *Tomicodon* from Ecuador. *Copeia*, (1):75–76, 1 fig.

———. 1969b. The clingfishes (Gobiesocidae) of Panama. *Copeia*, (4):774–778, 3 figs.

Chace, F. A., Jr. 1962. The non-brachyuran decapod crustaceans of Clipperton Island. *Proc. U. S. Nat. Museum*, 113(3466):605–635, 7 figs.

Coe, W. R. 1940. Revision of the nemertean fauna of the Pacific Coasts of North, Central and northern South America. *Allan Hancock Pacific Expedition*, 2(13):247–323, 8 plates.

Darwin, C. R. 1839. Journal and remarks. *in* P. P. King (editor), *Narrative of the surveying voyages of H.M.S. Adventure and Beagle between 1826 and 1836.* 3 vols. (in 4), illus., Colburn, London (this edition not seen).

———. (editor) 1839–1843. The zoology of the voyage of H.M.S. *Beagle*, under the command of Capt. Fitzroy, R.N., during 1832–36, in 5 parts. Smith, Elder, London.

Dawson, E. Y. 1959. Some algae from Clipperton Island and the Danger Islands. *Pacific Nat.*, 1(7):1–8, 1 fig.

Durham, J. W. 1963. Paleogeographic conclusions in light of biological data. *in* J. L. Gressitt (editor), *Pacific Basin biogeography, a symposium.* Bishop Museum Press, Honolulu, pp. 355–365, 4 figs.

———. 1966. Coelenterates, especially stony corals, from the Galápagos and

Cocos Islands. *in* R. I. Bowman (editor), *The Galápagos*. University of California Press, Berkeley, pp. 123–135.

Ekman, S. 1953. *Zoogeography of the sea*. Sidgwick & Jackson, London, xiv + 417 pp., 121 figs.

Emerson, W. K., and W. E. Old. 1965. New molluscan records for the Galápagos Islands. *The Nautilus*, 78(4):116–120.

Forbes, E. 1856. Map of the distribution of marine life. *in* Alexander K. Johnston, *The physical atlas of natural phenomena* (new edition). Edinburgh and London, W. and A. K. Johnston, plate no. 31.

Fournier, L. A. 1966. Botany of Cocos Island, Costa Rica. *in* R. I. Bowman (editor), *The Galápagos*. University of California Press, Berkeley, pp. 183–186.

Fowler, H. W. 1938. The fishes of the George Vanderbilt South Pacific Expedition, 1937. *Acad. Nat. Sci. Phila. Monograph No. 2*, v + 349 pp., 12 plates.

Garth, J. S. 1946. Distributional studies of Galápagos brachyura. *Allan Hancock Pacific Expedition*, 5(11):603–648, 10 charts.

———. 1955. A case for a warm-temperate marine fauna on the west coast of North America. *in Essays in the Natural Sciences in honor of Captain Allan Hancock on the occasion of his birthday, July 26, 1955*. University of S. California Press, pp. 19–27.

———. 1960. Distribution and affinities of the brachyuran Crustacea. Symposium: the biogeography of Baja California and adjacent seas. *Syst. Zool.*, 9(3–4):105–123, 3 figs.

———. 1965. The brachyuran decapod crustaceans of Clipperton Island. *Proc. Calif. Acad. Sci.*, 33(1):1–46, 26 figs.

Gilbert, C. H. 1890. A preliminary report on the fishes collected by the steamer Albatross on the Pacific coast of North America during the year 1889, with descriptions of twelve new genera and ninety-two new species. *Proc. U. S. Nat. Museum*, 13(797):49–126.

———. 1897. Descriptions of twenty-two new species of fishes collected by the steamer Albatross, of the United States Fish Commission. *Proc. U. S. Nat. Museum*, 19(1115):437–457, 7 plates.

Günther, A. C. L. G. 1880. *An introduction to the study of fishes*. Adam and Charles Black, Edinburgh, xvi + 720 pp., 320 figs.

Haig, J. 1960. The Porcellanidae (Crustacea Anomura) of the Eastern Pacific. *Allan Hancock Pacific Expedition*, 24:viii + 440, 12 figs., 42 plates.

Hedgpeth, J. W. 1957. Marine biogeography. *in Treatise on marine ecology and paleoecology*. vol. 1. *Mem. Geol. Soc. Amer.*, (67)359–382, 16 figs., 1 plate.

Hertlein, L. G. 1932. Mollusks and barnacles from Malpelo and Cocos Islands. *The Nautilus*, 46(2):43–45.

———. 1963. Contribution to the biogeography of Cocos Island, including a bibliography. *Proc. Calif. Acad. Sci.*. 32:219–289, 4 figs.

———, and E. C. Allison. 1960. Gastropods from Clipperton Island. *Veliger*, 3(1):13–16.

———, and ———. 1966. Additions to the molluscan fauna of Clipperton Island. *Veliger*, 9(2):138–140.

———, and W. K. Emerson. 1953. Mollusks from Clipperton Island (Eastern

Pacific) with the description of a new species of gastropod. *Trans. San Diego Soc. Nat. Hist.*, 11(13):345–364, 2 plates.

———, and ———. 1957. Additional notes on the invertebrate fauna of Clipperton Island. *Amer. Museum Novitates*, (1859):1–9, 1 fig.

———, and A. M. Strong. 1955. Marine mollusks collected at the Galápagos Islands during the voyage of the Velero III, 1931–32. *in Essays in the natural sciences in honor of Captain Allan Hancock on the occasion of his birthday, July 26, 1955.* University of S. California Press, pp. 111–145.

Hubbs, C. 1952. A contribution to the classification of the blennioid fishes of the family Clinidae, with a partial revision of the Eastern Pacific forms. *Stanford Ichthyol. Bull.*, 4(2):41–165, 64 figs.

Hubbs, C. L. 1952. Antitropical distribution of fishes and other organisms. Symposium on problems of bipolarity and of pan-temperate faunas. *Proc. 7th Pacific Sci. Congr.*, 3:324–329.

———. 1960. The marine vertebrates of the outer coast. Symposium: the biogeography of Baja California and adjacent seas. *Syst. Zool.*, 9(3–4):134–147.

Jordan, D. S. 1901. The fish fauna of Japan, with observations on the geographical distribution of fishes. *Science*, new series 14(354):545–567.

———, and R. C. McGregor. 1899. List of fishes collected at the Revillagigedo Archipelago and neighboring islands. *Rep. U. S. Commissioner Fish and Fisheries for 1898*, part 29:273–284, 4 plates.

Keen, A. M. 1958. *Sea shells of tropical west America.* Stanford University Press, Stanford, viii + 624 pp., text figs., 10 plates.

Mackay, D. C. G. 1943. Temperature and the world distribution of crabs of the genus *Cancer. Ecology*, 24(1):113–115.

Newell, I. M. 1948. Marine molluscan provinces of western North America: a critique and a new analysis. *Proc. Amer. Phil. Soc.*, 92(3):155–166, 7 figs.

Olsson, A. A. 1961. *Mollusks of the tropical Eastern Pacific, Panamic-Pacific Pelecypoda.* Paleontological Research Institution, Ithaca, N. Y., 574 pp., 86 plates.

Ortmann, A. E. 1896. *Grundzüge der marinen Tiergeographie.* Gustav Fischer, Jena, iv + 96 pp., 1 map.

Osburn, R. C. 1950–53. Bryozoa of the Pacific Coast of America. *Allan Hancock Pacific Expeditions*, 14(1–3):1–841, 82 plates.

Ricker, K. E. 1959. Fishes collected from the Revillagigedo Islands during the 1954–1958 cruises of the "Marijean." *Inst. Fish. Univ. Brit. Col., Museum Contrib.*, (4):1–10.

Rosenblatt, R. H. 1967. The zoogeographic relationships of the marine shore fishes of tropical America. *Studies Trop. Oceanog. Miami*, (5):579–592.

Sachet, M.-H. 1960. Histoire de l'île Clipperton. *Cahiers du Pacifique*, (2):1–32, 1 plate.

———. 1962. Monographie physique et biologique de l'île Clipperton. *Ann. Inst. Oceanog. (Paris)*, 40(1):1–107, 3 figs., 12 plates.

Salvat, B. 1967. Importance de la faune malacologique dans les atolls Polynésiens. *Cahiers du Pacifique*, (11):7–49, 7 figs., 6 plates.

Schenk, H. G., and A. M. Keen. 1936. Marine molluscan provinces of Western North America. *Proc. Amer. Phil. Soc.*, 76:921–938, 6 figs.

Silva, P. C. 1966. Status of our knowledge of the Galápagos benthic marine algal flora prior to the Galápagos International Scientific Project. R. I. Bowman (editor), *The Galápagos.* University of California Press, Berkeley, pp. 149–156.

Slevin, J. R. 1959. The Galápagos Islands. A history of their exploration. *Occas. Papers Calif. Acad. Sci.*, 25:1–150.

Snodgrass, R. E., and E. Heller. 1905. Papers from the Hopkins-Stanford Galápagos Expedition, 1898–1899. XVII. Shore fishes of the Revillagigedo, Clipperton, Cocos and Galápagos Islands. *Proc. Wash. Acad. Sci.*, 6:333–427.

Soule, J. D. 1960. The distribution and affinities of the littoral marine Bryozoa (Ectoprocta). Symposium: the biogeography of Baja California and adjacent seas. *Syst. Zool.*, 9(3–4):100–104, 1 fig.

Springer, V. G. 1958. Systematics and zoogeography of the clinid fishes of the subtribe Labrisomini Hubbs. *Publ. Inst. Marine Sci., Texas*, 5:417–492, 4 figs., 7 plates.

———. 1959. A new species of *Labrisomus* from the Caribbean Sea, with notes on other fishes of the Subtribe Labrisomini. *Copeia*, (4):289–292, 1 fig.

Stephens, J. S., Jr. 1963. A revised classification of the blennioid fishes of the American family Chaenopsidae. *Univ. Calif. Publ. Zool.*, 68:iv + 1–133, 11 figs., 15 plates.

Strong, A. M., and G. D. Hanna. 1930. Marine mollusca of the Revillagigedo Islands, Mexico. *Proc. Calif. Acad. Sci., Ser. 4*, 19(2):7–12.

Vinton, K. W. 1951. Origin of life on the Galápagos Islands. *Amer. J. Sci.*, 249:356–376, 2 figs., 2 plates.

Walker, B. W. 1960. The distribution and affinities of the marine fish fauna of the Gulf of California. Symposium: the biogeography of Baja California and adjacent seas. *Syst. Zool.*, 9(3–4):123–133, 1 fig.

———. 1966. The origins and affinities of the Galápagos shorefishes. *in* R. I. Bowman (editor), *The Galápagos.* University of California Press, Berkeley, pp. 172–174.

Wiggins, I. W. 1966. Origins and relationships of the flora of the Galápagos Islands. *in* R. I. Bowman (editor), *The Gálapagos.* University of California Press, Berkeley, pp. 175–182.

Williams, H. 1966. Geology of the Galápagos Islands. *in* R. I. Bowman (editor), *The Galápagos.* University of California Press, Berkeley, pp. 65–70.

Woodward, S. P. 1851–1856. *A manual of the mollusca.* John Weale, London, viii + 1–486 pp., 24 plates + 24 pp., 1 map.

Zullo, V. A. 1966. Zoogeographic affinities of the Balanomorpha (Cirripedia: Thoracica) of the Eastern Pacific. *in* R. I. Bowman (editor), *The Galápagos.* University of California Press, Berkeley, pp. 139–144, 1 fig.

The
Western
Atlantic Region

It is one of the most familiar facts in Natural History, that many countries possess a distinct Fauna and Flora, or assemblages of animals and plants peculiar to themselves; and it is equally true, though less understood, that the sea also has its provinces of animal and vegetable life.

S. P. Woodward in A Manual of the Mollusca, *1856.*

From Bermuda, southern Florida, and the southwestern Gulf of Mexico, a rich tropical fauna extends down the continental coast and through the West Indies. This tropical fauna is found along the South American mainland all the way to Cape Frio near Rio de Janeiro and is continued offshore to Fernando de Noronha and Trindade Islands.

Traditionally, the tropical Western Atlantic has been regarded as a single, homogeneous faunal area. This is understandable because many of the species are wide-ranging and may be found from one end of the region to the other. For example, when the distribution of the fishes of Florida was studied (Briggs 1958:237), it was found that 42.8 percent of the shore species had also been recorded from the Atlantic coast of South America.

As recently as 1953, Ekman regarded this region as "too little known in detail" to separate it into zoogeographic subdivisions. Yet now, there are indications from both revisional and faunal works that the Western Atlantic should be divided into three parts, each with its own characteristic assemblage.

At this date, the fauna of the northern end of the region is by far the best known—especially that of Bermuda, Florida and, to a certain extent, the Bahamas. The available systematic literature would lead one to believe

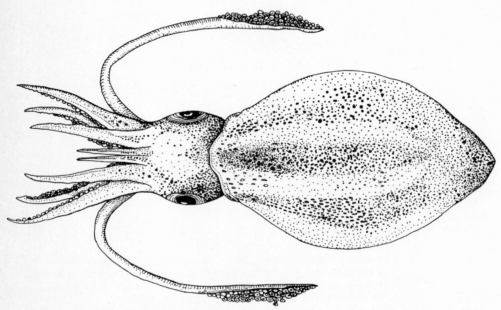

Figure 3-1   The squid (*Sepioteuthis sepioidea*). A shallow-water species apparently confined to the Western Atlantic Tropics (Caribbean and West Indian Provinces). After Boone (1933).

that the greatest concentration of species is in Florida waters. But theoretically, the greatest amount of speciation should occur in the area of the most stable temperature, provided that a sufficient variety of habitats were available. This means that the richest shore fauna should be found along the northeastern South American coast and perhaps up into the Lesser Antilles. In the winter months, the surface temperature of the ocean remains considerably higher in this area than in the region north of the Caribbean Sea.

The eastward projection or "hump" of Brazil splits the warm South Equatorial Current into two streams. One runs southward (the Brazil Current) and remains within the South Atlantic Gyre system. The other runs northwest parallel to the shore and picks up speed as it becomes reinforced by flow from the North Equatorial Current. It runs strongly into the southern Caribbean and on through to the Yucatán Channel. From there, the main stream makes a rather abrupt turn to the east and, now named the Florida Current, surges rapidly through the Florida Straits.

The Florida Current is called the Gulf Stream as it leaves the coast of the eastern United States to head across the North Atlantic. The Gulf Stream has a profound effect upon the distribution of shore animals in the Western Atlantic. Because of its tendency to transport larvae and occasional adults, many tropical forms are left stranded along the inhospitable shores of northeastern North America.

The position of Bermuda is unique, for nowhere else in the world does a tropical fauna occur at such a high latitude (32°15′). The swiftly flowing and often erratic Gulf Stream is Bermuda's lifeline, carrying warmth, food, and genetic reinforcement for a thriving tropical community across some 900 miles of ocean.

At the southern end of the region, the island of Fernando de Noronha is probably close enough to the continental shelf to account for its principal faunal relationship, but Trindade is well isolated and, apparently, is dependent on the Brazil Current.

The Provinces
Caribbean
Florida Peninsula

The northern boundary for the Caribbean Province on the east coast of the United States is also the dividing line between the warm-temperate and tropical marine faunas. Many opinions about the boundary location have been expressed. Forbes (1856:map) indicated that such a boundary, separating his "Caribbean" and "Carolinian" Provinces, was located at about the border between Florida and Georgia. Parr

(1933:62) suggested that we should expect to find a critical and perhaps limiting point for the winter migrations of subtropical species in the vicinity of Cape Kennedy, Florida. Other writers, who have also considered the fauna in general, appear to be in agreement (Stephenson and Stephenson 1950:398, Hedgpeth 1953:201, 1957:map).

Ekman (1953:46) considered the tropical boundary to be located at Cape Hatteras or a little to the south, and Schilder (1956:86) also indicated Cape Hatteras. More recently, Cerame-Vivas and Gray (1966), in their study on the benthic macroinvertebrates of the North Carolina shelf, noted that the inshore fauna was warm-temperate (Carolinian) but that the offshore fauna was tropical. The same pattern occurs in the benthic marine algae, for Humm (1969:48) wrote that the shelf of the open sea beyond a depth of 50 feet, in the area between Cape Kennedy and Cape Hatteras, was occupied by a tropical flora. He stated further that these plants represented a northward extension in waters that are not greatly cooled by winter atmospheric temperatures.

Other opinions, based on the distribution of certain animal groups, have been expressed. Woodward (1856), in his *Manual of the Mollusca*, showed a boundary at the tip of the Florida peninsula. Ortmann (1896:map), who worked on the decapod crustaceans, indicated a boundary at Cape Hatteras. Johnson (1934), who compiled a distributional list of the mollusks of the northwestern Atlantic, said that the older works of Fischer (1880–1887) and Tryon (1882–1884) indicated Cape Kennedy, but Johnson himself observed that the Caribbean Province should extend north to Jupiter Inlet, Florida, if not to Cape Kennedy. Clench (1945:32) indicated a preference for Jupiter Inlet.

More recent observations by malacologists include those of Abbott (1957:7), who considered his temperate "Appalachian Province" to extend down to the northern third of Florida; Warmke and Abbott (1961:319), who showed an onshore tropical fauna extending north to about the same place but an offshore fauna extending all the way to Cape Hatteras; and three others (Rehder 1954:472, Pulley 1952, Coomans 1962:98) who indicated boundaries at Cape Kennedy. Finally, Work (1969:24) felt that a better line of demarcation between the molluscan faunas could be identified at St. Lucie Inlet near Stuart, Florida. However, Work also noted that Cape Hatteras served as a northern limit for an offshore mixed Carolinian and West Indian fauna.

Bayer (1961:328) found a faunal break for octocorals just north of Palm Beach, Florida. Gray, Downey, and Cerame-Vivas (1968), who worked on the sea stars of North Carolina, found 13 species that occurred only

Figure 3-2   Caribbean, Brazilian, and West Indian Provinces.

in a northward extension of the Caribbean Province along the outer shelf. They also found that these tropical species tended to range slightly past Cape Hatteras northward to about Oregon Inlet.

In regard to fishes, it is apparent that a number of species breaks occur in northeastern Florida, especially in the area from the mouth of the St. John's River to St. Augustine (Briggs 1958). On the other hand, Christensen (1965), who conducted an ichthyological survey in the Jupiter Inlet area, identified several temperate species that extended this far south and also called attention to a large number of tropical species that apparently had reached their northern boundary.

The question of how far the warm-temperate fauna extends southward along the eastern coast of Florida or, conversely, how far north the tropical fauna extends is a complicated matter. That there is such a faunal change is clear, but the boundary is obviously not a sharp one. Some tropical species drop out quite far to the south at such places as Palm Beach, Jupiter Inlet, or St. Lucie Inlet while others reach Cape Kennedy, and some go even farther north. The situation is further complicated by the fact that many of the tropical species inhabiting the outer shelf range all the way to Cape Hatteras.

In regard to the fauna of the inner shelf, Cape Kennedy is an intermediate point in a rather lengthy area of change and is also the site of an upwelling and movement of cooler waters near shore. Since we know that many species terminate their ranges here, it seems reasonable to consider it as the most likely boundary. At the same time, we need to keep in mind that the fauna changes with depth and that tropical species can exist far to the north if the water temperature remains suitable. For example, Cerame-Vivas and Gray (1966) recorded a temperature of 19.5°C at 80 fathoms off Cape Hatteras in early March— the coldest time of the year.

The next boundary to be considered is also a temperate-tropical division that is probably located somewhere on the Gulf Coast of Florida. There is a general agreement among the students of animal distribution that the northern coast of the Gulf of Mexico comprises a warm-temperate rather than a tropical region. Of the modern workers, only Ekman (1953:53), Schilder (1956:86), Van Name (1954:495), and Taylor (1955:261, marine algae) considered the whole of the Gulf to be tropical.

Although the 1952 zoogeographic study by Pulley (Ph.D. thesis), based on bivalve distribution, divided the Gulf of Mexico into five different provinces—too many for the shore fauna in general—it is interesting to

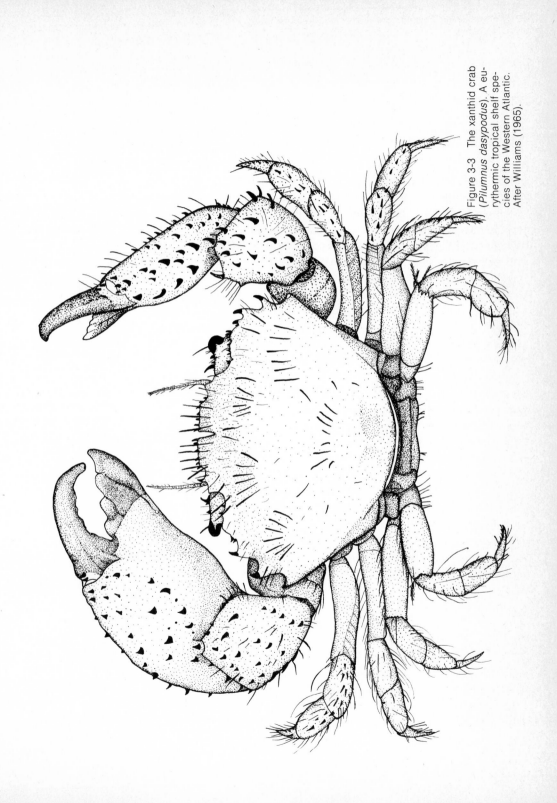

Figure 3-3 The xanthid crab (*Pilumnus dasypodus*). A eurythermic tropical shelf species of the Western Atlantic. After Williams (1965).

find that he (p. 45) considered Cape Romano on the Florida west coast to be the southern limit for species that, on the Atlantic coast, are restricted to the region north of Cape Kennedy. Hedgpeth (1953:201), in discussing the general distribution of invertebrates, noted that the Caribbean fauna must certainly extend as far north as Sanibel Island or Tampa.

On the other hand, Tabb and Manning (1962), who conducted an ecological survey of northern Florida Bay, at the very tip of mainland Florida, found that the majority of species were more common in the waters along the west coast of Florida and the northern Gulf of Mexico than they were in Biscayne Bay, only about 50 miles to the east. They compared the 239 species of invertebrates with the 116 species that had been reported from Soldier Key (in Biscayne Bay) by Voss and Voss (1955) and found that only 15 species, or 4.2 percent of the total, occurred in both areas.

Rehder (1954:472) considered Tampa Bay to be the northern limit for the tropical molluscan fauna, and Abbott (1957:7) considered the Tampa Bay area to be a special subregion (the Conradian) because of the presence of several endemic molluscan species. The map published by Warmke and Abbott (1961:319) shows the tropical molluscan fauna extending northward all the way to Cedar Key, and Coomans (1962:map) also indicated a boundary here. Finally, Work (1969), who also studied mollusks, called the stretch of coastline from just north of Clearwater to near the mouth of the Ancilla River (southeast of St. Marks) the "disjunct Astraea zone" and considered its fauna to be more tropical than that of southwest Florida.

Earle (1969), who published a detailed study of the Phaeophyta of the eastern Gulf of Mexico, observed that there was no clear-cut temperature break or series of distribution patterns along this coast comparable to those of the Cape Cod or Cape Hatteras–Beaufort area. She noted further that the Tampa Bay area is the northern known limit for eight tropical species and the southern known limit for seven temperate species.

The fish fauna of Cedar Key was first studied by Reid (1954) who was followed by others (Caldwell 1954, 1955, 1957; Berry 1958a, 1958b). The fishes of Alligator Harbor on Apalachee Bay, a locality some 95 miles north of Cedar Key, are fairly well known (Joseph and Yerger 1957, Yerger 1961). A thorough ecological study of the fishes of Tampa Bay was carried out by Springer and Woodburn (1960). These works all indicate that the resident shallow-water fishes demonstrate a warm-temperate facies similar to other localities in the northern Gulf of

Mexico. Furthermore, a number of the fishes taken in northern Florida Bay by Tabb and Manning (1961) proved to be warm-temperate species, although some were not common except during times of exceptionally cold weather.

As is demonstrated by the above information, the west coast of Florida supports a complex biotic assemblage. From north to south, warm-temperate species reach their range limits and tropical species make their appearance at various points. There is also considerable evidence of faunal change with depth, the tropical species being more numerous on the deeper parts of the shelf. In some places, there are considerable seasonal changes involving an inshore movement of tropical species during the warm months of the year.

In order to identify the area where the greatest number of range terminations takes place, and thus locate the zoogeographic boundary, distributional studies of the various marine animal groups that are common along the Florida west coast need to be made. At this time, and considering especially the Florida Bay studies of Tabb and Manning (1961, 1962), it appears that the boundary for the inshore fauna may occur quite far to the south. Accordingly, it is suggested that Cape Romano, just south of Naples, be considered the boundary—at least until more conclusive data becomes available.

Southern Portion The southern and longest portion of the Caribbean Province extends along the continental coastline of Central and South America from Cape Rojo (just below Tampico), Mexico to eastern Venezuela north of the Orinoco delta. Although the location of a northern boundary at Cape Rojo, in the western Gulf of Mexico, does not represent a universal agreement among marine biologists, there are persuasive reasons for its placement here. Hedgpeth (1953, 1957), in his general zooge-ographic works, depicted the boundary near the United States–Mexican border. But Pulley (1952:32), in addition to his studies on the distribu-tion of bivalve mollusks, observed that Cape Rojo is the northern limit for the shallow-water, West Indian corals and for several species of gastropods. He also remarked (p. 51) that there was apparently no great faunal change between Brownsville, Texas and Tampico, Mexico.

Rehder (1954:472), in his review of the molluscan fauna of the Gulf of Mexico, thought that the northern boundary for the Caribbean fauna should be placed at Corpus Christi Bay, Texas. The book on Caribbean seashells by Warmke and Abbott (1961) is helpful in this regard, for the authors took pains to illustrate a number of typical distribution patterns. From these, it can be seen that a large number of tropical species do extend northward to or near the Cape Rojo vicinity. It is also apparent that other species of tropical mollusks extend only as far as Veracruz or the tip of the Yucatán peninsula.

In regard to the shore fishes, both Gunter (1952:38) and Hildebrand (1955:225) emphasized that the fauna of the Mexican coast north of Tampico is about the same as that found along the Texas coast. Hildebrand, Chávez, and Compton (1964) reported on the fishes of Alacranes Reef just north of the Yucatán peninsula. They found this fauna to be very closely related to that of the Tortugas Islands, Florida. Also in this report, the authors described a collection taken from Cape Rojo and noted that the nearby Blanquilla Reef was the northernmost surface reef (with living corals) in the western Gulf of Mexico. The tropical nature of the fish fauna at Cape Rojo and southward appears to contrast sharply with the more temperate complex found at Tampico and northward.

In the deeper waters of the shelf off the Texas coast, where there is protection against the colder, winter surface temperatures, and where a firm substrate exists, living corals and a relatively rich associated fauna may be found. During the warm months of the year, I have observed that the shallow waters, particularly around piers and jetties, become temporarily colonized by tropical fishes (especially young) that probably originate from the permanent, offshore populations. So, the seasonal changes in the fauna that take place in the inshore waters of the Texas coast are very similar to those that occur along the northwest coast of Florida.

Brazilian

Beginning with the Orinoco River delta in Venezuela, almost the entire northeastern coast of South America as far as the vicinity of Fortaleza, Brazil, is virtually devoid of coral reefs (Wells 1957:map). Instead, there are vast stretches of mud bottom, and near the mouths of the great rivers such as the Orinoco, Amazon, Tocantins, and Parnaíba, the shallow waters have greatly reduced salinities. Thus, the coral reefs of Brazil and their associated faunas are separated from those of the Caribbean area by a markedly different environment occupying a straight-line distance of about 1,800 miles.

Although one may again find many familiar Caribbean species inhabiting the Brazilian reefs, a significant endemic element is encountered, and as a matter of fact, the coral species themselves are quite different. Vaughan and Wells (1943) reported the presence of 14 species of hermatypic corals between Alcantara (2°S) and Rio de Janeiro and

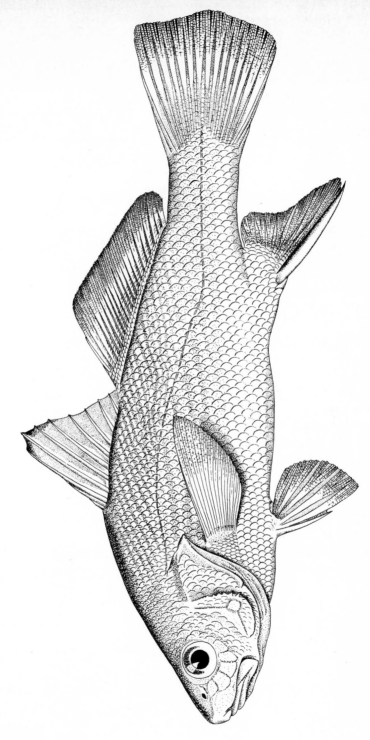

Figure 3-4 The croaker (*Bairdiella ronchus*). A tropical shelf species found in the Caribbean Province of the Western Atlantic.

noted that 10 of them, including 1 genus, were distinct from those of the Caribbean. More recently, Bayer (1961) recognized a "Brazilian" coral fauna that extended northward as far as Trinidad. Vannucci (1951:123) investigated the hydrozoans of the Brazilian coast and found that 26 of 116 species were endemic. Along this same line, Schilder (1956:75) recognized for the fauna in general a "Brasilianisch" subprovince occurring to the south of the "Antillisch" area. Work (1969:623), in his publication on the mollusks of Los Roques, Venezuela, observed that a fairly large percentage of the West Indian molluscan fauna extends southward beyond Trinidad but here it becomes intermingled with a remarkable, endemic molluscan fauna.

South of the last outpost of the Caribbean fauna—the coral reefs in the Gulf of Paria in Venezuela—the shore fish fauna undergoes a sudden transformation. The butterfly fishes (Chaetodontidae), parrotfishes (Scaridae), damselfishes (Pomacentridae), wrasses (Labridae), and other typical reef fishes become scarce, while there is a corresponding increase in the species of sea catfishes (Ariidae), croakers (Sciaenidae), and other groups that are better adapted to the changed environment.

Of the 20 species of sea catfishes that occur along the Brazilian coast, only 3 range as far north as the Gulf of Venezuela (Schultz 1944) and but 1 extends to the northern end of the tropical Atlantic. According to Travassos and Paiva (1957), there are 36 species of croakers known from Brazil, and apparently, most of these do not range north of the Brazilian Province. There is also a high percentage of endemism in the anchovies (Engraulidae), especially in the genera *Anchoa* (Paiva Carvalho 1950a) and *Anchoviella* (Paiva Carvalho 1951), and in the toadfishes (Batrachoididae, Collette 1966).

Despite the above-mentioned works, the fish fauna of the Brazilian Province is still very poorly known. It seems apparent that sufficient endemism exists so that there is little doubt of the need to consider it separate from the Caribbean fauna to the north. Certainly, the endemism amounts to more than 10 percent, and conceivably, it could be as high as 25 percent. There is a similar dearth of information on the main groups of invertebrates.

Ekman (1953:47) decided that the tropics ended "in the neighborhood of Rio or a little to the south," while Schilder's (1956:75) map indicates an area farther south (apparently near Port Alegre), as does the 1957 map of Hedgpeth. On the basis of molluscan distribution, Woodward (1856:378) found the boundary to be at Rio, Abbott (1957:7) made vague reference to "the southernmost shores of Brazil," and Pulley (1952:fig. 2) indicated a locality far to the north (Recife). Ortmann (1896:map), who investigated the decapod crustaceans, placed the boundary much farther south at the Río de la Plata. From ascidian distribution, Van Name (1945) considered the "West Indian" fauna to extend south to at least Santos, Brazil.

More recently, several authors who work in South America have published information indicating that the tropical fauna of the Brazilian coast extends southward to Cape Frio (a few miles to the west of Rio de Janeiro). At least, this appears to be true for such diverse groups as the hydrozoans (Vannucci 1951, 1964), echinoderms (Bernasconi 1964), mollusks (Stuardo 1964), penaeid shrimps (Boschi 1964), and fishes (Lopez 1964). The coral reefs of the southwestern Atlantic also terminate at about the same place (Wells 1957:map).

Cape Frio is selected as the southern boundary for the Brazilian Province and, at the same time, represents the southern edge of the Western Atlantic tropics. Its designation rests on our current knowledge about the distribution of several important groups of marine shore animals including the hermatypic corals, hydrozoans, echinoderms, mollusks, penaeid shrimps, and fishes.

Trindade The island of Trindade lies in the southwestern Atlantic about 700 miles off the coast of Brazil at latitude 20°28′S, longitude 31°46′W. Although small collections were reported on by Nichols and Murphy (1914), Miranda Ribeiro (1919), and Paiva Carvalho (1950), the fish fauna is still poorly known, and the extent of endemism difficult to assess. Two endemic species were described by Nichols and Murphy and two more by Miranda Ribeiro in the aforementioned works. More recently, a fifth endemic was discovered by Ypiranga Pinto (1957).

There is reason to cast doubt on most of the Trindade endemics that have so far been described. Springer (1959:291) expressed the opinion that Ypiranga Pinto's species (*Labrisomus trinidadensis*) was the same as the common transatlantic form *L. nuchipinnis* and also (1962:432) placed *Ophioblennius trinitatis* Miranda Ribeiro in the synonymy of *O. atlanticus atlanticus*. In addition, the figures and descriptions of the two new forms described by Nichols and Murphy (op. cit.) appear to lack distinctive characters in the light of information in the more recent literature. The invertebrate fauna is apparently very poorly known.

Other Islands Fernando de Noronha at 3°50′S, 32°25′W, about 200 miles off the Brazilian coast, and Saint Paul Rocks at 0°23′N, 29°23′W are both oceanic localities of considerable interest. Unfortunately, very little is known about their faunas.

Boulenger (1890) described a few fishes from Fernando de Noronha, and de Souza Lopes and Alvarenga (1955) discussed a mollusk collection. The latter authors found the great majority of species to be shared with the mainland coast of Brazil but also noted some endemism. In addition, they called attention to some evidence of relationship to St. Helena and Ascension Islands as a result of the influence of the South Equatorial Current.

Apparently, no more than an occasional specimen has ever been taken from the vicinity of Saint Paul Rocks. Its shore fauna should be similar to that of Ascension, for it appears that it would also be affected by the South Equatorial Current.

Although the West Indian Province includes an extensive geographic area, it consists entirely of islands. Bermuda is an isolated northern outpost while the main portion is an archipelago stretching from the Bahamas to Grenada—the southernmost of the Windward Island group. The southern Netherlands West Indies (Aruba, Curaçao, Bonaire) are still a distributional enigma; a number of Antillean species have been found there as well as some that are otherwise restricted to the mainland coastline.

West Indian

As the result of recent investigations, particularly work on the fishes of the Florida Keys and the Bahamas, it becomes increasingly apparent that the Straits of Florida—only 60 miles wide—are an important barrier to the dispersal of marine shore animals. Böhlke and Robins (1959:37) state, "In their respective field work in the shallow-water areas in the Bahamas and southern Florida, the writers repeatedly have found marked differences in the ichthyofaunas of the two regions."

In their revision of the clinid fish genus *Starksia*, Böhlke and Springer (1961) discussed eight West Indian species, four of them found in the Bahamas; of these, only one ranged across the Florida Straits. Böhlke and Robins (1960) discovered that in the gobioid genus *Lythrypnus* only one species, *L. spilus*, was common to the Florida Keys and the West Indies, while three others were found in the Bahamas and one other was a Florida endemic. Springer (1958) found seven species of the clinid genus *Malacoctenus* in the Bahamas, but only three of them reach Florida waters. Smith (1959:323), as the result of his extensive work with the large sea basses (*Epinephelus* and related genera), concluded that the Gulf Stream appears to have acted as an effective barrier, allowing two centers of differentiation, one in the Gulf of Mexico and the other in the West Indies.

Evidence for a barrier at the Florida Straits is by no means confined to the fishes. Following an ecological survey of Bimini, Bahamas, Voss and Voss (1960:113) stated,

"A careful study of these differences seems to indicate that while the Bahamas and south Florida lie directly opposite each other and both are tropical, the Florida Keys have been populated largely by fortuitous invasion and by a route around the west end of Cuba by way of the Yucatán Channel. This route appears to have limited the Florida Caribbean fauna in respect to certain of its shore species while the Bahamas have been populated around the east end of Cuba and directly from the north Cuban coast, and by easy stepping stones of the southern Bahamas."

The book of Warmke and Abbott (1961) shows that an impressive number of Caribbean mollusk species do not reach the Florida Keys but do range extensively through the West Indies and, some of them, north along the mainland coast to the Yucatán Peninsula or into the southwestern Gulf of Mexico. Mayr (1954:14) has remarked on the inability of the two sea urchin genera *Arbacia* and *Encope* to jump the gap from the American mainland to the Bahamas.

Why should a relatively insignificant distance of 60 miles present more of a dispersal problem to the shore faunas than, for example, the 900-mile stretch from the Bahamas to Bermuda? A study by Wennekens (1959) has shown that the Florida Current (the portion of the Gulf Stream system that passes through the Straits of Florida) is not only very fast flowing, but its water mass properties are such that very little mixing takes place. According to Wennekens, Caribbean or Yucatán water, identified by its well-defined salinity maximum, is found along the entire insular margin of the Florida Current. A continental edge water, also easily identified, becomes differentiated in the eastern Gulf and is found along the continental margin of the Florida Current throughout the length of the Straits.

Since the dispersal of many shore animals, especially the young stages, depends on transport by ocean currents, it can be seen why the gap between Florida and the Bahamas would be difficult to negotiate. Even if an organism were picked up by the current from the northern coast of Cuba, it would find itself in the Caribbean water mass and would tend to be carried along the insular margin of the Florida Current.

Heretofore, marine zoologists in general have been more impressed by the similarity of the faunas of either side of the Florida Straits rather than by any differences that may have been found. Consequently, the Florida tropics have almost invariably been considered an extension of a "West Indian" or "Caribbean" Province. However, it can now be said that considerable evidence exists for a zoogeographic barrier between Florida and the Bahamas. Its effectiveness depends not so much on a spatial separation as on the rapid flow of the Florida Current combined with a lack of mixing of the component water masses.

The West Indian Province extends close to the Caribbean Province in two other places. One is the Yucatán Channel, about 130 miles wide, and the other is apparently the passage between Grenada and Trinidad (or Tobago), about 75 miles across. As in the case of the Florida Straits, the effectiveness of these barriers probably depends primarily on the strong current movement.

The existence of a separate West Indian Province as defined here has not been previously recognized. Schilder (1956:75) depicted an "Antillisch Unterprovinz" but it covered all of the Western Atlantic tropics north of Brazil. Bayer (1961:343), in discussing coral distribution, referred to an "Antillean Region," but this included the Florida Keys. Rehder (1963:117), after studying the mollusks, recognized two general regions—one from southern Florida and the Bermudas to the island of Grenada, and the other from the Yucatán Peninsula (or possibly Veracruz) to eastern Brazil including Trinidad, Tobago, and the islands of the Venezuelan coast.

In zoogeography, the ultimate proof, as far as the determination of provinces is concerned, is in the extent of endemism. As more ichthyological research is accomplished, it becomes clearer that the West Indies are blessed with a fauna that has interesting indigenous characteristics. The work on the fishes of the Bahamas (Böhlke and Chaplin 1968), the most comprehensive ever done for a tropical shelf area, included descriptions of 466 shore species; of these, 87, or about 19 percent, have not been taken outside the West Indies. Twenty-four species of seven-spined gobies (Gobiidae) occur in the West Indies, but only 11 of them appear to be shared with the mainland (Böhlke and Robins 1968). There are 12 known species of Antillean clingfishes (Gobiesocidae), and 7 of them are probably endemic (although 2 do extend to Curaçao).

Some time ago, H. L. Clark (1919) made a detailed analysis of the distribution of the littoral echinoderms of the West Indies. According to his data, 18 (15.5 percent) of the total of 116 species may be considered endemics. The more recent publication on the brittle stars

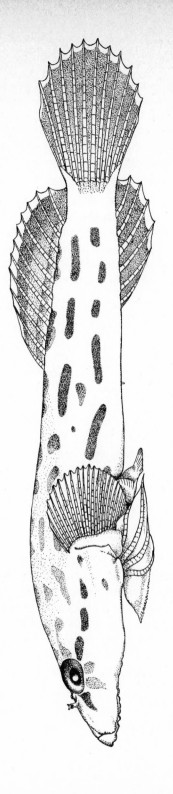

Figure 3-5 The clingfish (*Gobiesox lucayanus*). A tropical shelf species confined to the West Indian Province of the Western Atlantic.

(Ophiuroidea) of the Lesser Antilles by Parslow and Clark (1963) contains a table showing the distribution of all the West Indian species that have been taken at a depth of 10 fathoms or less. This demonstrates that 13 of 51 species, or 25.5 percent, are probably endemic. The extensive work of Warmke and Abbott (1961) covered 760 species of West Indian shallow-water mollusks, and judging from the ranges given, 185 (24.3 percent) of these should be classed as endemics.

Since the shore fauna of the southern part of the Caribbean Province is so poorly known, it is entirely possible that many species thus far considered to be West Indian endemics will turn up somewhere along the extensive continental coastline between eastern Venezuela and the southern Gulf of Mexico. However, it is difficult to imagine that this would occur to such a degree that the island-chain fauna could no longer be considered distinct.

Bermuda

Bermuda, actually a tiny archipelago consisting of about 360 small islands located at 32°14′ to 32°25′N and 64°38′ to 64°52′W, lies about 900 statute miles from the North Carolina coast. Its age has been estimated as probably Eocene or Oligocene (Wilson 1963). In general, the invertebrate fauna does not appear to be distinctive. Ekman (1953:54) observed that all of the 50 species of crabs and the 40 species of starfishes, brittle stars, and sea urchins are also found in the West Indies. Work (1969:619) noted that the mollusks of Bermuda, comprising a depauperate population, were West Indian in composition.

At this date, the shore fishes of Bermuda are reasonably well known. As nearly as can be determined, there is a total of 265 species; of these, 13, or about 5 percent, are endemic (Collette 1962). For a well-isolated oceanic island group that dates back probably to the early Tertiary, the existence of so little endemism is surprising.

It has been pointed out that Bermuda lies close to the northern boundary of the tropics and that its tropical fauna was probably decimated by the drop in surface temperature that took place during the most recent Pleistocene (Wisconsin) glaciation. The approximate 11,000 years that have passed since the glaciation have allowed the accumulation of a relatively rich fauna, but this amount of time has not been sufficient for much evolutionary change (Briggs 1966). Therefore, Bermuda, despite its great age and isolated location, shows a very low level of endemism. For this reason, it is considered an integral part of the West Indian Province.

Literature Cited

Abbott, R. 1957. The tropical western Atlantic province. *Proc. Phila. Shell Club,* 1(2):7–11.

Bayer, F. M. 1961. The shallow-water Octocorallia of the West Indian Region. *Studies on the fauna of Curacao and other Caribbean Islands,* 12:1–373, 101 figs., 28 plates (Martinus Nijhoff, The Hague).

Bernasconi, I. 1964. Distribución geografica de los equinoideos y austral de Sudamerica. *Bol. Inst. Biol. Mar., Mar del Plata, Argentina,* (7):43–49, 1 fig.

Berry, F. H. 1958a. Additions to the fishes known from the vicinity of Cedar Key, Florida. *Quart. J. Florida Acad. Sci.,* 20(4):232.

———. 1958b. Additions to the fishes of Cedar Key, and a list of Gulf of Mexico Carangidae. *Quart. J. Florida Acad. Sci.,* 21(2):190.

Böhlke, J. E., and C. C. G. Chaplin. 1968. *Fishes of the Bahamas and adjacent tropical waters,* Livingston, Wynnewood, Pa., xxxi + 771 pp., illus.

———, and C. R. Robins. 1959. Studies on the fishes of the family Ophidiidae. II. Three new species from the Bahamas. *Proc. Acad. Nat. Sci. Phila.,* 111:37–52, 2 figs., 1 plate.

———, and ———. 1960. Western Atlantic gobioid fishes of the genus *Lythrypnus* with notes on *Quisquilius hipoliti* and *Garmannia pallens. Proc. Acad. Nat. Sci. Phila.,* 112(4):73–101, 2 figs., 3 plates.

———, and ———. 1968. Western Atlantic seven-spined gobies, with descriptions of ten new species and a new genus, and comments on Pacific relatives. *Proc. Acad. Nat. Sci. Phila.,* 120(3):45–174, 21 figs.

———, and V. G. Springer. 1961. A review of the Atlantic species of the clinid fish genus *Starksia. Proc. Acad. Nat. Sci. Phila.,* 113(3):29–60, 16 figs.

Boschi. E. 1964. Los peneidos de Brasil, Uruguay y Argentina. *Bol. Inst. Biol. Mar., Mar del Plata, Argentina,* (7):37–42.

Boulenger, G. A. 1890. Pisces. *in* H. N. Ridley, *Notes on the zoology of Fernando Noronha. J. Linn. Soc. London, Zool.,* 20:483. (not seen).

Briggs, J. C. 1958. A list of Florida fishes and their distribution. *Bull. Florida State Museum, Biol. Sci.,* 2(8):223–318, 3 figs.

———. 1966. Oceanic islands, endemism, and marine paleotemperatures. *Syst. Zool.,* 15(2):153–163, 4 figs.

Caldwell, D. K. 1954. Additions to the known fish fauna in the vicinity of Cedar Key, Florida. *Quart. J. Florida Acad. Sci.,* 17(3):182–184.

———. 1955. Further additions to the known fish fauna in the vicinity of Cedar Key, Florida. *Quart. J. Florida Acad. Sci.,* 18(1):48.

———. 1957. Additonal records of marine fishes from the vicinity of Cedar Key, Florida. *Quart. J. Florida Acad. Sci.,* 20(2):126–128.

Cerame-Vivas, M. J., and I. E. Gray. 1966. The distributional pattern of the benthic invertebrates of the continental shelf off North Carolina. *Ecology,* 47(2):260–270, 6 figs.

Christensen, R. F. 1965. An ichthyological survey of Jupiter Inlet and Loxahatchee River, Florida. M.S. thesis, Florida State University, viii + 318 pp., 1 fig.

Clark, H. L. 1919. Distribution of the littoral echinoderms of the West Indies. *Papers Dept. Mar. Biol., Carnegie Inst.,* 13(3):51–74, 3 plates.

Clench, W. J. 1945. The West Indian fauna in southern Florida. *Nautilus*, 59(1):32–33.

Collette, B. B. 1962. *Hemiramphus bermudensis*, a new halfbeak from Bermuda, with a survey of endemism in Bermudian shore fishes. *Bull. Marine Sci. Gulf Carib.*, 12(3):432–449, 1 fig.

――――. 1966. A review of the venomous toadfishes, subfamily Thalassophyrninae. *Copeia*, (4):846–864, 12 figs.

Coomans, H. E. 1962. The marine mollusk fauna of the Virginian area as a basis for defining zoogeographical provinces. *Beaufortia*, 9(98):83–104.

Dahlberg, M. D. 1970. Atlantic and Gulf of Mexico menhadens, genus *Brevoortia* (Pisces: Clupeidae). *Bull. Florida State Museum, Biol. Sci.*, 15(3):91–162, 8 figs.

Earle, S. A. 1969. Phaeophyta of the eastern Gulf of Mexico. *Phycologia*, 7(2):71–254, 126 figs.

Ekman, S. 1953. *Zoogeography of the sea.* Sidgwick & Jackson, London, xiv + 417 pp., 121 figs.

Fischer, P. H. 1880–1887. *Manuel de Conchyliologie et de Paléontologie Conchyliologique.* Librairie F. Savy, Paris, xxiv + 1369 pp., 1138 figs., 24 plates, 1 map.

Forbes, E. 1856. Map of the distribution of marine life. *in* Alexander K. Johnston, *The physical atlas of natural phenomena* (new edition). Edinburgh and London, W. and A. K. Johnston, plate no. 31.

Gray, I. E., M. E. Downey, and M. J. Cerame-Vivas. 1968. Sea-stars of North Carolina. *Fish. Bull., U.S. Fish Wildlife Serv.*, 67(1):127–163, 40 figs.

Guest, W. C., and G. Gunter. 1958. The sea trout or weakfishes (genus *Cynoscion*) of the Gulf of Mexico. *Gulf States Marine Fish. Comm. Tech. Sum.*, (1):1–40, 2 figs.

Gunter, G. 1952. Records of fishes from the Gulf of Campeche, México. *Copeia*, (1):38–39.

Hedgpeth, J. W. 1953. An introduction to the zoogeography of the northwestern Gulf of Mexico with reference to the invertebrate fauna. *Publ. Inst. Marine Sci. Univ. Texas*, 3:107–224, 46 figs.

――――. 1957. Marine biogeography. *in* Treatise on marine ecology and paleoecology. vol. 1. *Mem. Geol. Soc. Am.*, (67):359–382, 16 figs., 1 plate.

Hildebrand, H. H. 1955. A study of the fauna of the pink shrimp (*Penaeus duorarum* Burkenroad) grounds in the Gulf of Campeche. *Pub. Inst. Marine Sci. Univ. Texas*, 4(1):169–232.

――――, H. Chávez, and H. Compton. 1964. Aporte al conocimiento de los peces del Arrecife Alacranes, Yucatán (México). *Ciencia, Méx.*, 23(3):107–134.

Humm, H. J. 1969. Distribution of marine algae along the Atlantic coast of North America. *Phycologia*, 7(1):43–53.

Johnson, C. W. 1934. List of the marine mollusca of the Atlantic coast from Labrador to Texas. *Proc. Boston Soc. Nat. Hist.*, 40(1):1–204.

Joseph, E. B., and Yerger, R. W. 1957. *The fishes of Alligator Harbor, Florida, with notes on their natural history.* Florida State University Studies No. 22, Oceanog. Inst. Papers No. 2, pp. 111–156.

Lopez, R. B. 1964. Problems de la distribucion geografica de los peces marinos suramericanos. *Bol. Inst. Biol. Mar., Mar del Plata, Argentina*, (7):57–62, 1 fig.

Mayr, E. 1954. Geographic speciation in tropical echinoids. *Evolution*, 8(1):1–18, 7 figs.

Miranda Ribeiro, A. de. 1919. A fauna da I. da Trindade. *Arch. Museum Nac. Rio de Janeiro*, 22:171–194 (not seen).

Nichols, J. T., and R. C. Murphy. 1914. Fishes from South Trinidad Islet. *Bull. Amer. Museum Nat. Hist.*, 33(20):261–266, 3 figs.

Ortmann, A. E. 1896. *Grundzüge der marinen Tiergeographie.* Gustav Fischer, Jena, iv + 96 pp., 1 map.

Paiva Carvalho, J. de. 1950a. Engraulídeos brasileiros, do gênero *Anchoa. Bol. Inst. Paulista Oceanog.*, 1(2):43–69, 2 plates.

———. 1950b. Resultados científicos do cruzeiro do "Baependí" e do "Vega" à I. da Trindade-Piexes. *Bol. Inst. Paulista Oceanog.*, 1(1):97–133, 4 figs.

———. 1951. Engraulídeos brasileiros do genero *Anchoviella. Bol. Inst. Paulista Oceanog.*, 2(1):41–66, 2 plates, 1 map.

Parr, A. E. 1933. A geographic-ecological analysis of the seasonal changes in temperature conditions in shallow water along the Atlantic coast of the United States. *Bull. Bingham Oceanog. Lab.*, 4(3):1–90, 28 figs.

Parslow, R. E., and A. M. Clark. 1963. Ophiuroidea of the Lesser Antilles. *Studies Fauna Curaçao Carib.*, 15(67):24–50, 11 figs.

Pulley, T. E. 1952. A zoogeographic study based on the bivalves of the Gulf of Mexico. Ph.D. thesis, Harvard University, 1–215 pp., 8 figs., 19 plates.

Rehder, H. A. 1954. Mollusks. Gulf of Mexico its origin, waters, and marine life. *Fish. Bull. U. S. Fish Wildlife Serv.*, 55(89):469–474.

———. 1963. Contribucion al conocimiento de los moluscos marinos del Archipielago de Los Roques y la Archila. *Mem. Soc. Cienc. Nat. La Salle, Caracas*, 22(62):116–138, 6 figs.

Reid, G. K. 1954. An ecological study of the Gulf of Mexico fishes in the vicinity of Cedar Key, Florida. *Bull. Marine Sci. Gulf Carib.*, 4:1–94.

Schilder, F. A. 1956. *Lehrbuch der Allgemeinen Zoogeographie.* Gustav Fischer, Jena, viii + 150 pp., 134 figs.

Schultz, L. P. 1944. The catfishes of Venezuela, with descriptions of thirty-eight new forms. *Proc. U. S. Nat. Museum*, 94(3172):173–338, 14 plates.

Smith, C. L. 1959. A revision of the American groupers (*Epinephelus* and allied genera). Ph.D. thesis, University of Michigan, xiv + 563 pp., 25 plates, 14 maps.

de Souza Lopes, H., and M. Alvarenga. 1955. Contribuicão ao conhecimento dos moluscos da Ilha Fernando de Noronha, Brasil. *Bol. Inst. Oceanograf., Univ. São Paulo*, 6(1–2):157–196, 3 plates.

Springer, V. G. 1958. Systematics and zoogeography of the clinid fishes of the subtribe Labrisomini Hubbs. *Publ. Inst. Marine Sci., Univ. Texas*, 5:417–492, 4 figs., 7 plates.

———. 1959. A new species of *Labrisomus* from the Caribbean Sea, with notes on other fishes of the subtribe Labrisomini. *Copeia*, (4):289–292, 1 fig.

————. 1962. A review of the blenniid fishes of the genus *Ophioblennius* Gill. *Copeia*, (2):426–433, 4 figs.

————, and K. D. Woodburn. 1960. An ecological study of the fishes of the Tampa Bay area. *Prof. Papers, Florida Board Conserv.*, (1):1–104, 18 figs.

Stephenson, T. A. and Stephenson, Anne. 1950. Life between the tide-marks in North America I. the Florida keys. *J. Ecol.*, 38(2):354–402, 10 figs., 7 plates.

Stuardo B. J. 1964. Distribucion de los moluscos marinos litorales en latinoamerica. *Bol. Inst. Biol. Mar. Mar del Plata, Argentina*, (7):79–91, 1 fig.

Tabb, D. C., and R. B. Manning. 1961. A checklist of the flora and fauna of northern Florida Bay and adjacent brackish waters of the Florida mainland collected through the period July, 1957 through September, 1960. *Bull. Marine Sci.*, 11(4):552–649, 6 figs.

————, and ————. 1962. Aspects of the biology of northern Florida Bay and adjacent estuaries. *Florida Board Conserv., Tech. Ser.*, (39):39–81, 5 figs.

Taylor, W. R. 1955. Marine algal flora of the Caribbean and its extension into neighboring seas. *in Essays in the natural sciences in honor of Captain Allan Hancock*. University of S. Calif. Press, Los Angeles, pp. 259–270, 8 figs.

Travassos, H., and M. P. Paiva. 1957. Lista dos *Sciaenidae* marinhos Brasilieros contendo chave de identificacão e proposta de "nomes vulgares oficiais." *Bol. Inst. Oceanograf. Univ. São Paulo*, 8(1–2):139–164, 4 plates.

Tryon, G. W. 1882–1884. *Structural and systematic conchology*, 3 vols. Tryon, Philadelphia, viii + 312, 430, and 453 pp., 140 plates, map.

Van Name, W. G. 1945. The North and South American ascidians. *Bull. Amer. Museum Nat. Hist.*, 84:1–476, 31 plates.

————. 1954. The Tunicata of the Gulf of Mexico. Gulf of Mexico its origin, waters, and marine life. *Fish. Bull. U. S. Fish Wildlife Serv.*, 55(89):495–497.

Vannucci, M. 1951. Distribiucão dos Hydrozoa até agora conhecidos nas costas do Brasil. *Bol. Inst. Paulista Oceanograf.*, 2(1):105–124.

————. 1964. Zoogeografia marinha do Brasil. *Bol. Inst. Biol. Mar., Mar del Plata, Argentina*, (7):113–121.

Vaughan, T. W., and J. W. Wells. 1943. Revision of the suborders, families, and genera of the Scleractinia. *Geol. Soc. Amer. Spec. Papers*, (44):xv + 1–363, 39 figs., 51 plates.

Voss, G. L., and N. A. Voss. 1955. An ecological survey of Soldier Key, Biscayne Bay, Florida. *Bull. Marine Sci. Gulf. Carib.*, 5(3):203–229, 4 figs.

————, and ————. 1960. An ecological survey of the marine invertebrates of Bimini, Bahamas, with a consideration of their zoogeographical relationships. *Bull. Marine Sci. Gulf Carib.*, 10(1):96–116.

Warmke, G. L., and R. T. Abbott. 1961. *Caribbean seashells*. Livingston Pub. Co., Narberth, Pa., x + 346 pp., 34 figs., 44 plates.

Wells, J. W. 1957. Coral reefs. *in* J. W. Hedgpeth (editor), *Treatise on marine ecology and paleoecology*, vol. 1. *Ecology. Geol. Soc. Amer., Mem.*, 67:609–631, 2 figs., 9 plates.

Wennekens, M. P. 1959. Water mass properties of the Straits of Florida and related waters. *Bull. Marine Sci. Gulf Carib.*, 9(1):1–52, 19 figs.

Wilson, J. T. 1963. Evidence from islands on the spreading of ocean floors. *Nature*, 197(4867):536–538, 4 figs.

Woodward, S. P. 1851–1856. *A manual of the mollusca*. John Weale, London, viii + 1–486 pp., 24 plates + 24 pp., 1 map.

Work, R. C. 1969. Systematics, ecology, and distribution of the mollusks of Los Roques, Venezuela. *Bull. Marine Sci.*, 19(3):614–711, 4 figs.

Yerger, R. W. 1961. Additional records of marine fishes from Alligator Harbor, Florida, and vicinity. *Quart. J. Florida Acad. Sci.*, 24:(2):111–116.

Ypiranga Pinto, S. 1957. Um novo Clinidae da Ilha da Trindade, Brasil. *Bol. Museum Nac. Rio de Janeiro, Zool.*, (163):1–13, 7 figs.

The
Eastern
Atlantic Region

The history of dispersal of animals seems to be primarily the history of successions of dominant groups, which in turn evolve, spread over the world, compete with and destroy and replace older groups. . . .
Philip J. Darlington, Jr. in Area, Climate, and Evolution, 1959

The Eastern Atlantic tropics occupies the smallest area, possesses the poorest fauna, and is the least known of the four tropical regions of the world. This lack of knowledge has been considerably alleviated in the past few years by means of scientific reports from several well-organized expeditions and by work done at coastal research laboratories.

The most important of the recent expeditions to the West African coast in terms of scientific results to date were the Danish *Atlantide* from 1945 to 1946, the Belgian Oceanographic Expedition from 1948 to 1949, the Danish *Galathea* from 1950 to 1952, and the French *Calypso* in 1956. The Institut Français d'Afrique Noire at Dakar, Sénégal, has produced an outstanding series of publications, and notable work has been done on the fauna of Angola by the Portuguese research organization, the Junta de Investigacões do Ultramar.

The location and relatively restricted latitudinal area of the region are due primarily to the effect of the two oceanic gyres of the North and South Atlantic. Of course these are, in turn, maintained by the surface wind systems of the two hemispheres. Along the western coast of Africa, two cool currents flow toward the equator, one from the north (Canary Current) and one from the south (Benguela Current). When these two currents leave the proximity of the coast and turn westward across the ocean, they leave between them an area that can be occupied by tropical water.

The water of the western coast of Africa between Cape Verde and Mossâmedes is tropical because it is heated by the sun and, to a certain extent, because it is fed by the warm Equatorial Countercurrent that flows eastward across the Atlantic between the North and South Equatorial Currents. As the countercurrent approaches the coast it is called the Guinea Current because it usually flows directly into the Gulf of Guinea.

As Longhurst (1962) pointed out in his review of the oceanography of the Gulf of Guinea, the tropical surface water layer above the thermocline is only about 30 to 40 meters deep in the Eastern Atlantic, while in the west, off the South American coast, it may be as deep as 130 to 150 meters. This state of affairs has considerable biological significance, for it means that along the African coast only a relatively small portion of the continental shelf can be occupied by tropical organisms. It also means that in this area some temperate species can, by submergence, rather easily transgress the tropics and achieve an antitropical (bipolar) distribution.

The tropics of the Eastern Atlantic have been referred to as a comparatively poor faunal area because the number of species (and higher taxa) is less than those found in any of the other three regions. The reasons for this situation are the latitudinal restriction of the area (to about 30°), the thin surface layer of tropical water, and the virtual absence of coral reefs. These factors and perhaps severe Pleistocene climatic changes have considerably limited opportunity for speciation.

The Eastern Atlantic Region extends from the Cape Verde Islands and Cape Verde in the north probably as far south as Mossâmedes, Angola. The oceanic islands of St. Helena and Ascension are also included. The region is conveniently divided into two provinces, one for the West African mainland and the Cape Verde Islands and the other for St. Helena and Ascension. Each of these demonstrates a high degree of endemism.

The Provinces
West African

Although some difference of opinion has existed, the northern (mainland) boundary of the West African Province is quite well defined. From the standpoint of the fauna in general, Stephenson (1947:221), Ekman (1953:56), and Schilder (1956:75) identified a major change at about Cape Verde. Much earlier, Forbes (1856:map) indicated a point farther north (approximately Cape Yubi, Spanish West Africa), but more recently, Hedgpeth (1957:map) was in favor of a locality more to the south, about the coast of Sierra Leone.

In discussing the distribution of the mollusca, Woodward (1856:361) and Tryon (1882–1884) placed the boundary at Cape Yubi, Fischer (1950) at Cape Blanco (Mauritania), and Coomans (1962:97) at Cape Verde. Ortmann (1896:map), on the basis of his studies of the decapod crustaceans, indicated the Straits of Gilbraltar and included the whole of the Mediterranean in the tropics.

As far as the fishes are concerned, Poll (1951:7) mentioned Sénégal (? Cape Verde) as a northern limit for the tropics. Two warm-temperate clingfish (Gobiesocidae) species (Briggs 1955:151, 1957:208) reach Cape Verde but extend no farther south, and two of the three tropical species are not found north of the Cape. The northwest African distributional list of Postel (1959–1960) showed that many tropical species have not been taken north of Cape Verde and that many warm-temperate forms have not succeeded in penetrating to the south past this point.

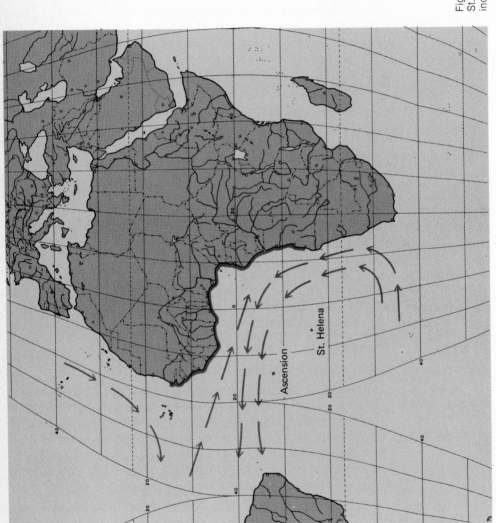

Figure 4-1   West African and St. Helena–Ascension Provinces.

According to Postel's list, *none* of the West African snappers (Lutjanidae) and only two of the eight known damselfishes (Pomacentridae), one of the four barracudas (Sphyraenidae), and one of the four parrotfishes (Scaridae) extend northward past Cape Verde. Also, Bauchot and Blanc (1961:93) presented a distributional map of the Eastern Atlantic butterfly fishes (genus *Chaetodon*), showing that only one of the four mainland species rounded the Cape to the north. Most recently, Longhurst (1962:695) stated a preference for a boundary at 14°N, a locality just a little to the south of Cape Verde.

As Postel (op. cit.) noted, many of the tropical species recorded from north of Cape Verde were captured only during the summer, indicating a temporary seasonal shift in their ranges. All in all, the evidence, particularly the distributional pattern of the shore fishes, is definitely in favor of Cape Verde as a northern (mainland) boundary for the province.

The designation of a southern boundary for the West African Province is considerably more difficult. The rapid change in the surface temperature that takes place from about Cape Santa Marta at 14°S to Cape Frio at 18°30'S presents an unusual situation that is greatly in need of study.

Forbes' early (1856) map showed the tropics extending all the way to the southern South-West African coast, but modern workers have suggested more northerly limits. Stephenson (1947:221) indicated Cape Frio as did Schilder (1956:75), Hedgpeth (1957:map), and Postel (1962:18). Ekman (1953:56) suggested Mossâmedes (15°S) or possibly Great Fish Bay (also called Baia dos Tigres, 16°30'S).

From the standpoint of the invertebrate distribution, Ortmann (1896: map, decapod Crustacea) indicated about Point Albino just south of Mossâmedes, and both Kramp (1955:311, neritic medusae) and Kirkegaard (1959:112, polychaetes) recognized a boundary at Great Fish Bay. Coomans (1962:97, mollusca) indicated a faunal change at 18°S just north of Cape Frio.

Longhurst (1962:659) discussed the distribution of the demersal fish faunas and concluded that the oceanographic frontal zone at approximately 14°S (about Cape Santa Marta, Angola) formed a very important boundary. Poll (1951:7) recognized a southern limit for the tropical fishes at Mossâmedes, and Matthews (1960:119) observed that the temperate pilchard, *Sardinops ocellata*, ranged north to Great Fish Bay. Blache (1962) utilized Mossâmedes as the southern boundary for a "Guineo-Equatoriale" Province.

In the absence of detailed knowledge about upwelling, the seasonal behavior of the Benguela Current, and other local conditions, it is difficult to decide upon a boundary with any real degree of confidence. In general, the most recent publications, particularly by those authors who have concentrated on specific animal groups, indicate a location to the north of Cape Frio. Perhaps as Longhurst (loc. cit.) observed, it should be at 14°S but for the present, Mossâmedes, located at about 15°20'S, seems the better choice.

Although West Africa has a relatively poorly developed tropical fauna, it is still distinctly richer than the warm-temperate fauna to the north or to the south, and there is evidence of a considerable amount of endemism. The fauna of hermatypic corals in the Gulf of Guinea was called a modified Caribbean fauna (Vaughan and Wells 1943:77); of a total of 18 species, 9 were considered to be endemic and 5 others closely related or identical to Caribbean species. In his 1953 account, Ekman (pp. 59–60) noted endemic levels of 40 percent among decapod crustaceans, 60 percent for ascidians, 63 percent for mollusks, and 57 percent for three echinoderm groups (starfishes, brittle stars, and sea urchins).

Analysis of some of the more recent invertebrate literature indicates that the endemic level may vary considerably from group to group. Kramp (1955) gave data that indicated only 25 percent of the neritic medusae were peculiar to West Africa, Knudsen (1956) indicated about 46 percent for the prosobranch mollusks (not including the families Conidae and Terebridae), Burton (1956:111) reported about 18 percent for the sponges, Monod (1956) about 41 percent for anomuran and brachyuran crabs, Kirkegaard (1959:112–113) about 17 percent for the polychaetes (both Monod and Kirkegaard included some material from the area between Cape Verde and Cape Blanco), and Dekeyser (1961) gave information that indicated about 54 percent for the ascidians. Finally, Stubbings (1967) found that about 26 percent of the barnacles were endemic.

Although ichthyological work in West Africa began in the 1830s, over 100 years passed before anyone attempted to bring together the widely scattered literature in order to view the fauna as a whole. This was accomplished by Fowler (1936), and the result was a classic work of 1,493 pages and 566 figures. Since 1936, many notable papers dealing with the West African ichthyofauna have been published. At present, the most useful for identification purposes is the nicely illustrated work of Blache, Cadenat, and Stauch (1970).

In general, the recent ichthyological literature indicates that a large

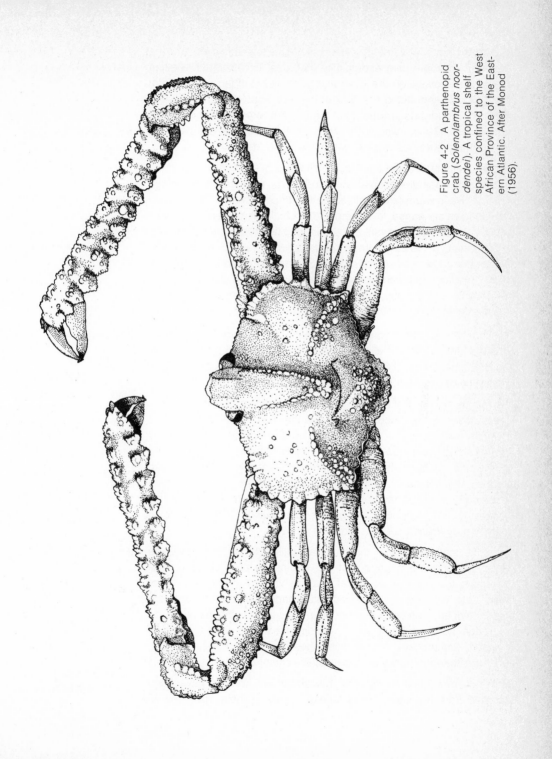

Figure 4-2 A parthenopid crab (*Solenolambrus noordendei*). A tropical shelf species confined to the West African Province of the Eastern Atlantic. After Monod (1956).

proportion (about 40 percent) of the shore fish species that occur in the West African tropics are endemic to that area. For example, Nielsen (1961), who worked on the pleuronectid flatfishes (Pleuronectoidea), provided data indicating that four of the nine species were probably endemic. The triglids (Triglidae) of the Eastern Atlantic were reviewed by Richards (1968), who found that three of the five species that occurred in the tropics were apparently confined to it.

A large component of the total shore fish fauna of West Africa, as in the other major tropical regions, is comprised of species that have the ability to live in warm-temperate as well as tropical waters. In this work, such species are called "eurythermic tropical," with the term "tropical" being reserved for those that are confined to the tropics. In the Eastern Atlantic, many of the eurythermic tropical species range northward into the Mediterranean and southward into South-West Africa or even South Africa. Another significant portion of the West African fish fauna (about 25 percent) is the group, both tropical and eurythermic tropical, that have transatlantic distributions.

The Cape Verde Islands comprise an oceanic archipelago consisting of 10 main islands located at 14°47' to 17°13'N and 22°52' to 25°22'W, about 320 miles west of Cape Verde, Sénégal. The islands are very old, apparently dating from the Lower Cretaceous (Wilson 1963:593).

Cape Verde Islands

The marine fishes of the Cape Verde Islands were, like those of the Galápagos and the Cocos-Keeling Islands, first made known to the scientific world (Jenyns 1842) as the result of collections made by Charles Darwin and crew during the voyage of the H.M.S. *Beagle.* Other early works on the fishes were written by Franz H. Troschel, João Cardoso, and Balthazar Osorio. A good indication of our current knowledge of the fish fauna may be gained from a provisional list by Cadenat (1951), later additions by the same author (1961), and a list published by Franca and Vasconcelos (1962).

Since the main portion of the archipelago lies to the north of Cape Verde, the northern boundary of the tropics on the mainland, it is interesting to note that its faunal relationships are primarily tropical rather than warm-temperate. Cadenat's papers (op. cit.) show the presence of an impressive number of species that, on the mainland, do not range north of Cape Verde. Among these are four species of snappers (Lutjanidae), five damselfishes (Pomacentridae), two parrotfishes (Scaridae), and two bigeyes (Priacanthidae).

As things now stand, a total of 130 shore fish species have been recorded from the Cape Verde Islands. Many of these (perhaps 40

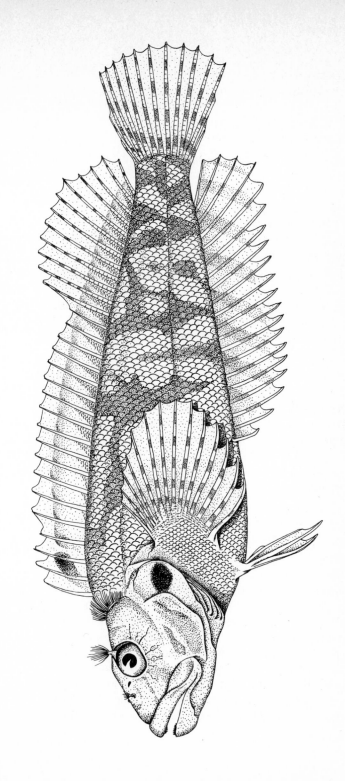

Figure 4-3 The hairy blenny (*Labrisomus nuchipinnis*). A tropical shelf species with a transatlantic distribution.

percent) are transatlantic species, and since most of them are relatively widespread in the Western Atlantic and many represent typical American genera, it seems reasonable to suppose that transport from the Western Atlantic by means of the Equatorial Countercurrent has been an important factor in the population of the islands.

Despite its great age and isolation, the Cape Verde archipelago does not display much endemism. At present, only 5 fish species out of the total of 130, or about 4 percent, may be considered as endemics. A total of 37 species are shared with the island of Madeira, and the same number with the Mediterranean. However, most of the latter group are eurythermic tropical species, so their presence does not indicate special faunal ties to these more northerly localities.

There seems to be little comparative information for the major invertebrate groups. Three species of hermatypic corals are known only from the Cape Verdes (Vaughan and Wells 1943:77), Knudsen (1956) recognized three endemic prosobranch mollusks out of a total of 27, and Postel (1966) showed that one of the three species of lobsters (Palinuridae) is endemic.

The surprisingly low rate of endemism is probably due to the geographic location of the archipelago. As in the case of Bermuda, it is very close to the northern boundary of the tropics so that a marked drop in sea-surface temperature of the North Atlantic Ocean, such as apparently occurred during the most recent glaciation, would have a disastrous effect upon its tropical fauna. It follows that if the islands had to be essentially repopulated since the Wisconsin glaciation, very little endemism would have had time to develop.

Taking into consideration the tropical nature of the marine fauna, the very low degree of endemism (at least among the fishes), and the obvious relationship to the West African coast, the Cape Verde archipelago is considered to be an integral part of the West African Province.

Although the oceanic islands of the St. Helena and Ascension both lie in the eastern part of the tropical Atlantic, they are separated by about 700 miles. Despite this relatively great distance from each other, these two islands demonstrate an interesting faunal relationship. Among the major groups of shore animals, the fishes are known the best. A number of recent publications have appeared (Cadenat and Marchal 1963, Bauchot 1966a and b, Blache 1967, Eschmeyer 1969) and I have had the opportunity to examine the good collections from Saint Helena and Ascension that are maintained in the British Museum of Natural History.

St. Helena–
Ascension

As nearly as I can determine at this time, a total of 80 shore fish species have been taken from the St. Helena–Ascension area, but the number of species common to both islands is only 29. St. Helena has 12 endemics, Ascension 2, and 6 other species are shared by the two islands but are found nowhere else. A comparison of the total number of endemics (20) to the total shore fish fauna (80) gives a 25 percent endemism for the province. Although the species complex at each island is quite characteristic, the existence of the 6 common endemics, and the very low degree of endemism for Ascension by itself, indicates a relationship that is probably best expressed by including both islands in the same province.

St. Helena St. Helena has the distinction of being the most isolated of all tropical islands. Its location at 14 °56'S and 5°42'W is 1,200 miles west of the African mainland and 700 miles southeast of Ascension Island. It also lies about 430 miles east of the Mid-Atlantic Ridge, and its age is possibly Miocene (Wilson 1963:536). Here, perhaps better than anywhere else, the biologist is presented with a natural laboratory where he can observe the effects of extreme isolation on small populations.

The littoral fish fauna of St. Helena, which consists of 55 species, is sparse compared to other tropical Atlantic islands such as the Cape Verdes (130 species) and Bermuda (265 species); 12 of the total, or about 22 percent, are endemics, and 6 others are shared only with Ascension. Of the nonendemics, 18 are transatlantic species, being found in the shelf waters on both sides of the ocean; of 19 St. Helena species which have a mainland distribution that is confined to one side or the other, 14 are shared with the Eastern Atlantic or the Indo-Pacific via the Cape of Good Hope, and 5 are shared with the Western Atlantic.

Four of the 14 species of eastern origin are of unusual interest in that they must have been transported around the tip of South Africa and almost directly to St. Helena. *Gonorhynchus gonorhynchus*, *Synanceija horrida*, and *Abalistes stellaris* are distinctive, widespread Indo-Pacific species, and none of them has been taken elsewhere in the Atlantic. The presence of the flatfish *Nematops macrochirus* at St. Helena was reported by Nielsen (1961). It was previously known only from the Bali Strait.

The most recent group of shallow-water invertebrates to be investigated was the decapod crustaceans (Chace 1966). A total of 23 species were identified and 4 of them (17 percent) were considered to be endemics. Of the nonendemics, 10 demonstrated eastern relationships being found along the West African coast (7) or shared only with the Indo-Pacific (3). Three species were shown to be shared only with the Western Atlantic.

Older works on the St. Helena invertebrates were published by Smith (1890a) on the mollusks, who found about 52 percent of the species to be endemic and the closest relationship to be with the West Indies, and by Mortensen (1933) on the echinoderms, who found that 13 of the 26 species examined were endemic to St. Helena and Ascension. Mortensen also concluded that at least one of the nonendemics (a starfish) must have been transported from South Africa by drifting seaweed.

The most recent general summary about the fauna of St. Helena and its geographic relationships was published by Colman (1946). In this work, he concluded that the marine fauna (based on information then available about the fishes, mollusks, and echinoderms) was related principally to that of the West Indies. Earlier, Norman (1935:57), who worked on the coastal fishes of the South Atlantic, had written that the fauna of both St. Helena and Ascension was predominately West Indian and Brazilian in character. Cadenat and Marchal (1963:1307) felt that the fish fauna of both islands was more closely related to that of the Antilles than to that of the West African coast. As far as St. Helena is concerned, I cannot agree with these conclusions, for it now seems clear, especially from the relationships of the fishes and the decapod crustaceans, that the shore fauna of this island was derived primarily from the Eastern Atlantic.

A hypothesis of an Eastern Atlantic origin for the majority of the littoral species fits in well with the general pattern of surface currents. St. Helena lies toward the northeastern portion of the great South Atlantic Gyre and, presumably, is mainly within the influence of the Benguela Current, with the water movement being in a westerly or northwesterly direction. Under these conditions, the transport of pelagic larvae from the southern part of the African shelf could take place quite easily.

Ascension Island is located in the southeastern Atlantic Ocean at 7°55′S and 14°25′W. It is well isolated, being about 1,000 miles from the west coast of Africa and about 700 miles northwest of St. Helena. It may be described as a small, volcanic island of apparently recent origin, probably Pleistocene (Wilson 1963:536). It lies only about 90 miles from the Mid-Atlantic Ridge, but St. Helena is located about 430 miles from this fracture. According to Wilson's hypothesis, the most recently formed oceanic islands should lie closest to the midocean ridges.

Ascension

To this date, I find that a total of 54 species of shore fishes have been taken at Ascension Island so that, in this animal group, the species diversity is about the same as that found at St. Helena. However, only 2 of the total (about 4 percent) are endemics; 6 others are shared only with St. Helena. Of the nonendemics, 27 are transatlantic species being found in the shelf waters on both sides of the ocean; 19 Ascension species have a mainland distribution that is confined to one side of the Atlantic or the other; 7 of them are shared with the Eastern Atlantic and 12 with the Western Atlantic.

The marine invertebrates of Ascension Island are very poorly known. Smith (1890b) found 33 species of mollusks, considered 10 of them to be endemics, and noted that 3 others were shared only with St. Helena. He assumed the fauna to be related to that of the West Indies.

Most surface current charts show Ascension to be located on the northern limb of the South Atlantic Gyre where the water movement is relatively rapid and the direction of flow almost due west. If this were consistently true, we would expect to find a strong representation of West African fishes and little, if any, indication of faunal ties with St. Helena. Instead, we are confronted with a situation in which the closest mainland relationships are to the west and there is good evidence of traffic between Ascension and St. Helena. It seems apparent that Ascension Island must come within the influence of the Equatorial Countercurrent for at least part of the year. By this means, pelagic larvae could be transported from the region of the Lesser Antilles. Otherwise, it would be difficult to account for the presence of the 11 Ascension–Western Atlantic shore fish species.

The existence of the six fish species that are Ascension–St. Helena endemics is best explained by assuming the existence of some current movement from one island to the other. It would be easiest to suppose that a portion of the Benguela Current, having passed close to St. Helena, would occasionally run as far north as Ascension. Sverdrup, Johnson, and Fleming (1946:633–636) wrote about the complicated nature of the currents near the equator and mentioned that the countercurrent is best developed during the summer. Perhaps during the summer, Ascension comes under the influence of the countercurrent, while in other seasons, water may move in from the southeast (past St. Helena) and from the east (near the western coast of Africa).

The contrast between Ascension and St. Helena, in regard to the endemism among the shore fish species, is remarkable. The 22 percent endemic level at St. Helena is far higher than at any other oceanic island in the tropical Atlantic. Furthermore, St. Helena is located quite close to the southern border of the tropics. It has been suggested that the survival of so many endemics may be attributed to a relatively stable climatic history (Briggs 1966). The low level of endemism at Ascension (4 percent) is probably due to its young age (Pleistocene) rather than climatic disturbance.

Since Ascension Island does not exhibit enough endemism to be considered a province by itself, it needs to be associated with the area to which it bears the strongest faunal resemblance. This poses a difficult problem. On one hand, most of the recruitment has probably come from the Western Atlantic, but on the other hand, the present of six fish species that are shared only with St. Helena indicates strong faunal ties in that direction. Although Ascension Island is here aligned with St. Helena and placed in the Eastern Atlantic Region, it can be argued that some justification exists for placing Ascension in the West Indian Province of the Western Atlantic Region.

In summary, we have considered, mainly on ichthyological evidence, the faunal relationships of two oceanic islands in the southeastern part of the tropical Atlantic, St. Helena and Ascension. St. Helena is relatively old (possibly Miocene), it lies about 430 miles from the Mid-Atlantic Ridge, and it demonstrates a high level of endemism. Most of its shore fauna was apparently derived from the east—the West African shelf or the Indo-West Pacific via the Cape of Good Hope. Ascension is relatively young (probably Pleistocene), it lies about 90 miles from the Mid-Atlantic Ridge, and it demonstrates a low level of endemism. Much of its shore fauna was apparently derived from the Western Atlantic, but there are also strong faunal ties to St. Helena.

Literature Cited

Bauchot, M. L. 1966a. Poissons marins de l'Est Atlantique Tropical. Téléostéens Perciformes. II. Percoidei (3ème partie). III. Acanthuroidei. IV. Balistoidei. *Atlantide Rep.*, (9):7–43, 8 figs.

———. 1966b. Poissons marins de l'Est Atlantique Tropical. Téléostéens Perciformes. V. Blennioidei, *Atlantide Rep.*, (9):63–91, 4 figs.

———, and M. Blanc. 1961. Poissons marins de l'Est Atlantique Tropical. I. Labroidei II. Percoidei. *Atlantide Rep.*, (6):43–100, 4 figs., 1 plate.

Blache, J. 1962. Lista de poissons signales dans l'Atlantique tropico-oriental Sud. *Trav. Cent. Oceanog. Pointe-Noire*, (2):13–102, 2 maps.

———. 1967. Contribution à la connaissance des poissons Anguilliformes de la côte occidentale d'Afrique. Quatrième note. Le genre *Lycodontis* McClelland, 1845. *Bull. Inst. Franç. d'Afrique Noire*, 29(3):1122–1187, 29 figs.

———, J. Cadenat, and A. Stauch. 1970. Clés de détermination des poissons de mer signalés dans l'Atlantique Oriental. *Faune Trop.*, (18):1–479, 1,152 figs. O.R.S.T.O.M., Paris.

Briggs, J. C. 1955. A monograph of the clingfishes (Order Xenopterygii). *Stanford Ichthyol. Bull.*, 6:1–224, 114 figs.

———. 1957. A new genus and two new species of Eastern Atlantic clingfishes. *Copeia*, (3):204–208, 3 figs., 1 plate.

———. 1966. Oceanic islands, endemism, and marine paleotemperatures. *Syst. Zool.*, 15(2):153–163, 4 figs.

Burton, M. 1956. The sponges of West Africa. *Atlantide Rep.*, (4):111–147, 4 figs.

Cadenat, J. 1951. Lista provisoria dos peixes observados nas Ilhas de Cabo Verde, de 1 de Maio a 24 de Junho de 1950 (published in the *Cabo Verde*, April 19, 1951).

———. 1961. Notes d'Ichtyologie ouest-africaine XXXIV. Liste complémentaire des espèces de poissons de mer (provenant des côtes de l'Afrique occidentale). *Bull. Inst. Franç. d'Afrique Noire, Ser. A,* 23(1):231–245.

———, and E. Marchal. 1963. Résultats des campagnes océanographiques de la Reine-Pokou aux iles Sainte Helene et Ascension. Poissons. *Bull. Inst. Franç. Afrique Noire, Ser. A,* 25(4):1235–1315, 48 figs.

Chace, F. A., Jr. 1966. Decapod crustaceans from St. Helena Island, South Atlantic. *Proc. U.S. Nat. Museum,* 118(3536):623–661, 14 figs.

Colman, J. 1946. Marine biology in St. Helena. *Proc. Zool. Soc. London,* 116(2):266–281, 3 plates.

Coomans, H. E. 1962. The marine mollusk fauna of the Virginian area as a basis for defining zoogeographical provinces. *Beaufortia,* 9(98):83–104.

Dekeyser, P. L. 1961. Liste provisoire des urocordés de la côte occidentale d'Afrique. *Bull. Inst. Franç. d'Afrique Noire, Ser. A,* 23(1):217–230.

Ekman, S. 1953. *Zoogeography of the sea.* Sidgwick & Jackson, London, xiv + 417 pp., 121 figs.

Eschmeyer, W. N. 1969. A systematic review of the scorpionfishes of the Atlantic Ocean (Pisces: Scorpaenidae). *Occas. Papers Calif. Acad. Sci.,* (79):iv + 1–143, 13 figs.

Fischer, P.-H. 1950. *Vie et moeurs des Mollusques.* Payot, Paris, 1–312 pp., 180 figs., 1 map.

Forbes, E. 1856. Map of the distribution of marine life. Alexander K. Johnston, *The physical atlas of natural phenomena* (new edition). W. and A. K. Johnston, Edinburgh and London, plate no. 31.

Fowler, H. W. 1936. The marine fishes of West Africa. *Bull. Amer. Museum Nat. Hist.,* 70:1–1493, 565 figs.

Franca, P., and M. S. Vasconcelos. 1962. Peixes do Arquipélago de Cabo Verde. Notas Mimeografadas, Centro Biologia Piscatória, Junta Invest. Ultramar, Lisboa (28):1–86, map.

Hedgpeth, J. W. (editor). 1957. Marine biogeography. *in Treatise on marine ecology and paleoecology.* vol. 1. *Geol. Soc. Amer., Mem.,* (67):359–382, 16 figs., 1 plate.

Jenyns, L. 1842. Fish. *in* C. Darwin, *The zoology of the voyage of H.M.S. Beagle.* part 4, xv + 172 pp., 29 plates.

Kirkegaard, J. B. 1959. The Polychaeta of West Africa. Part I. Sedentary species. *Atlantide Rep.*, (5):7–117, 25 figs.

Knudsen, J. 1956. Marine prosobranchs of tropical West Africa (Stenoglossa). *Atlantide Rep.*, (4):8–110, 2 figs., 4 plates.

Kramp, P. L. 1955. The Medusae of the tropical west coast of Africa. *Atlantide Rep.*, (3):239–324, 14 figs., 3 plates.

Longhurst, A. R. 1962. A review of the oceanography of the Gulf of Guinea. *Bull. Inst. Franç. d'Afrique Noire*, 24(3):633–663, 9 figs.

Matthews, J. P. 1960. Synopsis on the biology of the South African pilchard (*Sardinops ocellata* Pappé). *Fisheries Div., Biology Branch, F.A.O., Fishery Biol. Synopsis*, (8):vi + 115–133.

Monod, T. 1956. Hippidea et brachyura ouest-africains. *Mem. Inst. Franç. d'Afrique Noire*, (45):1–674, 884 figs., 1 map.

Mortensen, T. 1933. The echinoderms of St. Helena (other than crinoids). *Vidensk. Medd. Naturh. Foren.*, 93:401 (not seen).

Nielsen, J. 1961. Psettodoidea and Pleuronectoidea (Pisces, Heterosomata). *Atlantide Rep.*, (6):101–127, 8 figs., 1 plate.

Norman, J. R. 1935. Coast fishes. Part I. The South Atlantic. *Discovery Rep.*, 12:1–58, 15 figs.

Ortmann, A. E. 1896. *Grundzüge der marinen Tiergeographie*. Gustav Fischer, Jena, iv + 96 pp., 1 map.

Poll, M. 1951. *Expédition Océanographique Belge dans les Eaux Côtières Africaines de l'Atlantique Sud (1948–1949)*. Résultats Scientifiques, vol. 4, fasc. 1, Poissons. Institut Royal des Sciences Naturelles de Belgique, Bruxelles, 154 pp., 67 figs., 13 plates.

Postel, E. 1959–1960. Liste commentéc des poissons signalés dans l'Atlantique tropico-oriental nord, du Cap Spartel au Cap Roxo, suivie d'un bref aperçu sur leur répartition bathymétrique et géographique. *Bull. Soc. Sci. Bretagne*, 34(1–4):130–170, 241–281, 3 figs.

———. 1962. Survey of the natural resources of the African continent. Marine biology and biology applied to the fishing industry. UNESCO/NS/NR/2 Add. 2. (processed)

———. 1966. Langoustes de la zone intertropicale africaine. *Mem. l'Inst. Fondamental d'Afrique Noire*, (77):395–474, 15 figs.

Richards, W. J. 1968. Eastern Atlantic Triglidae (Pisces, Scorpaeniformes). *Atlantide Rep.*, (10):77–114, 13 figs., 1 plate.

Schilder. F. A. 1956. *Lehrbuch der Allgemeinen Zoogeographie*. Gustav Fischer, Jena, viii + 150 pp., 134 figs.

Smith, E. A. 1890a. Report on the marine molluscan fauna of the Island of St. Helena. *Proc. Zool. Soc. London*, pp. 247–317, 4 plates.

———. 1890b. On the marine Mollusca of Ascension Island. *Proc. Zool. Soc. London*, pp. 317–322.

Stephenson, T. A. 1947. The constitution of the intertidal fauna and flora of South Africa, part III. *Ann. Natal Museum*, 11(2):207–324, 11 figs., 2 plates.

Stubbings, H. G. 1967. The cirriped fauna of tropical West Africa. *Bull. Brit. Museum (Nat. Hist.) Zool.*, 15(6):229–319, 28 figs., 1 plate.

Sverdrup. H. U., M. W. Johnson, and R. H. Fleming. 1946. *The oceans.* Prentice-Hall, Englewood Cliffs, N.J., pp. 1–1087, 265 figs., 7 charts.

Tryon, G. W. 1882–1884. Structural and systematic conchology. 3 vols. Privately printed, Philadelphia, viii + 312, 430, and 453 pp., 140 plates, map.

Vaughan, T. W., and J. W. Wells. 1943. Revision of the suborders, families and genera of the Scleractinia. *Geol. Soc. Amer., Spec. Papers*, (44):xv + 1–363 pp., 39 figs., 51 plates.

Wilson, J. T. 1963. Evidence from islands on the spreading of ocean floors. *Nature*, 197(4867):536–538, 4 figs.

Woodward, S. P. 1851–1856. *A manual of the mollusca.* John Weale, London, viii + 1–486 pp., 24 plates + 24 pp., 1 map.

Relationships
of the Tropical
Shelf Regions

The distinguishing of the aboriginal from the invading population, and the determination of the causes which have produced and directed the invasion, are among the problems which the investigator of the distribution of animated creatures, has to endeavour to solve.

Professor Edward Forbes in The Natural History of European Seas, *1859.*

We have observed that the richest (most diverse) marine fauna is found in the shallow waters of the tropical oceans and that this environment may be divided into four great zoogeographic regions: the Indo-West Pacific, the Eastern Pacific, the Western Atlantic, and the Eastern Atlantic. Longitudinally, the tropical shelf regions are separated from one another by barriers that are very effective, since each region possesses, at the species level, a fauna that is highly endemic. By studying the operation of these longitudinal barriers, one can learn something about the interrelationship of the regions and, more important, obtain information leading to a better understanding of zooge-ography and evolution.

The East Pacific Barrier is the formidable stretch of deep water that lies between Polynesia and America. Its efficiency in regard to the migration of shore fishes has been investigated (Briggs 1961, 1964). The distributional patterns and evolutionary relationships of the 62 transpacific species[1] may be summarized as follows:

East Pacific Barrier

1 In the eastern Pacific, the ranges of many of the transpacific species were found to be quite restricted. In fact, the majority (35 out of the total of 62) are apparently confined to, or are most typical of, the offshore islands (principally the Galápagos, Cocos, Clipperton, or Revillagigedos).

2 In contrast, on the western side of the barrier almost all of the transpacific species are widespread, with the great majority (47 of 62) extending from Polynesia through the Indian Ocean and reaching the eastern coast of Africa or its vicinity.

3 Aside from the few monotypic genera represented, all the genera to which the transpacific species belong are better developed in the Indo-West Pacific.

4 There is not a single example of a species belonging to a typical New World genus gaining a foothold on the western side of the barrier, even at its outermost fringes.

[1] When this book was in press. Rosenblatt et al. (1972) published a paper which showed that some old records for Indo-West Pacific species in the Eastern Pacific should be discarded, but they also added some new records. Since the number of additions equalled the deletions, we can still recognize 62 trans-pacific shore species.

In view of the foregoing information, it has been concluded that there is good circumstantial evidence of a western origin for the great majority of the transpacific species. It can be said, therefore, that there is a recent (or current) invasion of the Eastern Pacific tropics from the west. This has been referred to as an "eastward colonization movement."

For fishes, a rough estimate of the degree of effectiveness of the East Pacific Barrier can be made by taking into consideration the total number of shore species that are theoretically available for transport

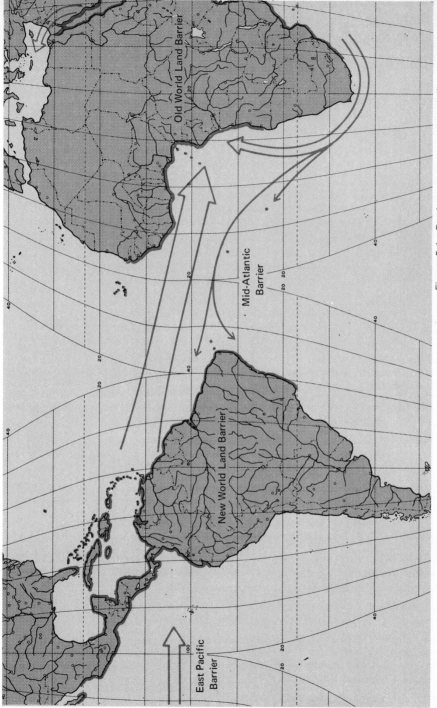

Figure 5-1 Barriers separating the tropical shelf regions of the world. Arrows indicate the direction and approximate relative amount of successful (colonizing) migration that has taken place across the barriers.

across the Pacific and comparing that total to the number that apparently have successfully made the journey.

The Line Islands are the easternmost Polynesian group that lie along the course of the Equatorial Countercurrent. Unfortunately, the fauna of this archipelago is very poorly known. However, one might expect to find at least as many shore fishes as have recently been reported from Hawaii by Gosline and Brock (1960)—387. There is probably a total of about 650 species inhabiting the coastal waters between southern Mexico and Peru. Since 62 (the shore fishes that are apparently transpacific in range) is about 6 percent of 1,037 (387 western species plus 650 on the eastern side), the efficiency of the barrier can be approximated at 94 percent.

When the general relationship of the tropical shelf regions was first discussed (Briggs, 1967a), comparable information on the major groups of the shallow-water invertebrates was not available. Emerson (1967) published a revealing analysis of the distribution of those Indo-West Pacific species of mollusks that have succeeded in penetrating across the barrier into the Eastern Pacific. He found that such transpacific species were largely restricted, in the Eastern Pacific, to the oceanic islands, the greatest numbers being found at Clipperton (33 species) and at the Galápagos (25 species). It was noted that the gastropods, which greatly outnumbered the bivalves, belonged to groups that were known to have relatively long-lived larval stages. Most important of all, Emerson pointed out that no molluscan species of apparent Eastern Pacific origin were known to occur in Polynesia.

Data on the other invertebrate groups are not as complete, but it is significant that some of the littoral echinoderms (Ekman 1946), holothurians (Deichmann 1963), decapod crustaceans (Chace 1962, Garth 1965), and hermatypic corals (Durham 1966) found in the Eastern Pacific (especially around the offshore islands) are transpacific species of apparent Indo-West Pacific origin. Therefore, it may now be said that for the tropical marine shore fauna in general, including both fishes and invertebrates, it seems likely that successful migration across the East Pacific Barrier takes place in one direction only—from west to east.

The New World Land Barrier, with the Isthmus of Panama forming its narrowest part, is virtually a complete block to the movement of tropical marine species between the Eastern Pacific and Western Atlantic. This state of affairs has existed since about the latest Pliocene or earliest

**New World
Land Barrier**

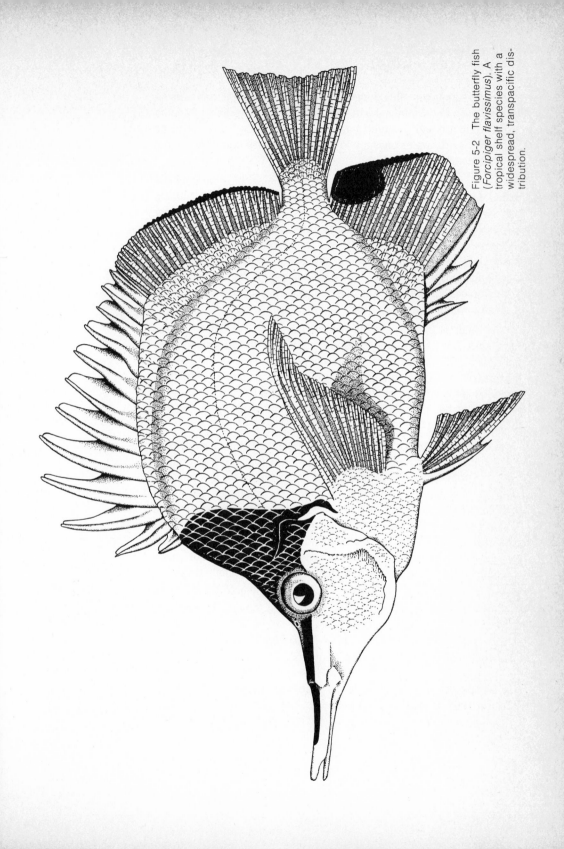

Figure 5-2   The butterfly fish (*Forcipiger flavissimus*). A tropical shelf species with a widespread, transpacific distribution.

Pleistocene (Simpson 1965, Patterson and Pascual 1968) so that, at the species level, the two faunas are well separated. The present Panama Canal has not notably altered this relationship since, for most of its length, it is a freshwater passage forming an effective barrier for all but a few euryhaline species.

Despite the obvious physical effect of the Isthmus in separating the marine populations of the American tropics, its evolutionary effects have not been fully investigated, and there is still an urgent need for research along this line. As recently as 1953, Ekman (p. 73) considered the East Pacific Barrier rather than the New World Land Barrier to be responsible for the most pronounced break in the circumtropical warm-water fauna of the shelf. The belief that a large portion of the New World tropical fauna is made up of amphi-American species apparently originated with the works of Albert C. L. G. Günther.

In 1861, Günther (p. 370) examined a small collection of fishes from the Pacific coast of Central America and commented on "the very strange fact" that 5 species out of a total of 14 were identical with Atlantic forms. In a later publication, Günther (1868:397–398) found that 59 of 193 (30.5 percent) marine species occurred on both sides of Central America. To explain this, he assumed the recent existence of an open channel across the Isthmus of Panama.

Finally in 1880, Günther published his popular text *An Introduction to the Study of Fishes* and by this time decided (p. 280) that "nearly one-half" of the species found on the Pacific side were the same as those on the Atlantic. The first to effectively challenge Günther's conclusions was Jordan (1885:393) who listed 407 species from the Pacific coast of Mexico and Central America and stated that only 71 species, or 17.5 percent, were also found on the Atlantic coasts. Also, Jordan noted that many of the 71 were cosmopolitan in tropical waters and that the fish faunas of the two shores were substantially distinct.

Evermann and Jenkins (1891:124–125) calculated that a total of 56 species were common to both coasts of the Americas (after the elimination of the cosmopolites), and this figure was reiterated by Jordan (1901:555). Gilbert and Starks (1904:206), in their study of the fishes of Panama Bay, recognized a total of 54 species as identical in the two faunas, and later, Meek and Hildebrand (1923:10) stated that only 24 species (cosmopolitan forms excepted) were common to both sides. In 1961, Briggs (p. 550) observed that, aside from the circum-tropical species and a few euryhaline forms, there are probably less than a dozen shore fishes common to the tropical waters on both sides of the Isthmus of Panama.

As more systematic work was done on the families and genera of amphi-American distribution, the species which could be recognized as common to both the Eastern Pacific and Western Atlantic became fewer. In fact, most recent studies along this line such as the revisions of the Antennariidae (Schultz 1957), the clinid genera *Labrisomus* and *Malacoctenus* (Springer 1958), Gobiesocidae (Briggs 1955), and Chaenopsidae (Stephens 1963) show no amphi-American species whatever!

Considering the total number of shore fish species now estimated to be present on both sides of Central America (roughly about 1,000), the fact that about 12 (aside from the circumtropical species and a few euryhaline forms) can still be considered identical indicates that the New World Land Barrier is approximately 99 percent effective. Apparently, evolutionary change has proceeded so slowly in a few species, such as the serranids *Paranthias furcifer* (Fig. 5-3) and *Epinephelus itajara*, that specific differences have not yet developed. However, for the great majority, it is interesting to see that about 3 million years has been sufficient to produce distinct morphological evidence of evolutionary change.

The New World Land Barrier is also important for the marine invertebrates, although in some groups, it does not seem to be quite as effective as it is with the fishes. Haig (1956, 1960) studied the crab family Porcellanidae in both the Western Atlantic and Eastern Pacific and found that 7 out of a total of 107 species (6.5 percent) are amphi-American; de Laubenfels (1936) found a similar distribution in 10.8 percent of the sponges he studied, and Ekman (1946) gave 2.3 percent and 2.5 percent (1953) for the echinoderms (Asteroidea, Ophiuroidea, and Echinoidea). In discussing speciation in tropical echinoids, Mayr (1954:13) noted that not a single species is identical on both sides of the Isthmus, and this fact was also stated by Hyman (1955:580).

The digenetic trematodes of marine fishes have been studied by Manter who has (1963:47) stated that the trematodes of the Tortugas-Caribbean area are similar to those of the American Pacific. Of 226 Caribbean species, 18.1 percent were found to occur in the Pacific with the hosts, in most cases, being different but related fishes. This suggests that the rate of evolution for such parasites may be slower than that of the free-living marine animals.

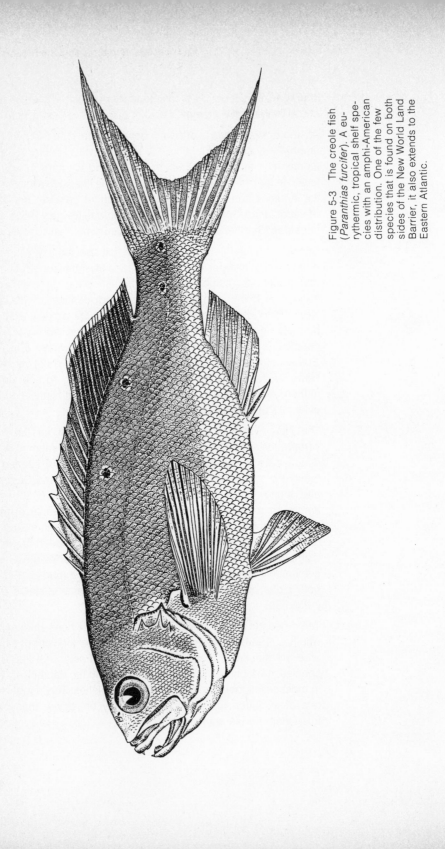

Figure 5-3 The creole fish (*Paranthias furcifer*). A eurythermic, tropical shelf species with an amphi-American distribution. One of the few species that is found on both sides of the New World Land Barrier, it also extends to the Eastern Atlantic.

Mid-Atlantic Barrier

The Mid-Atlantic Barrier is the broad, deep-water region that separates the Western Atlantic tropics from those of the West African coast. In regard to the fishes, it has been recognized for many years, particularly since the writings of the Portuguese zoologist Balthazar Osorio (a series from 1890 to 1911), that the tropical fauna of West Africa has many species in common with the American side of the Atlantic. Many additional transatlantic species have been noted in the more recent literature so that now an uncritical compilation of records would give a much greater total than is actually warranted.

In order to arrive at an assessment of the relationship between the shore fish faunas of the Western Atlantic and Eastern Atlantic, I compiled a manuscript list of transatlantic species. Fortunately, in the course of this work, it was possible to study Osorio's material in the Museu Bocage in Lisbon and to examine some critical specimens in the British Museum of Natural History. Although the list has undergone many changes over the years and will continue to change as new information becomes available, I believe it can be used now to give a reasonably dependable indication of relationship between the two sides of the tropical Atlantic.

In the light of present knowledge, it appears that a total of 120 species of shore fishes have transatlantic distributions. Approximately 900 species may be found in the Western Atlantic (Briggs 1967a) and about 434 species in the Eastern Atlantic (Blache, Cadenat, and Stauch 1970), giving a total of about 1,334 species. Since 120 is about 9 percent of 1,334, the efficiency of the Mid-Atlantic Barrier, insofar as the ichthyofauna is concerned, may be calculated at roughly 91 percent.

Perhaps more important than an attempt to discover the degree of faunal relationship is the determination of the direction of migration. Aside from a group of about 24 shore fishes that apparently make their way around the Cape of Good Hope and then westward across the Atlantic, the predominant migratory movement across the Mid-Atlantic Barrier seems to be from west to east. Many of the transatlantic species range broadly along the Western Atlantic shelf but have attained only limited purchase in the east. Others that have achieved broad distributions in the east are clearly representatives of American genera (such as *Mycteroperca*, *Rypticus*, *Eucinostomus*, *Sparisoma*, *Labrisomus*, etc.). None of the transatlantic species belong to genera that are typically Eastern Atlantic. Further, it is of interest to note that transatlantic species comprise about 25 percent of the shore fish fauna of tropical West Africa.

Works on some of the major groups of West African invertebrates also show that an appreciable number of the species are transatlantic: Dekeyser (1961) found that about 25 percent of the ascidians showed this distribution; Burton (1956), 18 percent of the sponges; Monod (1956), 16 percent of the anomuran and brachyuran crabs; Knudsen (1956), 6 percent of the prosobranch mollusks; Ekman (1953), 16 percent of the starfishes, brittle stars, and sea urchins; and Marcus and Marcus (1966), 29 percent of the opisthobranch mollusks. Furthermore, Chesher (1966), who found eight transatlantic species of sea urchins in the Gulf of Guinea, stated that gene flow appeared to take place from west to east.

It seems apparent that, in both the fishes and the invertebrates, the great majority of the transatlantic species originated in the Western Atlantic and then migrated eastward. The westward colonization traffic appears to be restricted to certain dominant species that originated in the Indo-West Pacific and then gained access to the Atlantic by rounding the Cape of Good Hope. So far, there are no indications that species originating in the Eastern Atlantic, and belonging to genera typical of that area, have been successful in becoming established on the western side.

Old World Land Barrier

The Old World Land Barrier, which was apparently established in the Lower Miocene (Ruggieri 1967), linked the continental masses of Eurasia and Africa and thus terminated the existence of the ancient Tethys Sea. The marine fauna of the Mediterranean underwent rapid changes, and African mammals used the new land bridge to invade Europe (Savage 1967). The Suez Canal has had interesting distributional repercussions, but its effects are just beginning to be felt. Currents of tropical water have extended around the Cape of Good Hope often enough to provide passage for some Indo-West Pacific shore species that have pelagic larvae or adults that can safely enter the pelagic environment.

The Suez Canal

Although water-borne traffic between the Red Sea and the Mediterranean was possible in the days of the Pharaohs, it was accomplished by means of canals supplied with fresh water from the Nile, so there was little or no effect on the distribution of marine animals. However, since the opening of the modern Suez Canal in 1869, some interesting changes have taken place.

It was once suggested that most of the migrations of marine animals through the Suez Canal took place during the first few years after its opening (Gohar 1954:27), but it now seems apparent that such movements are on the increase. At first, the Bitter Lakes portion of the canal, with an extremely high salinity of 68‰ (parts per thousand), provided an effective barrier to migrants. But by 1924 salinity had decreased to 52‰, and today it is about 41‰—not very different from that of the northern Red Sea or the southeastern Mediterranean (Oren 1969).

The movement of fishes through the Suez Canal has been studied more closely than that of other animal groups. Ben-Tuvia (1971a) recorded a total of 30 Red Sea species that have established themselves along the Mediterranean coast of Israel. There, they comprise about 11 percent of the total ichthyofauna. The same author (Ben-Tuvia 1971b) called attention to the presence of one Mediterranean species, the serranid *Dicentrarchus punctatus*, in the Gulf of Suez. It is significant that this species is very euryhaline and, in the Red Sea, has been found only in a high salinity (50 to 60‰) lagoon where there are apparently no native competitors.

Studies of three major invertebrate groups also indicate fairly large-scale movements of Red Sea species into the Mediterranean. No less than 41 species of mollusks (Barash and Danin 1971), 26 species of polychaetes (Ben-Eliahu 1971), and 16 species of decapod crustaceans (Holthuis and Gottlieb 1958) have undergone such a migration. A useful summary has been provided by Por (1971), who recorded a total of 140 animal species that have invaded the Mediterranean via the Suez Canal; of these, 21.3 percent are decapods, 17.8 percent fishes, 17.1 percent mollusks, 13.6 percent polychaetes, 5 percent ascidians, 5 percent sponges, and 15.2 percent other animal groups.

Por (1971), who described the movement of marine animals through the canal as "Lessepsian migration," emphasized that this was a one-way invasion from the Red Sea to the Mediterranean. He found that examples of successful migrations in the opposite direction that have occasionally appeared in the literature invariably turned out to be old Indo-Pacific residents, circumtropical species, or widespread fouling organisms.

The Cape of
Good Hope
Considering that the New World Land Barrier has been impassable to tropical marine animals, with the exception of a few euryhaline species since about the Pliocene-Pleistocene boundary, it is assumed that the circumtropical shore species have been able to preserve their genetic

homogeneity by means of migration around the Cape of Good Hope (in addition to crossing the open ocean barriers in the Pacific and Atlantic). Talbot and Penrith (1962:559) remarked that surface temperatures of 21°C are often present around the Cape outside a cold upwelling area, and under certain conditions there is evidence of mixing of Agulhas Current water and South Atlantic subtropical water.

In addition to the circumtropical species, there are a number of others that are common to the tropical Atlantic and the Indo-West Pacific also indicating the existence of gene flow around the Cape of Good Hope. Among the fishes, eight such species are found on both sides of the Atlantic: four are confined to the Western African fauna, and four have been taken only at the Island of St. Helena. These 16 species plus the 16 known circumtropical shore fishes (Briggs 1960), give a total of 32 that apparently transgress the Old World Land Barrier.[2]

²Leaving out of consideration the effects of the Suez Canal, since they are relatively short-term and, at this date, almost entirely confined to the eastern Mediterranean.

The successful migration of certain tropical shore species around the Cape of Good Hope has apparently taken place in one direction only. Of the total of 32 fish species, 8 are monotypic but the rest represent genera that are best developed in the Indo-West Pacific. There is, therefore, a similar situation to that seen in the case of the East Pacific Barrier (p. 104). That is, movement of dominant species has taken place from the major evolutionary center of the tropics into an area where the competition is apparently reduced. However, in this instance, the colonization movement is to the west instead of the east.

The efficiency of the Old World Land Barrier for shore fishes may be roughly calculated as follows: from the information given in Blache, Cadenat, and Stauch (1970) and other recent literature, the total shore fish fauna of tropical Western Africa can be estimated at about 434 species. The fish fauna of tropical southeast Africa is a good deal richer, with probably close to 1,000 species in the area between Mozambique and the mouth of the Kei River. If it can be assumed that some 1,434 species are immediately separated by the barrier in question and that only 32 have successfully crossed it, then it can be considered to be about 96 percent effective.

Apparently, only a few tropical invertebrate species have been able to migrate around the Cape of Good Hope. Monod (1956) in his monographic study of West African decapods (Hippidea and Brachyura), found that 10 out of 176 shore species also occurred in the Indo-West Pacific. Ekman (1953:60) noted that only 2 percent of the tropical Atlantic echinoderms (Asteroidea, Ophiuroidea, and Echinoidea) extended around the Cape. As yet, the evidence is sparse, but it looks as

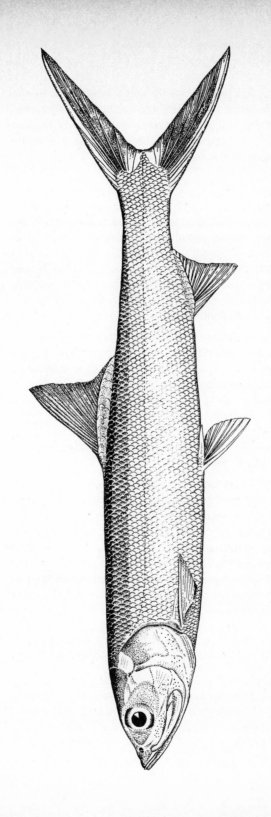

Figure 5-4 The ladyfish (*Elops saurus*); a euryth-ermic, tropical shelf species that ranges from the Indo-West Pacific to both sides of the Atlantic.

though the Old World Land Barrier may be about as effective for the invertebrates as it is for fishes. It appears, especially from the ichthyological evidence, that the limited colonization movement of tropical shore species around the Cape takes place entirely in a westerly direction, from the Indo-West Pacific to the Atlantic.

Conclusions

It may be said that the available information on the relationship of the tropical shelf regions permits the identification of three important facts: first, very few species have been able to migrate from one region to another and successfully establish themselves; second, the four zoogeographic barriers differ considerably in their efficiency or effectiveness; and third, the successful migratory traffic that does overcome the barriers tends to be unidirectional.

The data on fishes indicate that the New World Land Barrier is the most efficient, since it separates about 99 percent of the species found along the shores of Central America. The Old World Land Barrier is about 96 percent effective, since some migration around the Cape of Good Hope does occur (the relatively recent and apparently local effects of the Suez Canal were not considered in this calculation). The two deep-water barriers are certainly less efficient. The East Pacific Barrier separates about 94 percent of the species on both sides, and the Mid-Atlantic Barrier about 91 percent.

Successful (colonizing) migrations across the zoogeographic boundaries that delimit the Indo-West Pacific can apparently take place in one direction only, outward into areas where the fauna is poorer and the competition is less. The realization that the East Pacific and Old World Land Barriers operate as one-way filters enables us to understand better how the Indo-West Pacific Region serves as *the evolutionary and distributional center* for the tropical shore animals of the world. We can see that competitively dominant species continue to migrate, as they probably have for millions of years, from the Indo-West Pacific eastward across the open ocean to America and westward around the Cape of Good Hope into the Atlantic; since 1869, some of them have also been able to pass northward through the Suez Canal into the Mediterranean (Briggs 1970).

The Western Atlantic Region may be considered a secondary center of evolutionary radiation. Many species evolved in this area have proved capable of migrating eastward to colonize the tropical Eastern Atlantic. However, species originating in the Eastern Atlantic are apparently incapable of successfully invading the western side. Again, the advantage seems to lie with the area that possesses the richer fauna and the higher level of competition.

It can be seen that the completely eastward direction of successful migratory movements across the East Pacific Barrier and the predominantly eastward movements across the Mid-Atlantic Barrier take place in a direction opposite to that of the main flow of the surface waters via the North and South Equatorial Currents. In contrast, the surface and subsurface countercurrents in the tropical Pacific and Atlantic are weakly developed, but these smaller currents are obviously the principal means by which successful transport is achieved.

Fell (1967) noted that certain groups of shore species apparently demonstrated speciation gradients in which the number of species gradually diminished around the world in a westward direction. He interpreted this to mean that the direction of successful migrations had also been to the west and that such dispersals had been carried out by the North and South Equatorial Currents. Subsequently, it was pointed out that the existence of a gradient in numbers of species (or genera) across a major barrier did not necessarily indicate the direction of the original successful migration (Briggs 1967b). The fact that colonizations do take place in a direction opposite to that of the major currents is a good indication that biological competition rather than passive transport is probably the most important factor controlling the successful dispersal of tropical marine shore animals.

The Sea-Level Panama Canal

Man has undertaken major engineering projects for most of his civilized history, and the construction of such necessary facilities as canals, dams, and harbors will continue and expand as the human population grows larger. Although all such projects alter the natural environment to some extent, their effects on animal life are usually limited to local populations. Very seldom is the existence of an entire species threatened. However, the possible disruption of a major zoogeographic barrier, permitting an invasion of competitive species into new areas, poses a conservation problem of a new order of magnitude.

The Atlantic-Pacific Interoceanic Canal Commission has recommended to the President of the United States of America that a new, sea-level canal be excavated across the Isthmus of Panama. This would provide a continuous, saltwater passage for ships and, at the same time, would disrupt the New World Land Barrier, permitting animals to migrate between the Eastern Pacific and Western Atlantic. What would be the effects of such migrations?

Compared to the Western Atlantic, the Eastern Pacific Region is relatively small and its fauna is apparently less diverse. For example, there are probably about 600 species of shore fishes in the Western Caribbean but only about 400 in the Gulf of Panama and its adjacent waters. The Eastern Pacific tropics occupy almost the same amount of geographic area as do the Eastern Atlantic tropics, and the fauna of the latter is apparently only slightly poorer. Since the Mid-Atlantic Barrier has not been a complete block to migration by marine shore species (as has the New World Land Barrier), let us recall an important aspect of the faunal relationship between the two sides of the Atlantic.

Many species that apparently originally evolved in the Western Atlantic have migrated eastward across the Mid-Atlantic Barrier to successfully colonize the Eastern Atlantic, but species originating in the Eastern Atlantic have proved incapable of invading the western side. This may be considered a manifestation of a zoogeographic principle that operates wherever major barriers occur. When two faunally distinct but ecologically saturated areas are separated by a barrier that is partially passable, the one possessing the richer (more diverse) fauna will donate species to the poorer area but will seldom, if ever, accept species in return (p. 438).

If a sea-level canal is excavated across Panama, there is no reason to suspect that the foregoing principle would not apply. It has been argued that such a canal would not provide a very effective migratory route, yet we need only to consider that 80 to 85 percent of all tropical, benthic invertebrate species have planktonic larvae (Thorson 1966) and that most fish larvae and adults are relatively mobile. It is difficult to believe that, over the years, the majority of the shelf species in the vicinity would *not* migrate through a sea-level canal.

What happens to the fauna in an area that is apparently ecologically saturated (as most mainland shore areas probably are) when additional species are introduced? It seems reasonable to expect, if we can profit by records of similar happenings in the terrestrial environment, that the faunal enrichment would be temporary and, by means of the elimination of species through competition, the area would eventually return to its original level of diversity. Should a sea-level Panama canal be constructed, it has been predicted that the fauna of the tropical Eastern Pacific would be temporarily enriched by the arrival of species from the Western Atlantic, and since the Atlantic species come from a more diverse region, the invaders would tend to outcompete and replace the native species (Briggs 1968, 1969).

The tropical faunas on each side of Central America are not only very rich but, in general, are so poorly known that marine biologists simply cannot say how many species are present. As a very rough estimate, there are probably more than 8,000 species on the Atlantic side and nearly 6,000 species on the Pacific side. At least, it is evident that the disruption of the barrier would bring into contact for the first time very large numbers of distinct species. In the Eastern Pacific, this could result in the irrevocable loss of not one or two but hundreds or even thousands of species.

Because of its relatively diverse and stable ecosystem, the Western Atlantic would, in terms of numbers of species, remain relatively little affected by a sea-level canal. However, we may recall that a small but significant portion of the Eastern Pacific fauna consists of Indo-West Pacific species that have made their way across the East Pacific Barrier. Among these are the crown-of-thorns starfish (*Acanthaster planci*) and the poisonous sea snake (*Pelamis platurus*). The starfish feeds on living coral and is the species responsible for widespread destruction in the Great Barrier Reef and other parts of the Western Pacific. The sea snake is highly poisonous and is capable of causing human fatalities.

Should some of the Indo-West Pacific species that now also exist in the Eastern Pacific be able to migrate through a sea-level canal, it is expected that they would prove to be competitively dominant and would successfully establish themselves. Considering that, in the Western Atlantic, the prey of such species would consist of organisms that have had no opportunity to evolve defensive mechanisms, the result may well be a population explosion, with the invading species becoming far more numerous than in their native territories. This happened when the parasitic sea lamprey gained access to the western Great Lakes via the Welland Ship Canal and virtually destroyed the native populations of lake trout.

Aside from the possible economic and human welfare aspects of invasions by particular organisms, it is the prospect of the terrible loss of species in the Eastern Pacific that presents the real conservation problem. What is the value of a unique species? Do we have so many that we can afford to perform alterations to the earth's surface that are likely to result in large-scale extinctions? If one undertook the problem of devising a single construction project that would eliminate the greatest number of species from the face of the earth, none would be more effective than a sea-level passage through the New World Land Barrier—the most efficient of all the barriers separating the tropical marine faunas of the world.

The Suez Canal, although it has permitted a number of Red Sea species to invade the eastern Mediterranean, fortunately does not possess much of a potential for species extinction. Almost all of the Mediterranean Sea has winter surface temperatures that drop well below 20°C, so that it constitutes a warm-temperate rather than a tropical environment and is inhabited by a distinct warm-temperate fauna. Thus, the Mediterranean fauna is protected by its climate from extensive invasions by tropical animals. No such temperature barrier to migration would exist in the advent of a sea-level Panama canal.

In the United States, the controversy over the advisability of building a sea-level canal is likely to go on for years. The most attractive, and far more economical, alternative is to improve the present canal but maintain its basic structure. If this can be done, we would still have a high-level canal with a freshwater block to prevent the migrations of marine animals. For the sake of our biological heritage, let us hope that the latter alternative will be chosen.

Latitudinal (Temperature) Barriers

In determining the latitudinal boundaries of each of the four tropical regions, an objective attempt was made to gather all possible evidence in order that a boundary could be located where the rate of species change was the greatest, that is, where the most tropical forms dropped out and/or where the most temperate species began. As a result, boundaries were proposed as follows:

1 Indo-West Pacific—Amami Islands, Taiwan, and Hong Kong in the north. In the south, Fraser Island and Shark Bay in Australia and the mouth of the Kei River in South Africa.

2 Eastern Pacific—Magdalena Bay, Baja California, the mouth of the Gulf of California, and the southern Gulf of Guayaquil, Peru.

3 Western Atlantic—Cape Kennedy and Cape Romano, Florida, and Cape Rojo, Mexico, to Cape Frio, Brazil.

4 Eastern Atlantic—Cape Verde to Mossâmedes, Angola.

The important question is: What do these localities have in common that permits them to operate as major zoogeographic barriers? Biologists have been aware for a long time that the distributional patterns of most marine organisms reflect a high degree of temperature sensitivity. Beyond this, however, there has been little agreement. The most influential of the modern writers (Ekman 1953:57, Hedgpeth 1957:370, Coomans 1962) have utilized annual mean surface isotherms to help define zoogeographic regions. Others (Stechell 1920, Wells 1963) have used isotheres (lines of mean maximum surface temperature), and still others (Dana 1853, Forbes 1859, Gill 1884, Monod 1957) have used isocrymes (lines of mean minimum surface temperature).

It is possible to offer some solution to this problem, at least as far as the tropical fauna is concerned, for when the above-listed places are located on the monthly sea-surface temperature charts it can be seen that they fall on or very close to the 20°C isocryme line (calculated from the mean for the month of February in the Northern Hemisphere and for August in the Southern Hemisphere).

Many years ago, James D. Dana took part in the worldwide United States Exploring Expedition, 1838 to 1842. This trip did much the same thing for Dana as did the voyage of the *Beagle* for a young Englishman named Charles Darwin a short time earlier (1831 to 1836). It made Dana aware of some interesting facts about zoogeography and led to the publication of some classic works. Dana, after writing major reports on the corals and crustaceans collected by the expedition, concluded (1853), "The cause which limits the distribution of species northward or southward from the equator is the cold of winter rather than the heat of summer or even the mean temperature of the year." His remarks were accompanied by a chart relating his "isocrymal lines" to the distribution of marine animals.

The 20°C isotherm for the coldest month of the year (map inside the covers) is most important, since it appears to limit the tropical fauna regardless of whether it is rich (Indo-West Pacific) or poor (Eastern Atlantic) or whether the boundary is located to the north or south.

Higher Latitudes In addition to the tropics, the other major, shallow-water zones may be characterized by their temperature regimes. In most warm-temperate regions, the average temperatures for the coldest month range from 20°C down to about 13°C; however, in some places, such as the North Atlantic, a warm-temperate fauna may be found where the coldest month average reaches as low as 10°C. In general, the cold-temperate regions have winter averages ranging from about 13°C down to 2°C. The cold zones of the world (Arctic and Antarctic) have winter averages of about +2 to −2°C.

Literature Cited

Barash, A., and Z. Danin. 1971. *Indo-Pacific species of Mollusca in the Mediterranean.* Appendix to Progress Report, Hebrew University—Smithsonian Institution Joint Project, pp. 1–8.

Ben-Eliahu, M. 1971. *Polychaeta Errantia of the Suez Canal.* Appendix to Progress Report, Hebrew University—Smithsonian Institution Joint Project, pp. 1–18.

Ben-Tuvia, A. 1971a. *Revised list of the Mediterranean fishes of Israel.* Appendix to Progress Report, Hebrew University—Smithsonian Institution Joint Project, pp. 1–7.

———. 1971b. On the occurrence of the Mediterranean serranid fish *Dicentrarchus punctatus* (Bloch) in the Gulf of Suez. *Copeia*, (4):741–743.

Blache, J., J. Cadenat, and A. Stauch. 1970. Clés de détermination des poissons dans l'Atlantique oriental. *Faune Trop., XVIII, O.R.S.T.O.M.*, Paris, pp. 1–479, 1,152 figs.

Briggs, J. C. 1955. A monograph of the clingfishes (Order Xenopterygii). *Stanford Ichthyol. Bull.*, 6:1–224, 114 figs.

———. 1960. Fishes of worldwide (circumtropical) distribution. *Copeia*, (3):171–180.

———. 1961. The East Pacific Barrier and the distribution of marine shore fishes. *Evolution*, 15(4):545–554, 3 figs.

———. 1964. Additional transpacific shore fishes. *Copeia*, (4):706–708.

———. 1967a. Relationship of the tropical shelf regions. *Studies Trop. Oceanog. Miami*, 5:569–578.

———. 1967b. Dispersal of tropical marine shore animals: coriolis parameters or competition? *Nature*, 216(5113):350.

———. 1968. Panama's sea-level canal. *Science*, 169(3853):511–513.

———. 1969. The sea-level Panama canal: potential biological catastrophe. *BioScience*, 19(1):44–47, 1 fig.

———. 1970. Tropical shelf zoogeography. *Proc. Calif. Acad. Sci.*, 38(7):131–138.

Bruce, A. F. 1970. On the identity of *Periclimenes pusillus* Rathbun, 1906 (Decapoda, Pontonunae). *Crustaceana*, 19(3):308–310, 1 fig.

Burton, M. 1956. The sponges of West Africa. *Atlantide Rep.* (4):111–147, 4 figs.

Chace, F. A., Jr. 1962. The non-brachyuran decapod crustaceans of Clipperton Island. *Proc. U. S. Nat. Museum*, 113(3466):605–635, 7 figs.

Chesher, R. H. 1966. Report on the Echinoidea collected by R/V Pillsbury in the Gulf of Guinea. *Studies Trop. Oceanog. Miami*, 4(part 1):209–223.

Coomans, H. E. 1962. The marine mollusk fauna of the Virginian area as a basis for defining zoogeographical provinces. *Beaufortia*, 9(98):83–104.

Dana, James D. 1853. On an isothermal oceanic chart illustrating the geographical distribution of marine animals. *Amer. J. Sci.*, 16:314–327, 1 map.

Deichmann, E. 1963. The holothurians of Clipperton Island in the Eastern Tropical Pacific. *Breviora*, (179):1–5.

Dekeyser, P. L. 1961. Liste provisoire des urocordés de la côte occidentale d'Afrique. *Bull. Inst. Franç. d'Afrique Noire, Ser. A*, 23(1):217–230.

Durham, J. W. 1966. Coelenterates, especially stony corals, from the Galápagos and Cocos Islands. *in* R. I. Bowman (editor), The Galápagos. University of California Press, Berkeley, pp. 123–135.

Ekman, S. 1946. Zur Verbreitungsgeschichte der Warmwasserechinodermen in stillen Ozean (Asteroidea, Ophiuroidea, Echinoidea). *Nova Acta Regiae Soc. Sci. Upsaliensis*, 14(2):1–42, 1 map.

———. 1953. *Zoogeography of the sea*. Sidgwick & Jackson, London, xiv + 417 pp., 121 figs.

Emerson, W. K. 1967. Indo-Pacific faunal elements in the tropical eastern Pacific, with special reference to the mollusks. *Venus, Japan. J. Malacol.*, 25(3 and 4):85–93.

Evermann, B. W., and O. P. Jenkins. 1891. Report on a collection of fishes made at Guaymas, Sonora, Mexico, with descriptions of new species. *Proc. U. S. Nat. Museum*, 14(846):121–165, 2 plates.

Fell, H. B. 1967. Resolution of coriolis parameters for former epochs. *Nature*, 214:1192–1198, 7 figs.

Forbes, E. 1859. *The natural history of European seas* (edited and continued by Robert Godwin-Austen). John Van Voorst, London, viii + 306 pp., 1 map.

Garth, J. S. 1965. The brachyuran decapod crustaceans of Clipperton Island. *Proc. Calif. Acad. Sci.*, 33(1):1–46, 26 figs.

Gilbert, C. H., and E. C. Starks. 1904. The fishes of Panama Bay. *Mem. Calif. Acad. Sci.*, 4:1–304, 33 plates.

Gill, T. 1884. The principles of zoogeography. *Proc. Biol. Soc. Wash.*, 2:1–39.

Gohar, H. A. F. 1954. The place of the Red Sea between the Indian Ocean and the Mediterranean. *Pub. Hydrobiol. Res. Inst., Univ. Istanbul, Ser. B*, 2(2–3): 1–38, map.

Gosline, W. A., and V. E. Brock. 1960. *Handbook of Hawaiian fishes*. University of Hawaii Press, Honolulu, ix + 1–372 pp., 277 figs., 2 plates.

Günther, A. C. L. G. 1861. On a collection of fishes sent by Capt. Dow from the Pacific coast of Central America. *Proc. Zool. Soc. London*, (3):370–376.

———. 1868. An account of the fishes of the states of Central America, based on collections made by Capt. J. M. Dow, F. Godman, Esq., and O. Salvin, Esq. *Trans. Zool. Soc. London*, 6(7):377–494, 24 plates, 1 map.

———. 1880. *An introduction to the study of fishes*. Adam and Charles Black, Edinburgh, xvi + 720 pp., 320 figs.

Haig, J. 1956. The Galatheidea (Crustacea Anomura) of the Allan Hancock Atlantic Expedition with a review of the Porcellanidae of the western North Atlantic. *Allan Hancock Atlantic Expedition*, (8):1–44, 1 plate.

———. 1960. The Porcellanidae (Crustacea Anomura) of the Eastern Pacific. *Allan Hancock Pacific Expedition*, 24:viii + 440, 12 figs., 42 plates.

Hedgpeth, J. W. 1957. Marine biogeography. *in Treatise on marine ecology and paleoecology.* vol. 1. *Geol. Soc. Amer., Mem.,* (67):359–382, 16 figs., 1 plate.

Holthuis, L. B., and E. Gottlieb. 1958. An annotated list of the decapod Crustacea of the Mediterranean coast of Israel, with an appendix listing the decapoda of the eastern Mediterranean. *Bull. Sea. Fish. Res. Sta. Israel,* (18):1–126, 15 figs.

Hyman, L. H. 1955. *The invertebrates.* vol. 4. *Echinodermata.* McGraw-Hill, New York, 763 pp., 280 figs.

Jordan, D. S. 1885. A list of the fishes known from the Pacific coast of tropical America, from the Tropic of Cancer to Panama. *Proc. U. S. Nat. Museum,* 8(23):361–394.

———. 1901. The fish fauna of Japan, with observations on the geographical distribution of fishes. *Science,* new series 14(354):545–567.

Knudsen, J. 1956. Marine prosobranchs of tropical West Africa (Stenoglossa). *Atlantide Rep.,* (4):8–110, 2 figs., 4 plates.

de Laubenfels, M. W. 1936. A comparison of the shallow-water sponges near the Pacific end of the Panama Canal with those at the Caribbean end. *Proc. U. S. Nat. Museum,* 83(2993):441–466, 6 figs.

Manter, H. W. 1963. The zoogeographical affinities of trematodes of South American freshwater fishes. *Syst. Zool.,* 12(2):45–70, 12 figs.

Marcus, E., and E. Marcus. 1966. Opisthobranchs from tropical West Africa. *Studies Trop. Oceanog. Miami,* 4(part 1):152–208, 62 figs.

Mayr, E. 1954. Geographic speciation in tropical echinoids. *Evolution,* 8(1):1–18, 7 figs.

Meek, S. E., and S. F. Hildebrand. 1923–1928. The marine fishes of Panama. *Publ. Field Museum Nat. Hist., Zool. Ser.,* 15(215):xxx + 1045, 102 plates.

Monod, T. 1956. Hippidea et brachyura ouest-africains. *Mem. Inst. Franç. d'Afrique Noire,* (45):1–674, 884 figs., 1 map.

———. 1957. Scarides et milieu corallien: notes biogeographiques. *Proc. 8th Pacific Sci. Congr.,* 3:971–978, 2 plates.

Oren, O. H. 1969. Oceanographic and biological influence of the Suez Canal, the Nile and Aswan Dam on the Levant Basin. *Progr. Oceanog.,* 5:161–167.

Patterson, B., and R. Pascual. 1968. Evolution of mammals on southern continents. V. The fossil mammal fauna of South America. *Quart. Rev. Biol.,* 43(4):409–451, 13 figs.

Por, F. D. 1971. *The nature of the Lessepsian migration through the Suez Canal.* Progress Report, Hebrew University—Smithsonian Institution Joint Project, pp. 1–7.

Rosenblatt, R. H., J. E. McCosker, and I. Rubinoff. 1972. Indo-West Pacific fishes from the Gulf of Chiriqui, Panama. *Contrib. Sci., Nat. Hist. Museum Los Angeles,* (234):1–18, 3 figs.

Ruggieri, G. 1967. The Miocene and later evolution of the Mediterranean Sea. *Syst. Ass. Pub.,* (7):283–290, 2 figs.

Savage, R. J. G. 1967. Early Miocene mammal faunas of the Tethyan region. *in* C. G. Adams and D. V. Ager (editors), *Aspects of Tethyan biogeography. Syst. Ass. Pub.,* (7):247–282, 3 figs.

Schultz, L. P. 1958. Review of the parrotfishes family Scaridae. *Bull. U. S. Nat. Museum*, (214):v + 1–143, 31 figs., 27 plates.

————. 1957. The frogfishes of the family Antennariidae. *Proc. U. S. Nat. Museum*, 107(3383):47–105, 8 figs., 14 plates.

Simpson, G. G. 1965. *The geography of evolution.* Chilton Books, Philadelphia, x + 249 pp., 45 figs.

Springer, Victor G. 1958. Systematics and zoogeography of the clinid fishes of the subtribe Labrisomini Hubbs. *Publ. Inst. Marine Sci.*, 5:417–492, 4 figs., 7 plates.

Stechell. W. A. 1920. Stenothermy and zone-invasion. *Amer. Nat.*, 54(634):385–397.

Stephens, J. S., Jr. 1963. A revised classification of the blennioid fishes of the American family Chaenopsidae. *Univ. Calif. Pub. Zool.*, 68:iv + 1–133 pp., 11 figs., 15 plates.

Talbot, F. H., and M. J. Penrith. 1962. Tunnies and marlins of South Africa. *Nature*, 193(4815):558–559.

Thorson, G. 1966. Some factors influencing the recruitment and establishment of marine benthic communities. *Netherlands J. Sea Res.*, 3(2):267–293.

Wells, G. P. 1963. Barriers and speciation in lugworms. *Syst. Ass. Publ.*, (5):79–98, 6 figs.

The Southern Ocean

Southern
Hemisphere
Warm-Temperate
Regions

*The successive steps in the progress must appear first in some comparatively
limited region, and from that region the new forms must spread out, displacing
the old. . . . At any given period, the most advanced and progressive species of
the race will be those inhabiting that region; the most primitive and unprogres-
sive species will be those remote from this center.*

William Diller Matthew in Climate and Evolution, *1915*

The shallow, warm-temperate waters of the Southern Hemisphere may be divided into five distinct zoogeographic regions: Southern Australia, Northern New Zealand (including the Kermadec Islands), Western South America (with the Juan Fernández Islands), Eastern South America, and Southern Africa (with Amsterdam and Saint Paul Islands). The regions are well separated from one another both geographically and biologically. All except Eastern South America are subdivided into provinces.

Regional Relationships

A number of very interesting and peculiar shore fish families are confined mainly to the southern warm-temperate waters. Six of these occur in three or more regions in this part of the world. These are the Cheilodactylidae, Aplodactylidae, Latridae, Congiopodidae, Oplegnathidae, and Histiopteridae. Such families appear to be primarily phylogenetic and geographic relicts, and all but one (the Oplegnathidae with three species in southern Africa) appear to have undergone their main, recent development in southern Australia. For example, the Cheilodactylidae is represented by 12 species in southern Australia, 4 in New Zealand, 4 in western South America, 1 in eastern South America, and 4 in southern Africa. The family Latridae has 6 species in southern Australia, 5 in New Zealand, 3 in western South America, and none in eastern South America or mainland southern Africa.

The foregoing families were probably once widespread in the Pacific (species representing three of them may be found in warm-temperate Japan) and, in the Southern Hemisphere, may have been originally isolated in southern Australia. From there, a clockwise West Wind Drift dispersal, similar to that demonstrated for certain groups of echinoderms by Fell (1962), could have taken place. In the cases in which the species in the various regions are different from one another, the migration probably took place long ago, but where the species are the same, a recent migration can be assumed.

Holthuis (1963) has shown that the spiny lobster genus *Jasus* has essentially a warm-temperate, circumpolar distribution. Single, endemic species are found in South Africa, Amsterdam–St. Paul Islands, southeastern Australia-Tasmania, New Zealand, Juan Fernández Islands, and Tristan da Cunha Island. The ophidiid fish genus *Xiphiurus* also has a circumglobal distribution in the warm-temperate parts of the Southern Hemisphere. The same species, *X. blacodes*, is found in Australia, New Zealand, and on both sides of South America. A related species *X. capensis*, is found in southern Africa. Despite these indications of relationship among the southern regions, the general impression is one of extensive differences. The various faunas are comprised mainly of families and genera of tropical relationships and, except for Australia–New Zealand, their evolutionary derivation has proceeded independently.

Southern Australia (including Tasmania in many cases) is blessed with an old and highly peculiar fish fauna. In addition to serving as a distributional center for the families listed above (p. 126), it possesses six families that are found nowhere else: the Alabidae, Siphonognathidae, Peronedysidae, Dinolestidae, Gnathanacanthidae, and Pataecidae. In addition, it shares the family Leptoscopidae only with New Zealand, the Chironemidae with New Zealand and Lord Howe Island, and the Arripidae with New Zealand and Lord Howe and Norfolk Islands. This situation seems to parallel that in the marine algae, for in this area, there is probably more algal endemism than in any place in the world (Womersley 1961).

Warm-temperate New Zealand presents an interesting contrast to Australia. It shares three peculiar fish families that are found nowhere else except at an offshore island or two. However, with the possible exception of the Hemerocoetidae (which probably does not warrant separate family status), it has no endemic families of its own. Also, as Moreland (1959) has pointed out, the largest single element of the shore fish fauna is that shared with southeastern Australia. The mollusks (Powell 1962) and echinoderms (Fell 1949) also show a strong relationship of this kind.

Although Knox (1963b:361) gives credit to Fleming (1962) for picturing New Zealand and Australia in the lower Cretaceous as a connected land mass, Fleming's treatise does not show this. In fact, New Zealand could have been isolated throughout the Mesozoic. The present dual biogeographic pattern with a warm-temperate biota of tropical derivation in the north and a cold-temperate assemblage of circumpolar relationship to the south may have had its beginnings as long ago as the Permian (Fleming 1962:60).

Southern Africa gives ample evidence of long isolation. It has two strange endemic fish families (Halidesmidae and Parascorpididae), an endemic subfamily of clingfishes (Chorisochisminae), and many indigenous genera belonging to widespread tropical families. The two mainland provinces of the region occupy adjacent areas and meet at the Cape of Good Hope, yet one has a very rich fauna with a high incidence of endemism while the other has a relatively poor fauna with lesser endemism.

The rich area of southern Africa (Agulhas Province) has a fairly wide annual fluctuation in surface temperate from an August mean of about 14 to 16°C to a February mean of 20 to 22°C, while the poor area (South-West Africa just north of the Cape) has a lesser range from about 13 to 15°C to 16 to 18°C. It is probable that most of the Agulhas species, being of tropical derivation, still need relatively high temperatures for reproductive purposes. The lack of high temperatures in the summer could be an important factor in preventing migration around the Cape of Good Hope from east to west.

The two warm-temperate parts of South America share very few species, and the two faunas, for the most part, have been independently derived from tropical groups to the north. As would be expected, there is considerable similarity at the generic and family levels, since most of these higher categories have an amphi-American distribution. The Peru-Chilean fauna is a good deal better known that than of eastern South America and, perhaps for this reason, seems to be the richer. In contrast to southern Africa and southern Australia, there are no endemic shore fish families and few such genera.

The Southern Australia Region

The coast of southern Australia comes under the influence of three important ocean currents. To the west, the West Australian Current—the eastern arc of the main gyre of the south Indian Ocean—affects the coast from about Cape Leeuwin north as far as Shark Bay. The south coast from Cape Leeuwin to Victoria seems to receive contributions from the West Wind Drift. To the east, the East Australian Current comes close to the coast about in the area between Brisbane and Cape Howe. It is a southern branch of the South Equatorial Current of the Pacific Ocean and tends to form a small gyre of its own in the Tasman Sea between southeastern Australia and New Zealand.

The extreme southern part of the continent is also affected by a small, inshore branch of the East Australian Current that runs through Bass Strait between Tasmania and the mainland and then extends westward, paralleling the shore until it reaches the Indian Ocean. Although this current may be subject to seasonal reversal (Knox 1963b:346), it probably still serves to modify the influence of the cold water moving north from the West Wind Drift.

The early zoogeographers—Forbes (1856), who outlined the distribution of marine life in general, and Woodward (1856:376), who worked on the Mollusca—thought that southern Australia, Tasmania, and New Zealand should all be included in one faunal area. It was noted (p. 25) that Whitley (1932) summarized the findings of the early Australian zoologists in his marine zoology map of the Australian–New Zealand area. For southern Australia, Whitley recognized a "Flindersian" province from Geraldton, Western Australia to Wilson's Promontory, Victoria, and a "Peronian" province from the latter to about the northern tip of Fraser Island, Queensland. Tasmania was placed in a separate "Maugean" province.

Whitley's (1932) scheme was published in an obscure journal and was, at first, overlooked by biologists who worked in other countries. Clark (1946), on the basis of echinoderm distributions, recognized the Flindersian and Peronian faunas but extended their boundaries to southern Tasmania with a dividing line at Hobart. Kott (1952), from her work on the simple ascidians, added to Whitley's arrangement a "Baudinian" region for the area between Perth and Albany and an "Oxleyan" subregion for the coast between Sydney and Brisbane. Stephenson's (1947:plate 15) distributional map for the southern littoral faunas showed a single, warm-temperate zone for all of southern Australia and Tasmania. Ekman (1953:197) also recognized but one homogeneous, warm-temperate region including all of the area to the south of Perth on the west coast and Sydney on the east coast. Schilder's (1956:75) map indicated that he agreed with Stephenson and Ekman.

It was the well-documented study of the intertidal fauna and flora of Victoria by Bennett and Pope (1953) that first demonstrated the presence of a cold-temperate biota in this area and made it possible to see why two separate warm-temperate provinces could exist. The map of the Australian littoral provinces by Bennett and Pope (op. cit., 139) is somewhat similar to that offered by Whitley except that the shores of Victoria are linked with those of Tasmania in a Maugean "cool-temperate" province that serves, at least partially, to separate the south and east coast warm-temperate areas. Their arrangement was followed by Hedgpeth (1957:map). Womersley and Edmonds (1958) considered that southern Australia was essentially cold-temperate and belonged in the same province with Victoria and Tasmania. This suggestion was refuted by Bennett and Pope (1960) who, after field work in Tasmania, modified their original scheme only slightly. Knox (1960, 1963b) indicated agreement with the latter authors in regard to the extent of the cold-temperate area but considered the Flindersian province to the west of Victoria to harbor a transitional rather than a true warm-temperate biota.

The concept of a cold-temperate Tasmania-Victoria fauna is supported by the distributional patterns of both fishes and marine mammals. It is pointed out (p. 159) that the family Bovichthyidae is almost entirely restricted to southern cold-temperate waters. The presence of two species of this family in the area (McCulloch 1929–1930:336) is believed to have some significance. The herring, *Clupea bassensis*, is characteristic of Tasmanian waters (Blackburn 1941:53), and this pattern is consistent with the known arctic-boreal distribution of its relative *C. harengus* in the Northern Hemisphere. Davies (1958) showed that the area in question is occupied by the species of phocid and bladdernosed seals that, elsewhere, are good indicators of cold-temperate conditions.

The field investigations of Bennett and Pope (1953, 1960) and the biogeographic summaries of Knox (1960, 1963b) indicated that the cold-temperate zonation probably extends from Robe on the southern coast of Australia, just north of the Victorian border, eastward to Bermagui, New South Wales, also just north of the Victorian border. Bennett and Pope (1953) also noted the presence, in this area, of an extensive overlap between the two warm-temperate biotas and the cooler one of Victoria. It is now apparent that the Victorian coast acts as a barrier, more or less effectively separating two interesting, warm-temperate faunas. The proof of the barrier effect lies in the significant degree of endemism demonstrated by each.

The Provinces
Southwestern Australia

It was previously determined (p. 25) that the tropical-temperate boundary was probably located at Shark Bay, Western Australia. The Southwestern Australia Warm-Temperate Province can then be considered to occupy the coastline from that point to Robe, South Australia. So far, it has not been demonstrated that sufficient endemism occurs along the west coast portion of this area to consider designating it a separate province. The provisional recognition by Knox (1963b:377) of a "West Australian Province" seems to be based more on the presence of a different species combination in the intertidal community than on a truly distinctive biota.

Bennett and Pope (1953) mentioned several invertebrate species that are confined to the Southwestern Australia Province: the echinoderms *Holopneustes porossisimus* and *H. inflata*, the chitons *Clavarizona hirtosa* and *Onithochiton occidentalis*, and the gastropod *Ninella torquata whitleyi* and, perhaps, the gastropod *Dicathais aegrota*. Clark (1946:470), in his monograph of the echinoderm fauna of Australia, gave a total of 166 species for the area and listed 21 (12.6 percent) that

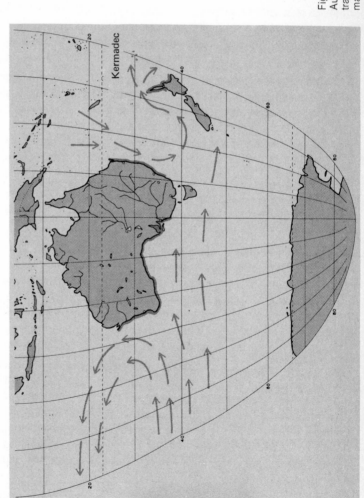

Figure 6-1 Southwestern Australia, Southeastern Australia, Auckland, and Kermadec Provinces.

are endemic. He also provided the names of 13 that are shared with the Southeastern Australia Province.

The work by Scott (1962) on the fishes of South Australia has been of great help in shedding some light on geographic relationships. A total of 253 shore fishes is listed, and 72 of these, or 28.4 percent, are apparently endemic to the Southwestern Province. Another group of 42 species is shared only with Tasmania indicating a surprisingly strong relationship with this area. Some 38 species have a regional warm-temperate distribution occurring in both the Southwestern and South-eastern Provinces but not elsewhere (except Victoria where there is evidence of considerable overlap). A larger group (63) seems to have a general temperate distribution being found in both the preceding provinces and in Tasmania-Victoria. The remaining 38 species are mainly eurythermic tropical forms that occur all around the continent.

Considering endemic rates of at least 12.6 percent in the echinoderms and 28.4 percent in the fishes, there is little doubt that Southwestern Australia should be considered a distinct province. Although the relationship to the Tasmania-Victoria fauna is strong, the latter is well characterized by the presence of distinctive cold-temperate elements and a significant amount of endemism (Bennett and Pope 1960).

Southeastern Australia

The Southeastern Australia Province includes the continental shelf from Sandy Cape on Fraser Island, Queensland, a comparatively well-established boundary (p. 24), south to Bermagui, New South Wales.

Our knowledge of the intertidal invertebrate fauna and its relationship is due mainly to the work of Dakin, Bennett, and Pope (1948) and Bennett and Pope (1953). In the latter work, lists of characteristic barnacles, echinoderms, chitons, gastropods, crustaceans, and ascidians were given. Although no estimate was made, it would appear that the rate of endemism is quite high, perhaps greater than in the Southwest Province. Clark (1946:468) noted that 107 species of echinoderms occurred in the province and gave a list of 21 that he considered to be "typical."

As would be expected of two provinces that are included in the same zoogeographic region, there is a basic distributional relationship. It has already been noted that 38 fish species and 13 echinoderms are shared. Bennett and Pope (1953) emphasized that a number of inter-tidal species ranged through both provinces but were absent along the Victoria coast. Examples given were the crabs *Leptograpsus variegatus* and *Cancellus typus*, the urchin *Heliocidaris erythrogramma*, and the asteroid *Patiriella gunnii*.

Further evidence of relationship between the two provinces may be found in the presence of closely related pairs of species or subspecies with one member being found in each area. Some of these are the urchins *Holopneustes porossismus* and *H. pycnotilus*, the chitons *Onithochiton quercinus* and *O. occidentalis*, and the gastropods *Ninella torquata torquata* and *N. t. whitleyi* (Bennett and Pope, op cit.). Similar pairs in the fishes are the sparids *Chrysophrys unicolor* and *C. guttulatus*, the Australian salmons *Arripis trutta esper* and *A. t. marginata*, and the velvet fishes *Aploactisoma milesii milesii* and *A. m. horrenda* (Scott 1962).

The most recent account of the fishes of New South Wales, the third edition of the book by McCulloch (1934), is so far out of date that one hesitates to rely on it for distributional data. However, according to Munro's (1956–) handbook on Australian fishes, many of the species that occur in the Southeastern Province seem to be confined to it, such as 3 of the 6 carpet sharks of the family Orectolobidae, 5 of the 7 soles (Solidae), and 11 of the 22 pipefishes and sea horses (Syngnathidae). At this time, only a very rough estimate of the extent of endemism in the shore fishes can be given—about 30 percent. At least, it seems that enough endemism exists in both fishes and invertebrates so that a Southeastern Province can be recognized.

The Northern New Zealand Region

The details of the New Zealand coastal current systems have been illustrated, mainly from Brodie's (1960) data, by Knox (1960:580, 1963a:12, 1963b:346). According to this interpretation, the Auckland Province (northernmost New Zealand) is under the influence of Trade Wind Drift water that comes in from the northeast. This water is depicted as giving rise to a West Auckland Current that flows southward along the west side of the Auckland Peninsula and an East Auckland Current that flows down the east side of the peninsula continuing as far as the East Cape vicinity.

The difficulty with the foregoing scheme is that it does not fit in with the faunal relationships. If the Auckland Province was consistently affected by surface water from the indicated area, one would expect to find a fauna (and flora) comprised mainly of eurythermic tropical species originating from the Samoa-Tonga area. Instead, such species are in the minority and there is a strong relationship to southeastern Australia (p. 135). This suggests that the area in question comes mainly under the influence of water from the Tasman Current, a continuation of the East Australian Current.

Although the tropical component is stronger, the Kermadec Islands also show faunal ties with northern New Zealand and with Australia. This situation too seems to require some influence by the East Australian–

Tasman Current system instead of simply the Trade Wind Drift from the northeast as was indicated by Knox (op. cit.).

The Northern New Zealand Region may be divided into two provinces: Auckland and the Kermadec Islands.

The Provinces
Auckland

Whitley (1932), in summarizing the findings of some of the early Australian zoologists, was one of the first to recognize that New Zealand belonged to a zoogeographic region separate from southern Australia. He delineated a "Neozelanic Region" that included all of New Zealand plus the Lord Howe, Norfolk, Kermadec, and Chatham Islands. The entire North Island of New Zealand comprised the "Cookian Province." Stephenson (1947:plate 15) gave no individual name to the area but indicated that all of New Zealand and the Chatham Islands were in the warm-temperate zone.

Although Ekman (1953) mentioned the molluscan faunal regions of Finlay (1925) and Powell (1937), he also considered all of New Zealand and the Chatham Islands to be warm-temperate. Schilder's (1956:75) map showed agreement with Stephenson and Ekman, but Hedgpeth's (1957) map illustrated a warm-temperate area extending only from the North Cape and its peninsula across to the East Cape. Knox (1960:611, 1963a) recognized the same limits using the name "Aupourian Province" proposed by Powell (1937). Later, Knox (1963b) utilized the same name but published a distributional map showing the west coast boundary shifted southward to about the Auckland vicinity. He also considered the biota to be a "transitional" rather than a true warm-temperate one.

As Knox (1963b) pointed out, the original schemes for subdivision of the New Zealand area were based on molluscan distribution. At first, Powell (1937) considered his Aupourian Province to be restricted to the northern tip of the Auckland Peninsula coming south only as far as Ahipara on the west coast and Whangaroa on the east coast. More recently, Powell (1962:11) extended the western boundary to a "division of some elasticity" between Ahipara and Manukau Heads and the eastern boundary to East Cape. On the basis of echinoderm distribution, Pawson (1961:13) would place the southern boundary in the west as far south as Cape Egmont. On the basis of algal distributions, Moore (1961) recognized an "Auckland Province" for the region north of Albatross Point and East Cape.

There is no doubt that, in general, the New Zealand invertebrate fauna is highly endemic, demonstrating the effects of prolonged isolation. For example, 80 percent of the echinoderms (Fell 1949), 46 percent of the thecate hydroids (Knox 1963b:359), and 31 percent of the polychaetes (Knox op. cit., 360) are indigenous. However, as far as the Auckland Province is concerned, it is more difficult to find detailed information. Powell (1940) listed 649 species of mollusks for the province and considered 261, or 40 percent, to be endemic, but some deep-water species were included. Knox (op. cit., 387) spoke of strong endemic elements in the major littoral biota and listed some characteristic algae, mollusks, barnacles, polychaetes, and echinoderms.

A thorough analysis of the marine fish fauna has never been made, although Moreland (1959) contributed a general discussion of the subject. According to his data, about 22 percent of the New Zealand fauna is made up of endemic species and the greatest degree of endemism is shown by the fishes of the littoral and sublittoral rocky shore, especially the families Acanthoclinidae, Gobiesocidae,[1] and Tripterygiidae. Moreland further stated that a "subtropical" fauna is confined to the far north, extending south to the eastern Bay of Plenty, and that none of the 50 shore fishes of this group are as yet known south of East Cape.

[1] The clingfishes (Gobiesocidae) are represented by eight species in seven genera, all of them endemic relicts (Briggs 1955:150).

The relationship to the Southeastern Australia Province is quite close. Powell (1940, 1962:11) pointed out that the Aupourian (Auckland) Province was characterized by a large percentage of "Peronian" (Southeastern Australia) organisms carried there by warm-water currents. A recent migratory movement of echinoderm species has taken place from Southeastern Australia to northern New Zealand (Fell 1949). Moreland (1959) remarked that the largest single element of the shore fish fauna is that shared with southeastern Australia and that it includes most, if not all, of those species restricted to the warm Bay of Plenty and North Auckland areas.

Apparently, a large proportion of those species capable of undergoing pelagic transport from southeastern Australia find suitable habitats in northern New Zealand. Since the Peronian (Southeastern Australia) Province was considered typically warm-temperate by Knox (1963b:371), this could be taken as an indication that the Auckland Province also supports a typical warm-temperate assemblage rather than a special *transitional* warm-temperate biota as interpreted by Knox.

Although there is very little information available about the exact amount of endemism, there is no doubt that an Auckland Province should be recognized. In his review of Australasian biogeography, Knox (1963b:367) observed that New Zealand possessed two main centers of distribution, a warm-water center in the north and a cold-water center in the south. He noted further that from each of these centers, characterized by endemic and nonendemic species confined to them, numbers of other species extended varying distances north and south. It may be concluded that the Auckland Province, extending south to about the Auckland vicinity and to East Cape, is a well-defined warm-temperate zone.

The Kermadec Islands are a small, volcanic group located in the South Pacific about 500 miles northeast of Auckland, New Zealand. A Kermadec Province was recognized by Whitley (1932) and, according to Powell (1937), was included as a separate province in Finlay's (1925) list. The fish fauna is very poorly known, the only published work having been done many years ago (Waite 1910, 1911). However, both the mollusks and echinoderms have recently been discussed, and these groups do reveal something about faunal relationships.

Kermadec

A total of 13 echinoderm species was recorded by Pawson (1961), 5 (39 percent) of them endemic, 6 Australian-Indo-Pacific, and 2 shared only with the New Zealand mainland. Of the latter 2, one ranges only to the Bay of Plenty in the Auckland Province, but the other is found throughout New Zealand. In addition, 2 of the endemics are very closely related to New Zealand species and one of the Australian-Indo-Pacific forms is also found in the Auckland Province. Pawson concluded that there was good evidence to include the Kermadecs in the same faunal region with New Zealand.

According to Dell (1957), the molluscan fauna is fairly rich with 85 species, or 34 percent, being endemic. The number of species shared only with New Zealand is small (8), but 16 are shared with both New Zealand and Australia, and 50 more occur in Australia but not in New Zealand. It is significant that a number of mollusk genera (*Endoxochiton*, *Maorichiton*, *Austronotoa*, *Haurakiopsis*, and *Pinnoctopus*) occurring at the Kermadecs are otherwise confined to New Zealand. Dell suggested that the Kermadecs and probably Lord Howe and Norfolk Islands[2] were distinct provinces.

[2]The Lord Howe-Norfolk Province is here considered to belong to the tropical Indo-West Pacific Region (p. 26).

In his review article, Knox (1963b) referred the Kermadec Islands to the "tropical-sub-tropical" coasts and apparently considered the temperature regime of the surface waters to be the same as that of tropical northern Australia. The 1944 U.S. Hydrographic Office Sea Surface Temperature Charts showed for the Kermadecs an August mean of about 17°C. The smaller isothermic map published by Sverdrup, Johnson, and Fleming (1946:chart III) indicated about 18°C for the same time of the year. In contrast, tropical Australia (north of Shark Bay in the west and Fraser Island in the east) has August mean temperatures of 20°C or more. So far, neither the temperature characteristics nor the relationship of the fauna demonstrate that this island group belongs to the tropics.

On the basis of present knowledge, the Kermadecs are considered to be a warm-temperate group with a distinct faunal relationship to New Zealand, particularly the Auckland Province. The high incidence of endemism (34 to 39 percent) shows the need to place these islands in a separate province.

The Western South America Region

The combined coastline of Peru and Chile, covering a straight-line distance of some 3,500 miles, is directly influenced by the Peru (Humboldt) Current. This important ocean stream has been described as the eastern limb of the South Pacific anticyclonic gyratory movement (Gunther 1936:113). However, it must also be noted that the Peru Current is reinforced by a portion of the circum-Antarctic current, the West Wind Drift, that is deflected to the north by the tip of the South American continent. The addition of this cold water is important, for on its way north, it is carried close to the shore and has profound effects on the biota of the shelf.

The prevailing winds, particularly off the Peruvian Coast, are southwest-erlies. These have two very important effects. First, by setting offshore currents in motion, the winds cause upwelling, bringing cooler water to the surface. This is the reason why a warm-temperate regime is found to extend all the way north to the Gulf of Guayaquil, a point only about 3° south of the equator. Second, the upwelling waters bring nutrients to the surface where they can again be cycled by living organisms. This helps make the coast of Peru one of the world's most productive marine areas.

The Juan Fernández Islands also come under the influence of the Peru Current, but they are located at its western edge where the water tends to be warmer than along the coast. The region may be divided in two provinces, one for the mainland and the other for the Juan Fernández Islands.

Perhaps because it occupies such an extensive area, the coastal fauna of Peru and northern Chile has been recognized as a distinct assemblage for well over 100 years. As far as general zoogeographic works are concerned, Forbes (1856) depicted a "Peruvian" fauna extending from the Gulf of Guayaquil to a point just south of Concepción, Chile. Among the modern writers, Stephenson (1947:plate 15), in his map showing the distribution of the southern littoral faunas, located a boundary at the north end of Chiloé Island; Ekman (1953:210) referred to a transitional region between the two temperate faunas approximately north of the island of Chiloé; Balech (1954:192) recognized a "Peruana" province extending to Valparaíso or a little to the south and, from there, a "Centrochilena" province to a second boundary at 40 to 41°S; Schilder (1956:75) placed a boundary somewhere near Concepción; and Hedgpeth's (1957) map indicates about the same place.

A modification of Balech's (loc. cit.) scheme was proposed by Knox (1960). He considered that the coast from Chiloé Island to approximately 30°S formed an extensive transition zone between the cold-temperate and warm-temperate faunas and called it the "Central Chilean" province. This arrangement was criticized by Dahl (1960:633), who stated that a positive characterization of regions and subregions by endemic elements was desirable and on this basis the coast of southern Chile was divisible into only two faunas but with a gradual change from one to the other.

Thanks mainly to a series of illuminating reports on the collections of the Lund University Chile Expedition, 1948 to 1949, several groups of the shore invertebrates have become relatively well known. Since Woodward's early (1856:376) book on the mollusca showed a faunal break at Valparaíso, Dall (1909:185), who also worked on mollusks, must receive the credit for being the first to recognize a boundary in the vicinity of Chiloé Island. Later, Soot-Ryen (1959:77), working on the Pelecypoda, decided that the true limit for the warm-temperate fauna was just to the north of Valparaíso but noted that a minor faunal border was also present at Chiloé Island. Stuardo (1964:85), in his work on the distribution of the littoral mollusks, recognized a transition zone from 37°37'S to the northern part of Chiloé Island.

In regard to the decapod crustaceans, Rathburn (1910:533) found a southern warm-temperate boundary at Chiloé; Haig (1955:4) agreed with this placement and noted that of 30 littoral and sublittoral species, 10 had their northern or southern limits of range close to the northern end of Chiloé Island; and Garth (1957:4) agreed with "Ekman's boundary" at the north end of this island. Madsen (1956:6), who studied the sea

The Provinces
Peru-Chilean

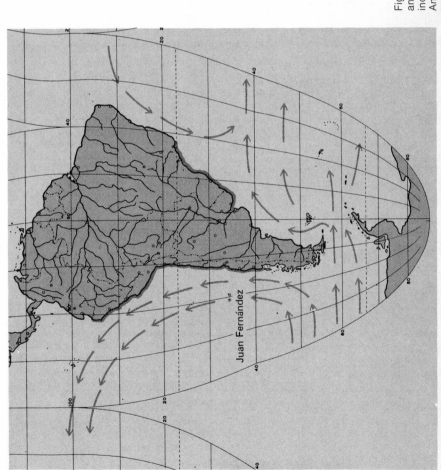

Figure 6-2 Peru-Chilean and Juan Fernández Provinces and Eastern South America Region.

Juan Fernández

stars (Asteroidea), found that the exclusively warm-temperate element had its southernmost limit north of Chiloé at about 42°S. Bernasconi (1964), from both sea urchin and sea star distributions, indicated the boundary at 41°30′S. Hartmann-Schroder and Hartmann (1962:43), on the basis of polychaete and ostracod distribution, recognized two regions for the warm-temperate part of the Chilean coast: a "Nordregion" from 18 to 30°S and a "Zentralregion" from 30 to 40° (42°) S.

From isopod studies, Menzies (1962:6) decided that the warm-temperate region should lie between 15 and 28°S. It is difficult to conceive of a well-defined fauna occupying such a restricted segment of this extensive coastline. Also, Menzies believed that for a warm-temperate region the "most frequent" average monthly temperature must lie between 17 and 20°C. Faunal evidence for this definition was not provided.

The great majority of fish species that have been reported from Peru and Chile are based on specimens taken at only a few convenient localities so that little is known about their latitudinal distribution along the coast. Mann (1954) would divide the warm-temperate area into two faunal aggregates or "Conjuntos": (1) a "Conjunto de Invasores septentrionales en la Corriente del Peru" extending south to Talcahuano at about 36°S and (2) a "Conjunto de Peces de las aguas frias Subantarcticus" from the latter point to 41°5′S. Norman (1937) considered the northern boundary of his "Patagonian Region" to lie at Chiloé Island and showed that, according to his data, an impressive faunal change took place, with only a few cold-temperate species extending northward past that locality. Lopez (1964) delineated a boundary farther north at about Valdivia.

Aside from some of the very early publications, it can be seen that a preponderance of opinion and evidence exists for the recognition of a southern warm-temperate boundary at Chiloé Island. Specifically, it can probably be placed at the north end of this large island (41°5′S). Notable exceptions are those of Schilder (1956:75), Hedgpeth (1957:map), Knox (1960), Menzies (1962:6), and Lopez (1964). In addition, Soot-Ryen (1959:77) recognized a minor but not a major boundary at Chiloé Island.

The pelecypod mollusks (Soot-Ryen, loc. cit.), the polychaetes and ostracods (Hartmann-Schroder and Hartmann 1962:43), and the fishes (Mann 1954) seem to offer some evidence for additional faunal breaks in such places as just north of Valparaíso, 30°S near La Serena, and 36°S near Concepción. Such apparent boundaries could mainly be due to the fact that a good many eurythermic cold-temperate species have managed to invade as far north as these places. The presence of these forms should not detract from the recognition of a well-established warm-temperate fauna with a southern boundary at Chiloé Island.

So far, there seems to be a rather high degree of endemism. Dall (1909) found that 315 of 566 species of littoral mollusks, or 55.5 percent, were indigenous, Haig (1955:4) discovered that 16 out of 30, or 53.3 percent, of the anomuran decapods were endemic, and Garth (1957:4) noted that the rate of endemism in the shorttail crabs was 14 out of about 60, or 23.3 percent. The shore fishes of Peru (Hildebrand 1946, Hildebrand and Barton 1949) are the best known, but a great deal remains to be learned about the fishes of both Peru and Chile. The rate of endemism in the fishes will probably prove to be 50 percent or more.

Juan Fernández   Aside from serving as the abode of the original Robinson Crusoe (Alexander Selkirk who stayed from October 1704 to February 1709), the Juan Fernández Islands are of unusual interest to the biologist. The group is fairly well isolated, lying about 400 miles west of the central Chilean coast, and are apparently quite old (between the upper Cretaceous aand Eocene according to Mann 1954), for an impressive amount of endemism has been found in both marine and terrestrial faunas.

The fishes of Juan Fernández were first collected for scientific purposes by Claudio Gay. Some of his specimens were described by Cuvier and Valenciennes (1828–1849) and others by Guichenot in Gay's monumental work *Historia Fisica y Politica de Chile* (1848). Since then, several small collections have been published, but the most complete work is still that of Rendahl (1921).

At the time of Rendahl's analysis, a total of 40 species were known. Of these, 34 may be considered shore fishes, with 19 of them endemic. Mann (1943:78), although recognizing substantial differences between the two, placed the fish faunas of Juan Fernández and Easter Island in the same province or "Conjunto" and considered the province to be different from those of the mainland. Juan Fernández endemics are still being described, and since an extensive collection has apparently never been made, it is not yet possible to make an accurate assessment of either the total fish fauna or the extent of endemism.

The early work of Odhner (1922) on the mollusks indicated a total of 39 species, and 20 of them (51 percent) were considered endemics. More recently, Madsen (1956:5), in his work on the sea stars (Asteroidea), found 5 species, 4 of them endemic and 1 shared with the Chilean mainland. Leloup (1956:86) listed 5 chitons, 3 endemic, and 1 Chilean mainland. Garth (1957:4), in his work on the brachyuran crabs, found 15 species with 4 endemics. Silva (1966:150) noted that the marine algal flora was characterized by a 32 percent endemism.

Although the biologist expects to find indigenous species in the shallow waters surrounding the oceanic islands, the foregoing works indicate that at Juan Fernández the incidence of endemism may be remarkably high, amounting in some groups to more than 50 percent of the species. According to present indications, the Juan Fernández Islands should certainly be considered a distinct province of the Peru-Chilean Region. So far, a strong relationship to the western South American coast has been demonstrated, with the ties to Easter Island and the Indo-West Pacific being very weak.

The very small island group of San Félix–San Ambrosio, about 400 miles to the north of Juan Fernández, is even farther from the mainland coast (600 miles), but its marine fauna is virtually unknown.

The Eastern South America Region

Below the vicinity of Maceió at about 10°S, the warm Brazil Current flows in a southwesterly direction paralleling the coastline. Its course, determined by prevailing onshore winds and coastal topography, is such that the entire southern coast of Brazil and eastern Uruguay comes under its influence. As it runs southward into higher latitudes, the current becomes cooler so that, below Rio de Janeiro, it can no longer support a tropical fauna.

Opposite the mouth of the Río de la Plata, the Brazil Current meets with the cold Falkland Current coming up the coast in the opposite direction. This, probably together with the latitudinal shift in the prevailing winds, causes an abrupt change in direction so that the Brazil Current leaves the coast and begins to flow eastward. As it extends out across the ocean, it forms the lower arc of the main South Atlantic Gyre.

The fauna of the east coast of South America below Rio de Janeiro is so poorly known that, as late as 1953, Ekman (p. 214) was unable to find evidence of an independent warm-temperate assemblage. Forbes (1856) recognized a "Urugavian" fauna extending from about Santa Catherina, Brazil to just below the Valdés Peninsula in Argentina. Woodward (1856:378), in discussing molluscan distribution, outlined the same area but called it the "Patagonian Province." Stephenson (1947:plate 15) showed the cold-temperate fauna extending north as far as the mouth of the Río de la Plata. Balech (1954:192) described a "Sudbrasilena" province extending from about Rio de Janeiro to a point from 30 to 32°S and, from there, an "Argentina" province to a boundary between 41 and 44°S. Hedgpeth (1957:map) depicted a warm-temperate area from about Port Alegre, Brazil to the Gulf of San Jorge, Argentina (46°S), and Knox (1960:614) accepted the subdivisions of Balech.

Works on the invertebrate fauna have proved useful for identification of the boundary area: Ortmann (1896:map), working with the decapod Crustacea, illustrated a faunal change at the Río de la Plata but thought that the tropics extended south to this point. Van Name (1945:8) spoke of a sub-Antarctic ascidian fauna extending along the Argentine coast north to 35 or 36°S; Bernasconi (1953), in her monograph of the sea urchins of Argentina, listed seven species that found their northern limits at about the Río de la Plata; and Madsen (1956), in his work on the Chilean sea stars (Asteroidea), noted that many (12) of the southern species ranged around the tip of the continent and north to the vicinity of the mouth of the Río de la Plata.

In regard to the fishes, the work of Lahille (1908) together with that of Fowler (1941) and de Buen (1950) showed that the Southern Hemisphere representatives of the family Zoarcidae apparently do not extend northward past the Río de la Plata. The various species of the cold-water families Nototheniidae and Bovichthyidae seem to be limited in the same way (Norman 1937, Fowler 1941, de Buen 1950). According to Svetovidov (1948:112), the southern gadid, *Urophycis brasiliensis*, does not extend north of 35°S. Also, Nani and Gneri (1951:219) found that the cold-temperate hagfish, *Notomyxine tridentiger*, did not range north of the Province of Buenos Aires. Finally, the Argentine checklist of Pozzi and Bordale (1935) showed that a multitude of eurythermic tropical species managed to reach the mouth of the Río de la Plata (35 to 37°S) but not beyond.

In October 1962, a seminar on the Biogeography of Marine Organisms was held at the Instituto de Biología Marine at Mar del Plata, Argentina. The papers presented were published in the *Boletin* of the *Instituto* in October 1964. Information bearing on the location of the warm-temperate versus cold-temperate boundary may be summarized as follows: the fauna in general at 35°S (Balech 1964:109), mollusks at the Valdés Peninsula about 42°S (Stuardo 1964:80), sea urchins and sea stars at 31°S (Bernasconi 1964:map), and fishes with an extensive overlap indicated between Uruguay and the Valdés Peninsula (Lopez 1964:map).

Although most of the foregoing works indicate quite well that the Río de la Plata serves as an important boundary, particularly for the cold-temperate species, very little information exists about the nature of the fauna that occupies the continental shelf north to the tropical limit at Cape Frio. Even so, there are indications in ichthyological works that at least some fish species may be confined to this area: for example, the lancelet, *Branchiostoma platae* (Bigelow and Farfante 1948:17); the anchovies *Anchoa marinii* and *A. howelli* (Paiva Carvalho 1950); several species of flatfishes (Paiva Carvalho, Tommasi, and Novelli 1968); the scorpaenid *Helicolenus dactylopterus lahillei* (Eschmeyer 1969:98); four rays belonging to the genus *Raja*, *R. agassizii*, *R. cyclophora*, *R. castelnaui*, and *R. platana* (Bigelow and Schroeder 1954); and the congrid eel, *Conger orbignyanus* (Kanazawa 1958:247). For invertebrates, it may be noted that the lists of Brazilian sea urchins and holothurians published by Tommasi (1966, 1969) include several species that appear to be warm-temperate endemics.

While the evidence is by no means conclusive, it indicates that a warm-temperate fauna probably does exist. Since the fauna of this coast is less known that that of any comparable area in the world, such things as the extent of endemism and the relative amount of invasion from the adjacent tropical and cold-temperate regions simply cannot be estimated.

In summary, an Eastern South America Warm-Temperate Region is recognized, but its salient features remain to be clarified. It is a most promising area for future investigation.

The Southern Africa Region

As the South Equatorial Current of the Indian Ocean approaches Madagascar, it becomes divided, with one part flowing around the north end of the island and the other flowing, at least to some extent, around the south end. A strong southward stream, sometimes called the Mozambique Current, parallels the African coast between Madagascar and the mainland. Apparently, it originates mostly from the north part of the South Equatorial Current. As the Mozambique Current enters the coastal area below 30°S latitude, it is joined by additional flow from around the south end of Madagascar. It then becomes a well-defined, swift, narrow stream called the Agulhas Current.

The Agulhas Current continues parallel to the coast and, as it nears the Cape of Good Hope region, flows almost due west. Although at times some Agulhas water rounds the Cape to become mixed with the Benguela Current (Talbot and Penrith 1962), the greater volume bends sharply to the south and then toward the east and is returned to the Indian Ocean. The warm Agulhas Current helps to maintain a warm-temperate fauna on the south coast of Africa and is responsible for the transport of many tropical species into these latitudes.

The southwestern African coast north of the Cape comes under the influence of the Benguela Current, the eastern arc of the main South Atlantic Gyre. In addition to the circulating South Atlantic water, the Benguela Current picks up reinforcement from the cold, circum-Antarctic, West Wind Drift. The prevailing winds come from the south-southeast so that their direction is either slightly offshore or parallel to the shore. This accounts for the continued upwelling that is characteristic of the coast, and it keeps the surface temperature remarkably constant from season to season. The Benguela Current is comparatively slow moving and becomes warmed by the sun until, just north of the Cape of Good Hope, the surface temperature reaches 13 to 15°C.

Both Forbes (1856) and Woodward (1856) depicted a "South African" province extending from about Bogenfels on the coast of South-West Africa around the Cape to a point on the Natal coast just north of Durban. Forty years later Ortmann (1896:map), who worked on decapod crustaceans, considered the region from about Point Albino, Angola around to a location about halfway between East London and Durban a part of his circumglobal "Antarktische Litoralregion."

More recently, the South African intertidal fauna and flora have become relatively well known, thanks mainly to the efforts of T. A. Stephenson and his colleagues at the University of Capetown (an extensive series of papers published from 1932 to 1947 listed in a bibliography by Stephenson 1947:309–323). At the time this work was being carried on, Ekman (1935) published his *Tiergeographie des Meeres*. Stephenson (1947) expressed disagreement with Ekman's concept of a general southern and southwestern African warm-temperate fauna with its center at the Cape of Good Hope and, instead, maintained that the area was divided into two faunas (and floras)—a warm-temperate existing in "purest form" between Cape Agulhas and Algoa Bay and a cold-temperate extending from about Kommetje just below Capetown north to tropical West Africa.

In his 1953 book, Ekman indicated considerable agreement with

Stephenson and recognized both a south coast fauna and a southwest coast or Namaqua fauna—the two being divided by the Cape of Good Hope. However, Ekman considered the southwest coast fauna to be warm-temperate rather than cold-temperate. Hedgpeth (1957:map) followed Stephenson in recognizing a west coast cold-temperate fauna for South Africa, and Knox (1960:615) did the same.

As Stephenson's work clearly shows, there is no doubt that the fauna to the east of the Cape is a warm-temperate one mainly of tropical derivation. But, is the assemblage from the Cape to Mossâmedes cold-temperate or should it be considered a separate warm-temperate province? Knox (1960) illustrated the distribution of the algal genus *Durvillea*, the mussel *Mytilus edulis*, and the siphonarid genus *Kerguelenella*. These were shown to have general cold-temperate distributions extending around the southern part of the globe, but they do not occur in southwestern Africa.

The fish fauna offers some good evidence along this line: such groups as the Nototheniidae, Bovichthyidae, and Zoarcidae are important components of the cold-temperate faunas in various parts of the Southern Hemisphere but are not represented in South Africa. The absence of the Bovichthyidae (except for one species at Amsterdam and Saint Paul Islands) is particularly indicative, since it is circumpolar and yet almost entirely confined to cold-temperate waters. Instead of these families, the shore fauna of southwestern Africa is characterized by species belonging to families of tropical relationship. Also, the seals of the family Phocidae and the bladdernosed seals of the subfamily Cystophorinae are widely distributed in Southern Hemisphere cold-temperate waters but do not reach South Africa (Davies 1958).

Despite the fact that the west coast of South Africa is somewhat cooler than the south coast, the surface temperatures for the coldest month are not as low as those found for cold-temperate areas in other parts of the world. According to the 1944 U. S. Hydrographic Office charts, the August mean is not ordinarily less than 14°C. Other published data (Ekman 1953:189) shows an August mean of 13 to 14°C.

If southwestern Africa were warm-temperate, one would expect to find a basic relationship to the south coast area. Stephenson's (1947:226) summary of the distribution of the geographic components of the intertidal fauna and flora does suggest that many species are probably endemic to the region as a whole. In southwestern Africa, this regional element may comprise 25 percent or more of the total fauna. A number of shore fishes can be considered characteristic of the entire region.

Some examples are *Merluccius capensis* (Svetovidov 1948:136), *Sardinops ocellata* (de Jager 1960), *Congiopodus spinifer*, *Xiphiurus capensis*, and others belonging to the families Clinidae, Gobiidae, and Blenniidae (Smith 1949).

In his monograph on the polychaetes of Southern Africa, Day (1967) observed that although Cape Point did represent a boundary between two different intertidal biotas, this was not true for the deeper waters of the shelf. In considering all the polychaete species taken from 0 to 200 meters, Day was able to identify only a single zoogeographic province for Southern Africa extending from the mouth of the Bashee River on the east, around the Cape of Good Hope, and then northward to Cape Frio or beyond. The endemic component was calculated to comprise about 36 percent of the total polychaete fauna.

It can be said, then, that a number of good reasons exist for considering all of nontropical Southern Africa to be occupied by a single, warm-temperate region. The majority of typical, Southern Hemisphere, cold-temperate groups of fishes, marine mammals, and invertebrates are missing; the endemic groups of the area are mainly tropical derivatives; many species are endemic to the area as a whole rather than to just a part; and, even in southwestern Africa, the minimum surface temperatures are not cold enough to support a cold-temperate fauna.

The Provinces
Southwestern
Africa

To the north, the province presumably extends as far as the tropical boundary at Mossâmedes, Angola, although a major faunal change here is not yet well documented (p. 88). The southern boundary is the Cape of Good Hope or, perhaps more exactly, at Kommetje on the west side of the Cape Peninsula (Stephenson 1947:229). Regrettably, Stephenson's investigations extended only as far north as Port Nolloth including about one-fifth of the total coastline of the province.

According to Ekman's (1953:194) interpretation of Stephenson's data, at most, 17 percent of the species of this fauna are endemic. Certainly, Stephenson's (op. cit., 226) distributional summary indicates only a small endemic element. This general impression is reinforced by our knowledge of the distribution of the fishes. As can be seen by Smith's (1949) book, there are many regional endemics, but relatively few of these are restricted to the Southwestern Africa Province. The Clinidae comprise the dominant group of shore fishes in the Southern Africa Region. A total of 33 species are known (Penrith 1969a); of these, 6 or 7

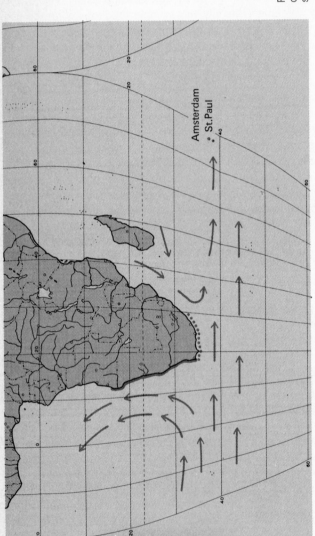

Figure 6-3 Southwest Africa, Agulhas, and Amsterdam–St. Paul Provinces.

appear to be confined to this province. In addition, 1 blenny (Blenniidae), and 3 gobies (Gobiidae) are apparently endemic (Smith 1949, 1960).

The fauna of this province is very poorly known compared to that of the neighboring Agulhas Province. The true amount of endemism and other basic features will not be revealed until extensive observations and collections are made in the area between Port Nolloth and Mossâmedes (or beyond).

Agulhas   Although Stephenson (1947:229) mentioned Cape Agulhas as a possible western boundary and regarded as "transitional" the area between that point and Kommetje, Cape Agulhas is here regarded as belonging to the Agulhas Province. Stephenson's (op cit., 226) diagrammatic summary shows that the majority of the provincial endemics extend westward past Cape Alguhas to the Cape of Good Hope or its vicinity. Available information about the local distribution of the endemic shore fishes (Smith 1949, 1960, Penrith 1969a) indicates that many of them do the same.

In contrast to the Southwestern Africa Province, Stephenson's works show that the Agulhas fauna and flora are highly endemic. According to Ekman's (1953:189) interpretation, this amounts to about 34 percent of the species. Again, the distribution of shore fishes shows agreement: of the 33 known Southern Africa members of the family Clinidae, 12 are confined to the Agulhas Province (Penrith 1969a) and 8 of the 13 gobies (Gobiidae) are similarly confined (Smith 1960). There are also peculiar genera and species of clingfishes (Smith 1964, Penrith 1969b) and toadfishes (Batrachoididae—Smith 1952). Millar (1962), in his work on Southern Africa ascidians, described a total of 68 species and considered more than half of them to be endemic. He further observed that most of the endemics were found in waters east of the Cape.

The Agulhas fauna is markedly richer than that of Southwestern Africa due not only to its superiority in numbers of endemic species but to an invasion by many eurythermic tropical species that belong primarily to the Indo-West Pacific fauna. This phenomenon has also been demonstrated nicely by Stephenson (op. cit.).

This area has been called both a "South Coast" and a "Cape" Province. Ekman (1953:187) indicated his objections to the latter name noting that both Capetown and the Cape Peninsula were located on the boundary of another province. Since the name "South Coast" is also objectionable because it is not particularly applicable to South Africa, the designation "Agulhas Province" seems appropriate. The Agulhas Current exerts an important influence upon the fauna, and Cape Agulhas is a major geographic feature within the area.

Amsterdam and Saint Paul are small, isolated, volcanic islands located in the south Indian Ocean 2,800 miles from southeastern Africa and about 1,200 miles northeast of the Kerguelen Islands. They seem to be influenced by both warm water from the main gyre of the south Indian Ocean and cold water from the West Wind Drift.

Amsterdam–
Saint Paul

Apparently the first fishes collected from the islands were taken by the Austrian Frigate *Novara* during its circumnavigation of the globe from 1857 to 1859. The specimens were described by Kner (1865). The famous French ichthyologist Henri Emile Sauvage contributed three papers on this fauna between 1875 and 1880, and further observations were added by Angot (1951). The most recent work was accomplished by Blanc and Paulian (1957), who provided a complete list of the fishes, both shore and pelagic.

A total of 14 species of shore fishes are now known from Amsterdam–Saint Paul with 4 of them (about 28 percent) possibly endemic. Three others are shared only with Southern Africa, 4 occur in other Southern Hemisphere localities as well as Southern Africa, and 3 are found elsewhere but not in Southern Africa—2 of the latter probably occur in South America and 1 in New Zealand.

Other forms shared with Southern Africa are: the algae *Macrocystis* (Blanc and Paulian 1957:331) and *Splachnidium rugosum* (Knox 1960:591), the echinoderm genus *Spoladaster* (Fell 1962:760), the crustacean genus *Jasus* (Holthius 1963), and the ascidian *Corella eumyola* (Ekman 1953:196). It seems likely that most, if not all of these forms, were transported from Southern Africa to the islands by the West Wind Drift.

In his account of the littoral ecology and biogeography of the southern oceans, Knox (1960:611) considered Amsterdam–Saint Paul to constitute a separate, cold-temperate province with, apparently, no special relationship to Southern Africa. Since 7 out of the 10 nonendemic shore fishes and at least several of the algal and invertebrate species occur in Southern Africa, and since Southern Africa does not possess a cold-temperate biota, it would be inconsistent to place these islands in such a category. Also, the mean surface temperature of about 12.5°C for winter and 17.5°C for summer (U. S. Hydrographic Office Charts, 1944) is quite comparable to the range in other warm-temperate localities.

Although the invertebrate fauna is still poorly known, the fishes demonstrate such a definite relationship to the Cape of Good Hope region of Southern Africa that it seems best to recognize this by including Amsterdam-Saint Paul as a province within the Southern Africa Warm-Temperate Region.

Literature Cited

Angot, M. 1951. Observations sur le faune marine et la pêche aux îles Saint-Paul et Amsterdam. *Mem. Inst. Sci. Madagascar, Ser. A,* 6(1):1–51, 25 figs. (not seen).

Balech, E. 1954. Division zoogeografica del litoral sudamericano. *Rev. Biol. Marina,* 4(1–3):184–195, 2 figs.

———. 1964. Caracteres biogeographicos de la Argentina y Uruguay. *Bol. Inst. Biol. Mar., Mar del Plata, Argentina,* (7):107–112.

Bennett, I., and E. C. Pope. 1953. Intertidal zonation of the exposed rocky shores of Victoria, together with a rearrangement of the biogeographical provinces of temperate Australian shores. *Australian J. Marine Freshwater Res.,* 4(1):105–159, 5 figs., 6 plates.

———, and ———. 1960. Intertidal zonation of the exposed rocky shores of Tasmania and its relationship with the rest of Australia. *Australian J. Marine Freshwater Res.,* 11(2):182–221, 7 plates.

Bernasconi, I. 1953. Monografia de los equinoideos Argentinos. *Anales Museum Hist. Nat., Montevideo, Ser. 2,* 6(2):1–58, 32 plates.

———. 1964. Distribucion geografica de los equinoideos y asteroideos de la extremidad austral de Sudamerica. *Bol. Inst. Biol. Mar., Mar del Plata, Argentina,* (7):43–49, 1 fig.

Bigelow, H. B., and I. P. Farfante. 1948. Lancelets. *in* John Tee-Van (editor), *Fishes of the Western North Atlantic.* Mem. Sears Foundation. Mar. Res., no. 1:1–28, 3 figs.

———, and W. C. Schroeder. 1954. *Fishes of the Western Atlantic.* Part two. *Sawfishes, guitarfishes, skates, rays, chimaeroids.* Sears Foundation, Bingham Oceanographic Laboratory, New Haven, xv + 588 pp., 127 figs.

Blackburn, M. 1941. The economic biology of some Australian clupeoid fish. *C. S. I. R., Commonw. Australia, Bull.,* (138):1–135, 29 figs., 1 plate.

Blanc, M., and P. Paulian. 1957. Poissons des îles Saint-Paul et Amsterdam. *Mem. Inst. Sci. Madagascar, Ser. F,* 1:325–335, 3 plates.

Briggs, J. C. 1955. A monograph of the clingfishes (Order Xenopterygii). *Stanford Ichthyol. Bull.,* 6:ix + 224 pp., 114 figs., 15 maps.

Brodie, J. W. 1960. Coastal surface currents around New Zealand. *New Zealand J. Geol. Geophysics,* 3(2):235–252, 7 figs.

de Buen, F. 1950. La fauna de peces del Uruguay. El Mar de Solís y su fauna de peces. II parte, Servicio Occanografico y de Pesca, Publ. Sci., (2):pp. 46–144.

Clark, H. L. 1946. The echinoderm fauna of Australia. *Carnegie Inst. Wash.*, publ. 566:iv + 567 pp.

Cuvier, G. L. C. F. D., and A. Valenciennes. 1828–1849. *Histoire naturelle des poissons.* 22 vols. Levrault, Paris.

Dahl, E. 1960. The cold temperate zone in Chilean seas. *Proc. Roy. Soc., Ser. B*, 152(949):631–633.

Dakin. W. J., I. Bennett, and E. Pope. 1948. A study of certain aspects of the ecology of the intertidal zone of the New South Wales coast. *Australian J. Sci. Res., Ser. B*, 1(2):176–230 (not seen).

Dall, W. H. 1909. Report on a collection of shells from Peru, with a summary of the littoral marine Mollusca of the Peruvian zoological province. *Proc. U. S. Nat. Museum*, 37(1704):147–294, 9 plates.

Davies, J. L. 1958. The Pinnipedia: an essay in zoogeography. *Geograph. Rev.*, 48(4):474–493, 10 figs.

Day, J. H. 1967. *A monograph on the Polychaeta of southern Africa.* Brit. Museum of Natural History, London, pp. viii + 878.

Dell, R. K. 1957. The marine mollusca of the Kermadec Islands in relation to the molluscan faunas in the southwest Pacific. *Proc. 8th Pacific Sci. Congr.*, 3:499–503, map.

Ekman, S. 1935. *Tiergeographie des Meeres.* Akademische Verlagsgesellschaft, Leipzig, xii + 512 pp.

———. 1953. *Zoogeography of the sea.* Sidgwick & Jackson, London, xiv + 417 pp., 121 figs.

Eschmeyer, W. N. 1969. A systematic review of the scorpionfishes of the Atlantic Ocean (Pisces: Scorpaenidae). *Occas. Papers, Calif. Acad. Sci.*, (79):1–143, 13 figs.

Fell, H. B. 1949. The constitutions and relations of the New Zealand echinoderm fauna. *Trans. Roy. Soc. N. Z.*, 77:201–212 (not seen).

———. 1962. West Wind Drift dispersal of echinoderms in the Southern Hemisphere. *Nature*, 193(4817):759–761, 2 figs.

Finlay, H. J. 1925. Some modern concepts applied to the study of Cainozoic Mollusca from New Zealand. *Verbeek Mem. Birthday Vol.*, pp. 161–172 (not seen).

Fleming, C. A. 1962. New Zealand biogeography. A paleontologist's approach. *Tuatara*, 10:53–108, 15 figs.

Forbes, E. 1856. Map of the distribution of marine life. *in* Alexander K. Johnston, *The physical atlas of natural phenomena* (new edition). W. and A. K. Johnston, Edinburgh and London, plate no. 31.

Fowler, H. W. 1941. A list of the fishes known from the coast of Brazil. *Arquivos Zool. Estado São Paulo*, 3(6):115–184.

Garth, J. S. 1957. The Crustacea Decapoda Brachyura of Chile. Rept. Lund University Chile Exped. 1948–49. *Lunds Univ. Årsskrift, N.F. Avd. 2*, 53(7):1–130, 11 figs., 4 plates.

Gay, C. 1848. *Historia fisica y politica de Chile.* Zoologia. Pisces. Paris and Santiago. 2(part 2):137–372. Atlas ictiologica, 17 plates.

Ginsburg, I. 1951. Western Atlantic tonguefishes with description of six new species. *Zoologica*, 36(3):185–201, 3 plates.

Gunther, M. A. 1936. A report on oceanographical investigations in the Peru coastal current. *Discovery Rep.*, 13:107–276, 3 plates.

Haig, J. 1955. The Crustacea Anomura of Chile. Rep. Lund Univ. Chile Exped. 1948–49. *Lunds Univ. Årsskrift, N.F. Avd. 2*, 51(12):1–68, 13 figs.

Hartmann-Schroder, G., and G. Hartmann. 1962. Zur Kenntnis des Eulitorals der chilenischen Pazifikküst und der argentinischen Küste Südpatagoniens unter besonderer Berücksichtigung der Polychaeten and Ostracoden. *Mitt. Hamburgischen Zool. Museum Inst.*, 60:1–270, 451 figs.

Hedgpeth, J. W. 1957. Marine biogeography. *in Treatise on marine ecology and paleoecology.* vol. 1. *Geol. Soc. Amer., Mem.*, (67):359–382, 16 figs., 1 plate.

Hildebrand, S. F. 1946. A descriptive catalog of the shore fishes of Peru. *Bull. U. S. Nat. Museum*, No. 189, 530 pp., 95 figs.

———, and O. Barton. 1949. A collection of fishes from Talara, Perú. *Smithsonian Misc. Coll.*, 3(10):1–36, 9 figs.

Holthius, L. B. 1963. Preliminary description of some new species of Palinuridea. *Koninkl. Nederl. Akad. Van Wetenschappen, Ser. C*, 66(1):54–60.

de Jager, B. v. D. 1960. Synopsis on the biology of the South African pilchard (*Sardinops ocellata* Pappé). U. N. Food and Agricultural Organization, *Fish. Biol. Synopsis*, (7):vi + 97–114.

Kanazawa, R. H. 1958. A revision of the eels of the genus *Conger* with descriptions of four new species. *Proc. U. S. Nat. Museum*, 108(3400):219–267, 7 figs., 4 plates.

Kner, R. 1865. *Reise der österreichischen Fregatte "Novara" um die Erde in den Jahren 1857–1859.* Zoologischer Theil. Fische, Wien, 433 pp., 16 plates.

Knox, G. A. 1960. Littoral ecology and biogeography of the southern oceans. *Proc. Roy. Soc., Ser. B*, 152(949):577–624, 73 figs.

———. 1963a. Problems of speciation in intertidal animals with special reference to New Zealand shores. *Syst. Ass. Publ.*, (5):7–29, 10 figs.

———. 1963b. The biogeography and intertidal ecology of the Australasian coast. *Oceanog. Marine Biol. Ann. Rev.*, 1:341–404, 5 figs.

Kott, P. 1952. Ascidians of Australia. I. Stolidobranchiata Lahille and Phlebobranchiata Lahille. *Australian J. Marine Freshwater Res.*, 3(3):205–333, 183 figs.

Lahille, F. 1908. Nota sobre los zoarcidos Argentinos. *Anales Mus. Nac. Buenos Aires, Ser. 3*, 9:403–441, 9 figs., 2 plates.

Leloup, E. 1956. Polyplacophora. Rep. Lund Univ. Chile Exped. 1948–49. *Lunds Univ. Årsskrift. N.F. Avd. 2*, 52(15):1–94, 53 figs.

Lopez, Rogelio B. 1964. Problems de la distribucion geografica de los peces marinos. *Bol. Inst. Biol. Mar., Mar del Plata, Argentina*, (7):57–62, 1 fig.

Madsen, F. J. 1956. Asteroidea. Rept. Lund Univ. Chile Exped. 1948–49. *Lunds Univ. Årsskrift., N.F. Avd. 2*, 52(2):1–53, 1 fig., 6 plates.

Mann, G. 1954. *La vida de los peces en aguas Chilenas.* Ministerio de Agricultura y Universidad de Chile, Santiago, 342 pp., illus.

McCulloch, A. R. 1929–1930. A check-list of the fishes recorded from Australia. *Australian Museum Mem.*, 5(1929–1930), 4 parts., 534 pp.

———. 1934. *The fishes and fish-like animals of New South Wales.* 3d edition. Royal Zoological Society of New South Wales, Sydney, xxv + 104 pp., 42 plates.

Menzies, R. J. 1962. The zoogeography, ecology, and systematics of the Chilean marine isopods. Rep. Lund. Univ. Chile Exped. 1948–49. *Lunds. Univ. Årsskrift., Avd. 2*, 57(11):1–162, 51 figs.

Millar, R. H. 1962. Further descriptions of South African ascidians. *Ann. S. African Museum*, 46(7):113–121, 45 figs.

Moore, L. B. 1961. Distribution patterns in New Zealand seaweeds. *Tuatara*, 9:18–23.

Moreland, J. 1959. The composition, distribution and origin of the New Zealand fish fauna. *Proc. New Zealand Ecol. Soc.*, (6):28–30.

Munro, I. S. R. 1956–. *Handbook of Australian fishes.* vol. 15. Fisheries Newsletter, Sydney (series still in course of publication).

Nani, A., and F. S. Gneri. 1951. Introduccion al estudio de los mixinoideos sudamericanos I. Un nuevo genero de "barbosa de mar," "Notomyxine." *Rev. Mus. Cienc. Nat. "Bernardino Rivadavia,"* 2(4):183–224, 6 figs., 3 plates.

Norman, J. R. 1937. Coast fishes. Part II. The Patagonian region. *Discovery Rep.*, 16:1–150, 76 figs., 5 plates.

Odhner, N. H. 1922. Mollusca from Juan Fernandez and Easter Island. *in* Carl Skottsberg, *The natural history of Juan Fernandez and Easter Island.* Almqvist and Wiksells, Uppsala, 3:219–254.

Ortmann, A. E. 1896. *Grundzüge der marinen Tiergeographie.* Gustav Fischer, Jena, iv + 96 pp., 1 map.

Paiva Carvalho, J. 1950. Engraulídeos brasileiros, do gênero *Anchoa. Bol. Inst. Paulista Oceanog.*, 1(2):43–69. 2 plates.

———, L. R. Tommasi, and M. D. Novelli. 1968. Lista dos linguados do Brasil. *Contr. Inst. Oceanog. Univ. S. Paulo, Ser. Oceanog. Biol.*, (14):1–26, 18 figs.

Pawson, D. L. 1961. Distribution patterns of New Zealand echinoderms. *Tuatara*, 9:9–18.

Penrith, M.-L. 1969a. The systematics of the fishes of the family Clinidae. *Ann. S. African Museum*, 55(1):1–121, 48 figs.

———. 1969b. *Apletodon pellegrini* (Chabanaud) and other clingfishes (Pisces: Gobiesocidae) from South West Africa. *Ann. S. African Museum*, 55(2):123–134, 5 figs.

Powell, A. W. B. 1937. New species of marine Mollusca from New Zealand. *Discovery Rep.*, 15:153–222.

———. 1940. The marine mollusca of the Aupourian Province, New Zealand. *Trans. Roy. Soc. N. Z.*, 70:205–248 (not seen).

———. 1962. *Shells of New Zealand.* 4th edition. Whitcomb and Tombs, Christchurch, 203 pp., 36 plates.

Pozzi, A. J., and L. F. Bordale. 1935. Cuadro sistematico de los peces marinos de la Republica Argentina. *Anales Soc. Cient. Argentina.* 120(4):145–189, 1 map.

Rathbun, M. J. 1910. The stalk-eyed Crustacea of Peru and the adjacent coast. *Proc. U. S. Nat. Museum*, 38:531–620, 21 plates (not seen).

Rendahl, H. 1921. The fishes of the Juan Fernandez Islands. *in* Carl Skottsberg, *The natural history of Juan Fernandez and Easter Island*. Almqvist and Wiksells, Uppsala, 3(10):49–58.

Schilder, F. A. 1956. *Lehrbuch der Allgemeinen Zoogeographie*. Gustav Fischer, Jena, viii + 150 pp., 134 figs.

Scott, T. D. 1962. *The marine and fresh water fishes of South Australia*. Govt. Printer, Adelaide, pp. 1–338, illus.

Silva, P. C. 1966. Status of our knowledge of the Galápagos benthic marine algal flora prior to the Galápagos International Scientific project. *in* R. I. Bowman (editor), *The Galápagos*. University of California Press, Berkeley, pp. 149–156.

Smith, J. L. B. 1949. *The sea fishes of southern Africa*. Central News Agency, Cape Town, xvi + 550 pp., 1,100 figs., 103 plates.

———. 1952. The fishes of the family Batrachoididae from South and East Africa. *Ann. Mag. Nat. Hist., Ser. 12*, 5:313–339, 3 figs.

———. 1960. Fishes of the family Gobiidae in South Africa. *Ichthyol. Bull. Rhodes Univ.*, (18):299–314, 13 figs.

———. 1964. The clingfishes of the western Indian Ocean and the Red Sea. *Ichthyol. Bull. Rhodes Univ.*, (30):581–596, 1 fig., 6 plates.

Soot-Ryen, T. 1959. Pelecypoda. Rept. Lund Univ. Chile Exped. 1948–49. *Lunds Univ. Årsskrift N. F. Avd. 2*, 55(6):1–86, 6 figs., 4 plates.

Stephenson, T. A. 1947. The constitution of the intertidal flora and fauna of South Africa. Part III. *Ann. Natal Museum*, 11(2):207–324, 11 figs., 2 plates.

Stuardo B., Jose. 1964. Distribucion de los moluscos marinos litorales en latinoamerica. *Bol. Inst. Biol. Mar., Mar del Plata, Argentina*, (7):79–91, 1 fig.

Sverdrup, H. V., M. W. Johnson, and R. H. Fleming. 1946. *The oceans*. Prentice-Hall, Englewood Cliffs, N.J., pp. 1–1087, 265 figs., 7 charts.

Svetovidov, A. N. 1948. Fauna of the U.S.S.R. Fishes. vol. IX, no. 4. Gadiformes. *Zool. Inst. Akad. Nauk. U.S.S.R.* (34):1–222, 39 figs., 71 plates.

Talbot, F. H., and M. J. Penrith. 1962. Tunnies and marlins of South Africa. *Nature*, 193(4815):558–559.

Tommasi, L. R. 1966. Lista dos equinóides recentes do Brasil. *Contr. Inst. Oceanog. Univ. S. Paulo, Ser. Ocean. Biol.*, (11):1–50, 72 figs., 9 plates.

———. 1969. Lista dos Holothurioidea recentes do Brasil. *Contr. Inst. Oceanog. Univ. S. Paulo, Ser. Ocean. Biol.*, (15):1–21, 4 plates.

Van Name, W. G. 1945. The North and South American ascidians. *Bull. Amer. Museum Nat. Hist.*, 84:1–476, 31 plates.

Waite, E. R. 1910. A list of the known fishes of Kermadec and Norfolk Islands, and a comparison with those of Lord Howe Island. *Trans. New Zealand Inst.*, 42:370–383.

———. 1911. Additions to the fish fauna of the Kermadec Islands. *Proc. New Zealand Inst.*, 1:28.

Whitley, G. P. 1932. Marine zoogeographical regions of Australasia. *Australia Nat.*, 8(8):166–167, map.

Womersley, H. B. S. 1961. The marine algae of Australia. *Bot. Rev.*, 25:545–614, 2 figs.

———, and S. J. Edmonds. 1958. A general account of the intertidal ecology of South Australian coasts. *Australian J. Marine Freshwater Res.*, 9(2):217–260, 2 figs., 12 plates.

Woodward, S. P. 1851–1856. *A manual of the mollusca.* John Weale, London, viii + 1–486 pp., 24 plates, + 24 pp., 1 map.

Southern
Hemisphere
Cold-Temperate
and Antarctic Regions

Thus he who views only the produce of his own country, may be said to inhabit a single world; while those who see and consider the productions of other climes, bring many worlds, as it were, in review before them.

Carl Linnaeus in Museum S.R.M. Adolphi Friderici Regis, *1754*

The shore faunas of the cold-temperate and Antarctic waters of the Southern Hemisphere were poorly known until recent years. Most of the available information was in the form of scientific reports issued as the results of a series of expeditions that took place mainly during the first half of the present century.

Although expeditions sent by northern nations still continue, a significant portion of our recent knowledge has been provided by marine biologists who reside in the Southern Hemisphere. This sudden growth of interest in marine ecology and zoogeography probably received its original impetus from the work of T. A. Stephenson and his colleagues in South Africa. By 1947, their contributions had become widely recognized.

In 1948, Isobel Bennett and Elizabeth Pope began their important work in the southern Australian-Tasmanian area. They were soon followed by others such as H. B. S. Womersley and T. D. Scott. In New Zealand, although A. W. B. Powell's significant work goes back to the 1930s, a new and prolific group has arisen (G. A. Knox, R. K. Dell, H. B. Fell, C. A. Fleming, J. A. F. Garrick, D. L. Pawson, and others). The work of these people has not only been important for marine biology in their respective areas but has stimulated similar investigations in other parts of the world.

The cold-temperate shores of the Southern Hemisphere may be divided into four zoogeographic regions: Southern South America (including the Tristan-Gough island group), Tasmania, Southern New Zealand (with Chatham and the islands of the Antipodean Province), and the Sub-Antarctic (with the Kerguelen and Macquarie areas). The Antarctic Region occupies an area where the surface temperatures are markedly colder (-2 to $+3.7°C$) and includes South Georgia and Bouvet Islands.

Regional Relationships

The importance of the Antarctic continent as an evolutionary and distributional center is nicely illustrated by its shore fish fauna. The suborder Notothenoidea is made up of five families that are entirely confined to Antarctic and south temperate waters. About 82 percent of the shore fish species of Antarctica belong to this group. Dollo (1904) and Regan (1914:29) discussed the relationship of these families and concluded that the Bovichthyidae was the most primitive. From the Bovichthyidae arose the Nototheniidae which, in turn, gave rise to the Harpagiferidae, Chaenichthyidae, and Bathydraconidae. This evolutionary sequence fits very well the present distribution patterns.

The Bovichthyidae is now entirely peripheral to the Antarctic, being found in Southern South America (plus the Juan Fernández Islands), the Tristan-Gough island group, St. Paul–Amsterdam Islands, Tasmania-Victoria, and southern New Zealand (with Chatham and the Antipodes area). The most primitive bovichthyid genus, *Pseudaphritis*, is confined to freshwater streams in Victoria and Tasmania. The Nototheniidae, Harpagiferidae, and Chaenichthyidae are mainly Antarctic but have some species that extend to cold-temperate localities, while the Bathy-draconidae is wholly restricted to the Antarctic Region.

The Bovichthyidae apparently provides a good example of a group that has been pushed out of its center of origin presumably by competition from its more advanced descendents. It probably retreated via the islands of the Scotian Arc or else bridged the deep-water gap between the Palmer Archipelago and Cape Horn. Then, a gradual clockwise dispersal via the West Wind Drift could have taken place.

The proliferation of four distinct families from a common ancestor must surely have taken a very long time, perhaps back into the Mesozoic. Since it now seems apparent that a cooling of the polar seas began to take place in the Middle Jurassic (p. 408), it is possible, considering that the tropical fish fauna was probably decimated by falling tempera-tures, that an ancestral bovichthyid could have become established and begun its evolutionary expansion by the Upper Jurassic. Regan (1914:40) was of the opinion that the Antarctic continent may have been isolated and its coasts washed by a cold sea probably throughout the Tertiary. More recently, Fell (1961:75), on the basis of echinoderm (ophiuroid) distribution, said that it was reasonable to infer that the peculiar Antarctic genera are the result of a long Tertiary evolution in relative isolation.

Other elements of the Antarctic and southern cold-temperate faunas are of comparatively recent northern derivation. In fishes, an important example is the family Zoarcidae—the eelpouts. This group is almost certainly of North Pacific origin. It probably made its way south along the west coast of the Americas, apparently managing to traverse the tropics by means of isothermic submergence. There are now 10 genera and about 15 species in the Magellan Province of South America and 3 genera with 5 species in the Antarctic. Powell (1951:65) depicted a similar situation in the Mollusca whereby 3 genera (*Acantina*, *Fusitriton*, and *Aforia*) reached the Magellan Province by the same route and from there the latter 2 became dispersed eastward.

Another important faunal source for the southern cold-temperate regions has been the Indo-West Pacific tropical fauna. Such fish families as the Labridae, Syngnathidae, Tripterygiidae, Gobiidae, and Gobiesocidae are well represented. In many cases, cold-temperate species belonging to these families are apparently phylogenetic relicts that are remnants of successive invasions that took place over a long period of time. Such tropical groups comprise large portions of the Tasmanian and southern New Zealand faunas but are scarce in Southern South America and virtually absent in the Sub-Antarctic and Antarctic Regions. On the other hand, those of North Pacific origin are numerous in South America, fairly well represented in the Sub-Antarctic and Antarctic, but scarce in Tasmania and New Zealand.

Although they have only a few species in common, the four cold-temperate regions demonstrate an interesting and marked relationship at the generic and family levels. This may be attributed to the influence of the West Wind Drift, and it is similar to the relationship which was noted for certain faunal elements of the southern warm-temperate regions (p. 126). This has been well illustrated for penguins by Mackintosh (1960) and for some of the echinoderm genera by Fell (1962:760).

For groups of Antarctic origin, the islands of the Scotian Arc, the tip of South America, and the Falkland Islands have apparently served as important "passenger boarding stations" for the West Wind Drift. Some of the groups of North Pacific origin that made their way to the Magellan Province may have also been able to take advantage of the latter two stations. As would be expected, most of those of tropical origin seem to have been dispersed from the Tasmanian–Southern New Zealand regions. The West Wind Drift effect is most pronounced in the Sub-Antarctic Region where a number of forms have been carried from Kerguelen Island some 3,000 miles to Macquarie Island.

In summary, it can be said that by far the greater portion of the shore fish fauna of the Antarctic Region belongs to families that have originated *in situ* and that their history probably extends back into the Mesozoic. Almost all the remaining fish species represent groups of North Pacific origin that apparently migrated south along the American west coast. They probably reached the Antarctic by jumping the gap from Tierra del Fuego to the Palmer Archipelago.

The shelf fauna of the southern cold-temperate regions apparently was derived from three primary sources: (1) an old evolutionary center in the Antarctic, (2) northern cold waters (particularly the North Pacific), and (3) the tropics (mainly the Indo-West Pacific). Although each of the regional faunas is highly distinct, generic and family relationships indicate that the West Wind Drift has served as an important dispersal agent. This effect is especially noticeable in the Sub-Antarctic.

The Southern South America Region

The tip of South America extends almost 20° farther south than any of the other continental masses reaching the circum-Antarctic region. This means that a considerable portion of its west coast is directly exposed to the West Wind Drift. A minor portion of this current is deflected northward and contributes to the formation of the Peru Current. By this means, the West Wind Drift is responsible for the cold-temperate conditions that exist along some 1,000 miles of Chilean coastline. The major portion of the West Wind Drift flows through Drake's Passage between South America and Antarctica. After this, another portion called the Falkland Current turns northward to run between Tierra del Fuego and the Falkland Islands.

The Falkland Current flows almost due north, paralleling the Argentine coast until it reaches the mouth of the Río de la Plata. The current then takes a sharp turn eastward and again joins the broad West Wind Drift. Quite possibly, it is Falkland Current water that is responsible for transporting cold-temperate organisms from South America to the islands of Tristan da Cunha and Gough. Although these islands are extremely isolated, some 2,500 miles from South America, they show definite faunal ties to the Magellan-Falkland area.

The Provinces
Magellan

The problem of the location of the boundaries between the warm-temperate and cold-temperate faunas of South America has already been discussed (pp. 141 and 144), and it was concluded that the areas of greatest faunal change were probably the northern tip of Chiloé Island on the west coast and the mouth of the Río de la Plata on the east coast. Cold-temperate South America has not always been considered to harbor a homogeneous fauna. For example, Forbes (1856), who was the first to give a detailed description of the distribution of marine animals, divided the area into three provinces: on the west coast he recognized an "Araucanian Province" from just south of Concepción to the Taitao Peninsula, on the east coast an "East Patagonian Province" from about the Valdés Peninsula to the northern part of Grande Bay, and a "Fuegian Province" for the rest of the southern tip of the continent.

More recently, Balech (1954:192) proposed a "Provencia Magellanica"

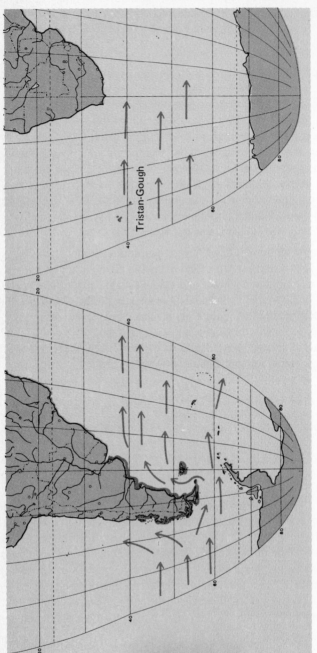

Figure 7-1 Magellan and Tristan-Gough Provinces.

that was subdivided into the following districts: to the west a "Distrito Valdiviano" from just south of Valparaíso to 40 to 41°S, below that a "Distrito Chiloense" to about 51°S, to the east a "Distrito Chubutiano" from just below the Valdés Peninsula to Cape Tres Puntas, below that a "Distrito Santacruceño" to the eastern entrance to the Strait of Magellan, and finally a "Distrito Fueguino" for the rest of the continent including Tierra del Fuego. Schilder (1956:75) recognized a "Südamerikan" province divided into two subprovinces: the first an "Argentinisch" subprovince on the east coast extending from about Port Alegre, Brazil to Necochea, Argentina, and the second a "Magellanisch" subprovince to include the rest of the continent between the latter point and about Valdivia, Chile.

Knox (1960:611) described three provinces for cold-temperate South America: to the west a "Central Chilean" province from about 30°S to Chiloé Island, to the east an "Argentinian" province from about 30 to 32°S to 41 to 44°S, and a "Magellanic" province for the remaining area to the south. Apparently, Knox attempted to map the distribution of certain rocky-shore biotic associations instead of true provinces in the biogeographic sense. The invertebrate fauna of southern Chile is now becoming much better known mainly because of the series of publications resulting from the Lund University Chile Expedition, 1948 to 1949. In general, these reports substantiate the concept of a single, cold-temperate fauna that extends around to the Atlantic and the Falkland Islands.

If one includes within the Magellan Province all of South America (and the Falkland Islands) below the boundaries with the adjoining warm-temperate regions, it becomes apparent that the area displays an interesting distinctiveness. Ekman (1953:215) gave an estimate of a 52 percent endemism for echinoderms (Asteroidea, Echinoidea, Ophiuroidea), Haig (1955) found 2 of 6 anomuran crabs to be endemic, Madsen (1956) indicated that 13 of 29 (about 45 percent) of the sea stars (Asteroidea) were endemic, the data presented by Soot-Ryen (1959) showed that 72 of 117 (about 61 percent) of the pelecypod mollusks should be considered indigenous, and Pawson (1969a:37) indicated that about 50 percent of the holothurians were endemic.

Menzies (1962), as the result of studies on the marine isopods, decided that a cold-temperate region is where the "most frequent" monthly temperatures lie between 12 and 16°C. On the South American west coast this means, according to Menzies, that the cold-temperate region must be located between 25 and 45°S. The coast south of 45°S was considered to be in a still colder "subpolar" region. As far as Chilean isopods are concerned, Menzies' (op. cit.) data could be interpreted to show that only one cold-temperate fauna exists and that it is confined to that area south of 40°S. About 30 of 66 (45 percent) of the isopod species found there seem to be cold-temperate endemics.

Both Dollo (1904) and Norman (1937a) have written interesting accounts of the history of ichthyological work in the Magellan area. By far the most extensive collections were those made by the R.R.S. *William Scoresby* during trawling surveys made from 1927 to 1928 and from 1931 to 1932. These specimens plus some from other sources provided a total of more than 3,000 individuals, representing 84 species. They were utilized by Norman for his 1937 *Discovery Report*.

Regan (1914), in his report on the fishes of the British Antarctic ("Terra Nova") Expedition, recognized within the South Temperate Zone a "Magellan District" for South America (and the Falkland Islands) south of Chiloé Island to the west and the Gulf of San Jorge on the east. Norman (1937a:137) outlined about the same area for his "Patagonian Region" except that the eastern boundary was moved north to the Valdés Peninsula. From this region, Norman recorded 128 species, with 63 (52 percent) apparently endemic. Mann (1954) recognized for Chilean waters a separate "Conjunto de Peces de los Canales Patagonicos" to include all of Chiloé Island and southward. DeWitt (1966) observed that the genus *Notothenia* had its center of abundance in the Magellan Province and that 12 of the 14 species were apparently endemic.

In summary, there seems to be but one cold-temperate fauna for Southern South America. It includes all of the mainland below the warm-temperate boundary areas and also the Falkland Islands. Endemic rates such as about 45 to 50 percent for echinoderms, 61 percent for pelecypod mollusks, 45 percent for marine isopods, and 52 percent or more for fishes all indicate the presence of a highly distinctive assemblage of animals.

Tristan-Gough

The three small, volcanic islands of Tristan da Cunha, Nightingale, and Inaccessible are situated in the South Atlantic roughly at 37°S, 13°W. Gough Island is quite well separated, lying about 200 miles to the southeast at 40°20'S, 9°56'W. The entire group is extremely isolated, being about 2,500 miles from the South American mainland and 1,800 miles from South Africa. Our current knowledge of both the terrestrial and marine biota is due mainly to the publications resulting from the Norwegian Scientific Expedition to Tristan da Cunha in 1937 to 1938.

One of the more recent Norwegian Expedition reports was that on the polychaete worms by Day (1954). He found this group to be essentially sub-Antarctic in distribution with particular affinity to the Falkland

Island–Magellan area. Five (14 percent) of the 36 species that were positively identified can be considered as endemics. Holdgate (1960) summarized the data for all the marine species and found a total of 107; of these, 35 were shared with the Magellan area, 25 (23 percent) were endemic, and 21 extended to South Africa.

Knox (1960) depicted Tristan-Gough as a cold-temperate area and offered some distributional evidence: at Gough Island, a siphonarid *Kerguelenella lateralis* and a brown algae *Adenocystis utricularis* occur that, elsewhere, are found only in sub-Antarctic, cold-temperate waters. At the same island, another algae *Durvillea antarctica* occurs that is mainly a cold-temperate species. The giant algae *Macrocystis pyrifera* is found on both Tristan and Gough Islands, but it can grow in either cold-temperate or warm-temperate waters.

Although the first report on the fishes of Tristan da Cunha was made by Carmichael (1817), the largest collection was that obtained by the Norwegian Expedition which stayed on the island from December 7, 1937 to March 29, 1938. The ichthyology report by Sivertsen (1945) contains accounts of 15 species, 5 of them pelagic and the rest belonging to the shore fauna. Rowan (1955:129) noted the presence of one additional shore species, the wreckfish (*Polyprion americanus*), a form with a transatlantic distribution. The most recent report on the fishes of Tristan da Cunha, Gough, and the Vema Seamount is that of Penrith (1967). He found that none of the shore species were confined to Tristan but that one endemic apparently occurs at each of the other places. In addition, one species found at all three localities apparently occurs nowhere else.

There is some indication, from the findings of both Knox and Penrith, that there may be considerable faunal differences between the Tristan da Cunha group and Gough Island which lies in somewhat cooler water. The fishes of the latter island are not well known, but those of Tristan da Cunha exhibit mainly a warm-temperate facies. However, at present, it seems best to rely on Holdgate's (1960) summary. Since this suggests a strong relationship to the Magellan Province, Tristan-Gough is considered to be a part of the Southern South America Region. An overall endemic rate of about 23 percent requires that it be placed in a province by itself. The probable age of Tristan da Cunha, estimated at 18 million years (Wilson 1963:536), indicates sufficient time for the development of this relatively high degree of endemism.

The Tasmanian Region

The west coast of Tasmania is directly exposed to the West Wind Drift. Also, currents diverging from the West Wind Drift extend around the southern tip of the island and up the east coast. In the Bass Strait between Tasmania and Victoria the circulation is quite complex. Apparently, in this area, there is some mixing with waters of tropical origin and a seasonal reversal of flow (Knox 1963b:346). The Victoria coast biota shows, to a considerable extent, an overlap of warm-temperate and cold-temperate species (Bennett and Pope 1953). This biogeographic pattern seems to fit well with the mixed origin of the surface currents.

Although Whitley (1932) placed Tasmania in a separate "Maugean" province, the area has been, until very recently, generally considered simply a continuation of the Southern Australia Warm-Temperate Region (Forbes 1856, Stephenson 1947:plate 15, Ekman 1953:197, Schilder 1956:75). The well-documented contributions of Bennett and Pope (1953, 1960) provided the first convincing demonstration that both Tasmania and the coast of Victoria possessed a distinctive cold-temperate biota. Their conclusions were adopted by Hedgpeth (1957:map) and Knox (1960, 1963b).

In discussing the location of boundaries between the warm-temperate

Figure 7-2 Tasmanian Region and the Cookian, Chatham, and Antipodean Provinces of the Southern New Zealand Region.

and cold-temperate faunas of Southern Australia (p. 130), it was suggested that these were probably situated at Bermagui, New South Wales, and at Robe, South Australia. This places the entire coast of Victoria in the cold-temperate zone as well as all of Tasmania. Since the area has not been well studied, there is available only a small amount of information to indicate the amount of endemism.

Bennett and Pope (1960:212) listed a number of common intertidal species that are cold-temperate endemics. Included were 14 mollusks, 4 algae, and 3 echinoderms. As Knox (1960:589, 1963b:380) has illustrated, the endemic giant brown algae *Durvillea potatorum* is a conspicuous indicator species. Womersley and Edmonds (1958:252) discussed the rich marine algal flora of Southern Australia and concluded that about 17 percent of the species were confined to Victoria and Tasmania. Knox (1963b:table IV) analyzed the distribution of limpets and found a total of 14 cold-temperate species. Of these, 4 (28.6 percent) were shown to be endemic.

Although there is no recent comprehensive account of Tasmanian fishes, there are indications that, in at least several families, significant numbers of species are regional endemics. This is probably true for 4 of the 7 species in the family Rajidae (Munro 1956–), 10 of the 13 species in the Clinidae, 1 of the 2 in the Bovichthyidae, and 3 of the 5 in the Ostraciontidae (McCulloch 1929–1930, Fraser-Brunner 1941).

It is apparent that Victoria and Tasmania have an unusually large complement of species that extends also into warm-temperate waters, particularly to the west. Kangaroo Island, South Australia, lies close to the mainland shore about 125 miles north of the cold-temperate boundary. A preliminary analysis of the algal flora of this island by Womersley and Edmonds (1958:253) showed that 78 percent of the species occurred on the Victorian coast and 54 percent in Tasmania. The great majority of Victorian mollusks range into warm-temperate waters (MacPherson and Gabriel 1962). The account by Scott (1962) of the fishes of South Australia showed that about 16.6 percent of the shore species of that state are shared with Tasmania (with many more ranging only as far as Victoria).

Due to its extreme isolation and rather small area, the cold-temperate biota of Tasmania-Victoria is relatively sparse, but that of the adjoining warm-temperate provinces is very rich. It is probable that many of the warm-temperate species have been able to invade these colder waters because of a lack of indigenous competition. The Southwestern Australia Province has a little cooler temperature regime than the Southeastern Australia Province, so it seems logical that more species from the former would be able to enter the Tasmanian Region.

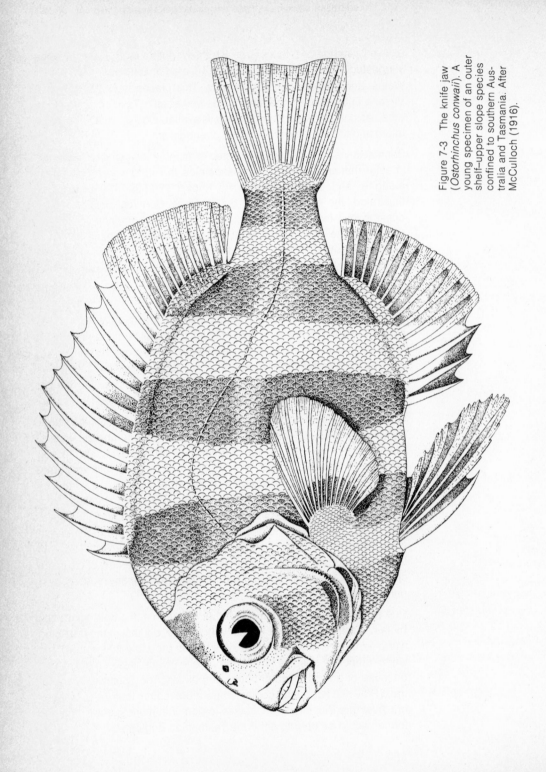

Figure 7-3 The knife jaw (*Ostorhinchus conwaii*). A young specimen of an outer shelf–upper slope species confined to southern Australia and Tasmania. After McCulloch (1916).

In general, it can be said that the regional biota is sparsely developed and not well known. As nearly as can be determined at this time—mainly through studies made on the algae, mollusks, echinoderms, and fishes—a significant amount of endemism does exist. This seems to vary, depending on the group, from about 10 to 30 percent.

The Southern New Zealand Region

As was noted in discussing the hydrography of Northern New Zealand (p. 133), the coastal current systems have been illustrated, mainly from Brodie's (1960) data, by Knox (1960:580, 1963a:12, 1963b:346). For New Zealand south of the Auckland Province, Knox described the following important components: (1) a Southland Current of modified subtropical water with a varying proportion of cold West Wind Drift water; the current comes in from the west, flows around the lower end of South Island between Stewart Island and the mainland, and then extends up the east coast as far as Banks Peninsula; (2) a cold Canterbury Current that runs up the east coast from Banks Peninsula to about Mahia Peninsula; (3) a Westland Current that flows northward along the west coast from the "Sounds" region to the Auckland vicinity; and (4) a D'Urville Current that apparently branches from the Westland Current to run eastward through the Cook Strait.

In the most recent of the publications mentioned above, Knox (1963b:346) depicted the Tasman Current as a narrow stream running almost directly from the southeastern corner of Australia to southwestern New Zealand. Close to the latter point, the Tasman Current is shown to bifurcate, one branch giving rise to the Southland Current and the other forming the Westland Current. This representation of surface current conditions in the Tasman Sea is quite different from the pattern in the standard ocean charts. Most such maps illustrate a very broad flow of warm water coming toward New Zealand from the west and, in fact, do not attempt to identify a discrete Tasman Current.

From the standpoint of Australian–New Zealand biotic relationships, it appears that most of the transport of organisms eastward across the Tasman Sea occurs considerably to the north of Knox's placement of the Tasman Current. The majority of the species shared by the two countries are warm-temperate and probably could not survive a trip via the cold waters of the South Island coast.

Knox (1963a:13) noted that, at times, the southern part of South Island and Stewart Island are under the direct influence of the cold West Wind Drift. Apparently, the New Zealand sub-Antarctic Islands (Auckland, Campbell, Antipodes, and Bounty) are almost constantly exposed to the West Wind Drift. The Chatham Islands are influenced by somewhat warmer water that appears to come directly from the central part of the New Zealand mainland.

As was noted elsewhere (p. 134), those who wrote about general marine zoogeography usually considered all of New Zealand proper to belong to a single warm-temperate area (Stephenson 1947, Ekman 1953, Schilder 1956). However, earlier New Zealand and Australian zoologists working mainly with mollusks had already subdivided the two main islands, indicating considerable latitudinal change in the fauna. Both Finlay (1925) and Whitley (1932) placed the entire North Island in one province (Cookian) and the South and Stewart Islands in another (Forsterian). In 1937, when Powell defined the "Aupourian Province" for the Cape Peninsula area, he also moved the southern boundaries of the Cookian Province down to Westport on the west coast of the South Island and to Banks Peninsula on the east coast.

Powell's latest (1962:11) arrangement included the "northern half" of the South Island in the Cookian area, leaving the southern half (with Stewart Island and the Snares) in the Forsterian. Knox (1960:619), who considered the biota in general, placed the Cookian-Forsterian boundary north of Dunedin on the east coast and said that the west coast limits still needed to be determined. A detailed map in a later work (Knox 1963a:12) showed the east coast boundary at about Oamaru and the western boundary at about George Sound. On the basis of algal distribution, Moore (1961) recognized, below the Auckland Province, an "Intermediate Province" for the remainder of the North Island excluding the Cook Strait region; a "Central Province" from Cook Strait to about the middle of South Island; and the Forsterian Province for the rest of South Island and Stewart Island. For fishes, Moreland (1959:30) discussed a northern fauna extending from North Cape to the Kaikoura-Banks Peninsula area and, from there, a southern fauna to Stewart Island.

A puzzling aspect of New Zealand biogeography is that, on one hand, a splitting of the mainland cold-temperate area into at least two provinces continues to be recognized while, on the other hand, it has been clear for several years that only one cold-temperate evolutionary center exists. Knox (1960, 1963a, 1963b) has given detailed accounts of the distributions of animals and plants along the east coasts of both islands. These showed very nicely that for all of New Zealand there are only two main evolutionary (and distributional) centers, a warm-water center in the north and a cold-water center to the south. In fact, Knox

The Provinces
Cookian

(1960:619) made a definite statement about the restricted Cookian Province: "It does not possess a distinctive flora and fauna of its own, but an overlapping to varying degree of northern warm-water and southern cold-water species." Powell's (1962) molluscan checklist shows but very little endemism for an intermediate Cookian area.

Since the most important criterion for the recognition of a zoogeographic province is evidence of independent operation as an evolutionary center, only one such area can be designated for Southern New Zealand. This would include all of the region south of the warm-temperate boundaries at Auckland and at East Cape (p. 136) and also Stewart Island and the Snares (Dell 1963). Of the possible names for the province, Cookian seems to be the most appropriate. Cook Strait is a prominent geographic feature, and of course, Captain Cook was the first to conduct a detailed exploration of the country.

The extent of endemism in the expanded Cookian Province is not easily determined. According to Powell's (1962) list, about 33 percent of the shallow-water mollusks seem to be endemic. Knox (1963b:table III) analyzed the distribution of limpets, and his data indicate 5 of 23, or 21.7 percent, are confined to the province. Of the 8 species of New Zealand clingfishes (Gobiesocidae), 5 are cold-temperate and the other 3 have general temperate distributions (Briggs 1955).

Chatham The Chathams are a cluster of small oceanic islands that lie due east of the Banks Peninsula and about 400 miles from the southern tip of North Island, New Zealand. Apparently of volcanic origin, the islands are quite old, dating from the Cretaceous or even earlier (Fleming 1962:67). However, they were submerged during parts of the Tertiary.

The marine fauna of the Chatham Islands had not been well collected until 1954 when an expedition was sent out under the sponsorship of the New Zealand Department of Scientific and Industrial Research and several kindred organizations (Knox 1957). The material taken at that time has provided data for a number of significant publications.

The first naturalist to visit the Chathams was Ernst Dieffenbach, who made collections in 1840 and later (1843) published his findings. Apparently the early visitors were mainly interested in the Mollusca. In 1925, Finlay, a malacologist, recognized the islands as a distinct zoogeographic area, the "Moriorian Province." This name and designation have been adopted by most of the modern workers.

Reporting on material from the 1954 expedition, Dell (1960a:6) identified 18 species of shorttail crabs but no endemics. Fell (1960:57) stated
that he was no longer able to recognize a distinctive Chatham Island
echinoderm fauna either on the shelf or the continental slope. He felt
that the fauna was best regarded as a part of the Cook Strait subregion,
i.e., central New Zealand between 38 and 46°S latitude. In contrast to
these reports, Dell (1960b:156) described a rich marine molluscan
fauna of 320 species with 49, or 15.3 percent, endemic and considered
the province as distinct as any in the New Zealand region. Bergquist
(1961:169) found 5 (23 percent) of the 22 siliceous sponges to be
endemic.

According to Knox (1960:596), Moore (1949) recorded 190 species of
algae from the Chathams and found that 10, or 5.3 percent, were
endemic. Recently, Knox (1963b) published a detailed account of the
littoral ecology and also discussed biogeographic relationships. He
recognized the Moriorian Province and observed that the present fauna
and flora was obviously derived from the New Zealand mainland.
Moreland (1957:34) reported that the entire marine fish fauna was New
Zealand in character and was probably continually reinforced by
drifting larval stages from the mainland.

In view of such conflicting evidence, it is difficult to make any positive
statement about the general relationship of the Chatham biota to
mainland New Zealand. It is most surprising to have recent reports of
little or no endemism in the algae, crabs, echinoderms, and fishes, but
appreciable amounts in the sponges and mollusks. Of the foregoing
groups, it appears that the mollusks have been the best collected and
have been the object of the most systematic work. Since they indicate a
15.3 percent endemism, the Chatham Islands are considered to comprise a separate province.

The sub-Antarctic islands of New Zealand comprise the Antipodean   Antipodean
Province. The islands are located to the south and east of the mainland
as follows: Auckland Islands, 190 miles; Antipodes Islands, 490 miles;
Campbell Island, 320 miles; and Bounty Islands, 490 miles. These
islands project from an extensive submerged plateau that stretches
from the southern tip of the mainland. During the Triassic to the
Cretaceous, most of the plateau was apparently above the surface and
continuous with New Zealand proper. The plateau and, at times, the
islands were submerged in the Tertiary. Auckland and Campbell

Islands were built up largely in the Pliocene when they were active volcanic sites (Fleming 1962).

The earliest interpretation of faunal relationship was made by the ichthyologist Regan (1914:37), who included the New Zealand sub-Antarctic islands along with the Snares, Stewart Island, and the southern tip of the mainland, in an "Antipodes District" of his "Subantarctic Zone." The New Zealand sub-Antarctic islands alone plus Macquarie Island were then placed in a "Rossian Province" by Finlay (1925) and by Whitley (1932). Later, Macquarie Island was determined to have other affinities, and the remaining four islands were placed in an Antipodean Province (Knox 1960:611, 1963b:371; Powell 1962:11).

From Powell's (1962) checklist it appears that about 50 percent of the Mollusca are endemic and that many of the nonendemics are shared with the mainland. Knox (1963b:395) compiled the available information about the echinoderms of the province: of 27 species, 6 (22 percent) are endemic, 18 are shared with the New Zealand mainland, 2 or 3 are circumpolar, and 1 is cosmopolitan. In the same work, Knox (table III) also analyzed the distribution of New Zealand limpets: of the 8 species known from the Antipodean Province, 6 are endemics and the other 2 extend to the mainland. Of the 6 species of shore fishes reported by Parrott (1958), 2 are possibly endemic.

Although, in general, the fauna is poorly known, there seems to be sufficient endemism so that a separate province can be recognized. Since strong external ties with mainland New Zealand are indicated so far, the province is considered to be a part of the Southern New Zealand Cold-Temperate Region.

The Sub-Antarctic Region

The Sub-Antarctic is the one zoogeographic region in the world that is entirely made up of small oceanic islands. Although they are scattered over a broad expanse of southern ocean, the islands have many genera and species in common. The West Wind Drift affects the region throughout the year and acts as an important dispersal agent. The surface temperatures are cold and show only a minor amount of seasonal fluctuation—from about 2°C in winter to 5°C in summer.

The independent character of the fauna, particularly the shore fishes, has been recognized for some time, but controversy has existed over the question of relationship to the Antarctic shelf as opposed to the cold-temperate parts of the Southern Hemisphere. Regan (1914) referred to the area as a "District" within the Antarctic Zone in order to express its affinity to Antarctica. Norman (1937b, 1938) agreed with this arrangement, but Nybelin (1947, 1951) presented evidence to demonstrate that the strongest relationship lay with the sub-Antarctic (cold-temperate) areas. Andriashev (1965:539) also recognized this distinction by identifying a Kerguelen Subregion (including Macquarie Island) as distinct from his Glacial Subregion for the rest of Antarctica.

For the shore fauna in general, Ekman (1953:219) regarded the Kerguelen Island group as belonging to a "transitional and mixed" region, but Hedgpeth (1969:fig. 10), in defining the biogeographic subdivisions for the Antarctic Map Folio Series, recognized a separate "Subantarctic Region" to include not only Kerguelen and its nearby islands but Macquarie, Tristan da Cunha, and the southern part of South America.

When one considers the distribution of the nonendemic species that are found in the region, the situation becomes somewhat clarified. Of the 15 species of nonendemic shore fishes known from Kerguelen Island, only 4 are also known from the shelf of the Antarctic continent (Dr. H. H. DeWitt, from a personal communication). The Kerguelen Province has 4 penguin species that are found elsewhere to the west (Falkland, South Georgia, and South Shetland Islands), but none of them occur on the Antarctic continent (Mackintosh 1960:629). Knox (1960:613) provided a list of the marine algae (9 species) that are also found at the tip of South America.

The Sub-Antarctic Region exhibits a highly endemic cold-temperate biota that was probably accumulated gradually as the result of a clockwise transport by the West Wind Drift. The predominate place of origin is Southern South America including the Falkland Islands. A few species were apparently picked up from South Georgia and the South Sandwich Islands. From the Kerguelen group, the current has carried a number of species some 3,000 miles eastward to Macquarie Island.

The Provinces
Kerguelen

The Kerguelen Province includes not only the Kerguelen group itself but the McDonald, Heard, Marion, Prince Edward, and Crozet Islands. The Kerguelen Islands lie in the southern Indian Ocean about 1,300 miles north of the Antarctic continent and directly on the mean position of the Antarctic Convergence. The McDonald-Heard Islands are located 300 miles southwest of the Kerguelens, and the two groups are linked by the submerged Kerguelen-Gaussberg Ridge which has a maximum depth

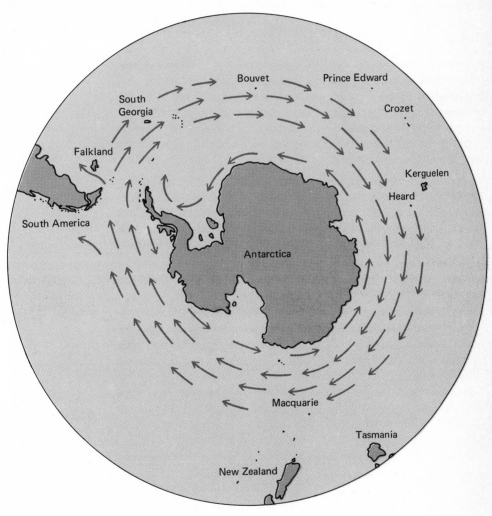

Figure 7-4   South polar projection depicting geographic relationship of areas included in Kerguelen, Macquarie, South Polar, South Georgia, and Bouvet Provinces.

of about 400 fathoms. The Crozet Islands are 1,000 miles west of Kerguelen and the Prince Edward–Marion group is another 300 miles farther west.

The first natural history collections from the province were made at Kerguelen Island by a British expedition from H.M.S. *Erebus* and *Terror*, probably in late 1841. The zoology of the voyage was published in parts by Richardson and Gray (1844 to 1875). During 1874 to 1875, three expeditions established camps at the Kerguelen Islands for the primary purpose of observing the transit of Venus in the Southern Hemisphere. The biological results of the American expedition entitled *Contributions to the Natural History of Kerguelen Island* were published in 1875 (J. H. Kidder, editor). The zoological results of the concurrent German expedition from S.M.S. *Gazelle* were published by Studer in 1879 and the report on the fishes taken by the British expedition appeared in the same year (Günther 1879).

The famous *Challenger* expedition made collections at Prince Edward and Kerguelen Islands in January 1874. Later, Kerguelen was visited by the German South Polar Expedition (1901 to 1903) with the *Gauss* and the British, Australian, New Zealand Antarctic Research Expedition (1929 to 1931) with the *Discovery*. The collecting activity of all these expeditions, plus some additional recent work, has served to make the fauna of Kerguelen itself quite well known. It is possible, therefore, to give a reasonably dependable analysis of its relationships.

The shore fishes of Kerguelen have been worked on more than the other groups of marine animals. Regan (1914:36) was the first to recognize that the various island groups listed above were characterized by a common fauna that was dissimilar to other areas. He proposed the name "Kerguelen District." At that time, 9 species were known with 7 of them apparently endemic. Norman (1937b:67), in his British, Australian, New Zealand Expedition report, recognized 17 species, 10 of them considered to be endemic. His 1938 *Discovery* Expedition report gave the same information.

In recent years, additional ichthyological material was reported on by Nybelin (1947, 1951) and by Blanc (1951, 1954, 1958). A new assessment of the fish fauna, kindly made by Dr. H. H. DeWitt (personal communication), indicates that it consists of 15 species and that 10 of them, or 66 percent, are endemic to the Kerguelen Province. Invertebrate endemism also appears to be high. Five of the 18 species of holothurians (Pawson 1969a:37), 5 of the 9 sea urchins (Pawson

1969b:39), and 6 of the 20 ascidians (Kott 1969) are apparently confined to the province.

Although its fauna is still poorly known, it is possible that the Prince Edward–Marion group (with probably the Crozets) constitutes a separate province within the region. Ekman (1953:217) noted that, on the basis of a total of 26 echinoderm species and 2 fishes, 21 percent of the shallow-water species are endemic and 57 percent are shared with Kerguelen, 39 percent with Southern South America, 32 percent with the Antarctic, and 21 percent with other temperate islands. For Kerguelen itself, Ekman (p. 219) indicated the presence of 69 echinoderm species with 32 (47 percent) considered to be endemic.

The first checklist of the Mollusca of Heard Island was published by Dell in 1964. A total of 19 species were recorded, with only 2 considered to be endemic. Of the rest, 13 are shared with Kerguelen and 2 with Macquarie Island. This gives strong evidence that the McDonald-Heard group is a part of the Kerguelen Province rather than belonging to the Antarctic Province (Knox, 1960:611).

Macquarie Macquarie Island is surrounded by deep water and is well isolated, being situated about 400 miles southwest of the Auckland Islands and some 650 miles from the New Zealand mainland. A collection of mollusks was made by Augustus Hamilton in 1894 (Dell 1964:267), but the first general biological survey was accomplished by the *Aurora* under the direction of Sir Douglas Mawson as a part of the activity of the Australasian Antarctic Expedition (1911 to 1914).

Both Finlay (1925) and Whitley (1932) included Macquarie Island in a "Rossian Province" with the New Zealand Sub-Antarctic Islands (Auckland, Campbell, Bounty, Antipodes). In writing about the fishes collected at Macquarie by the B.A.N.Z. Antarctic Research Expedition, Norman (1937b) quoted Regan's (1916a) observation to the effect that it was very interesting to find that the relationship appeared to be closer to Kerguelen than to the Sub-Antarctic Islands of New Zealand. A year later, Norman (1938) gave formal recognition to this relationship by recognizing a "Kerguelen-Macquarie District."

Although Ekman (1953:208) was uncertain about the faunal relationship, other modern workers have followed Norman in uniting Macquarie with Kerguelen and its associated islands. Nybelin (1947, 1951) utilized Norman's name, Andriashev (1959) referred to a "Kerguelen Transi-

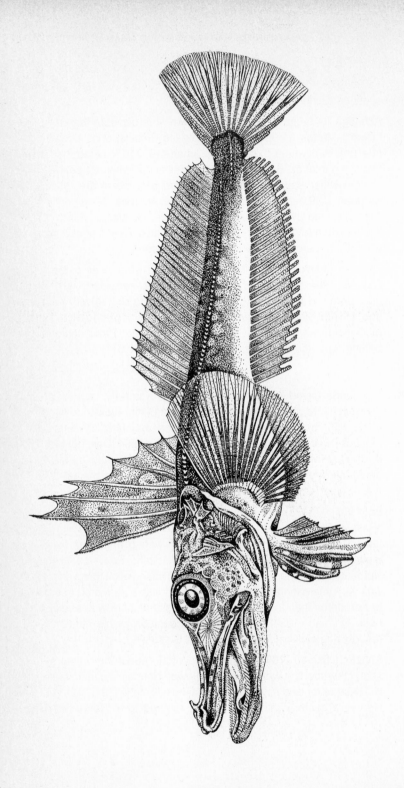

Figure 7-5 A channichthyid (*Channichthys rhinoceratus*); a cold-temperate shelf species confined to Kerguelen Island in the Sub-Antarctic. After Dollo (1904).

tional Province," and Knox (1960, 1963b) and Powell (1962) used the name "Kerguelenian Province." Hedgpeth (1969:fig. 10) included Macquarie in his "Kerguelen Subregion or Province."

Despite its spatial isolation and some good biological evidence of independent evolutionary development, no one has suggested that Macquarie might be placed in a province by itself. Ekman (loc. cit.) reviewed some of the data published as the result of collections made by the Australasian Antarctic Expedition and noted that 5 of the 6 echinoderms, 11 of 16 amphipods, and 30 percent of the mollusks were endemic. Later, Knox (1960:596) stated that 18, or 36.7 percent, of the 49 mollusks were endemic; and still more recently, Dell (1964:300) found that 27, or 64 percent, of a total of 42 molluscan species were endemic, with the great majority (12) of the nonendemics shared with other islands in the region. The fishes are still very poorly known. Waite (1916), in reporting on the ichthyological results of the Australasian Antarctic Expedition, identified 5 shore species; 4 of them are known to be present at Kerguelen, the other at the Antipodes. One of the 4 penguin species (the Royal eudyptes) is an endemic (Mackintosh 1960) but the others are found at Kerguelen (and elsewhere).

Knox (1963b) indicated agreement with Dell (1952) in recognizing the presence of a well-marked endemic element, a strong sub-Antarctic influence, and a very weak Neozelanic link. Knox further stated that the sub-Antarctic elements appeared to be largely derived from the west and that the relationship with Kerguelen was strongly marked. The existence of notable external ties to the Kerguelen Province together with a high degree of endemism (especially in the echinoderms, mollusks, and amphipods) can be expressed best by the establishment of a Macquarie Province within the Sub-Antarctic Region.

The Antarctic Region

The West Wind Drift is a very broad, circumpolar current that in many places covers some 20° of latitude. However, at Drake Passage between the South Shetland Islands and Cape Horn it must flow through a relatively narrow gap of about 400 miles. Within the West Wind Drift area is a circumpolar zone in which the cold surface water tends to sink, causing the surface temperature to alter rapidly with a change in latitude. This is called the Antarctic Convergence.

The Antarctic surface water, lying between the Convergence and the edge of the continents, is usually described as a cold, low salinity layer that varies in thickness from 100 to 250 meters. Much of it is covered by pack ice which forms a belt that expands and contracts with the seasons. The continental edge and its pack ice come under the influence of a countercurrent called the East Wind Drift. This current is relatively weak, irregular, and, being interrupted by the Palmer Peninsula, is not truly circumpolar.

The Antarctic continent and the South Shetland, South Orkney, and South Sandwich Islands all lie below the February 1°C isotherm and have winter temperatures as low as −2°C. The temperature regime at South Georgia is notably warmer, ranging from a winter mean of 1.3°C to a summer mean of 3.7°C. That of Bouvet Island is about −1 to + 1°C.

The Provinces
South Polar

In general, the South Polar shore animals demonstrate an unusual ability to cope with variation in depth. For example, the common nototheniid fish, *Trematomus bernacchii*, has been taken from near the surface to over 600 meters (Norman 1938:32, DeWitt and Tyler 1960) and many ophiuroid starfishes range from very shallow water down to 700 meters and, in some cases, as deep as 900 meters (Fell 1961).

It is known that a sinking of cold water takes place at the edge of the Antarctic continent so that fairly uniform temperature and salinity conditions exist from the surface down to considerable depths (Dietrick 1963:351). This, plus the fact that at the edge of the pack ice the water is quite deep, averaging about 400 meters, is apparently responsible for the development of many species with broad bathymetric tolerances.

In most parts of the world, the fauna of the continental shelf (down to 200 meters) is distinct from that of the slopes and the deep benthic regions. Apparently, this rule does not apply in the continental Antarctic, for the fauna from the surface down to about 500 meters seems, so far, to be rather uniform. For the purpose of distributional analysis, the shore fauna of this area is considered to extend to this depth.

Only in recent years has the biota of the Antarctic become well enough known so that meaningful estimates about relationship could be made. Ekman (1935, 1953:220) proposed a "Low-Antarctic" subdivision for the island of South Georgia and designated the rest of the area "High-Antarctic," which included "West Antarctic" and "East Antarctic" subregions. He also called attention to a very high degree of endemism for the Antarctic Region as a whole (fishes about 90 percent, echinoderms 73 percent, polychaetes 50 percent, amphipods 70 to 75 percent, and isopods 75 percent). Stephenson (1947:plate 15) depicted the Antarctic littoral fauna as occupying all of the periphery of the continent plus the island groups of the Scotian Arc and Bouvet Island.

In 1957, Hedgpeth (map) indicated the same general distribution of Antarctic fauna as Stephenson. However, Knox (1960) utilized the following arrangement: (1) an Antarctic Province with two subprovinces, a "Scotian" for Graham Land and the South Shetland, South Orkney, and South Sandwich Islands and a "Rossian" for the Victoria Land and Ross Sea area of east Antarctica; (2) a South Georgia Province for the island of the same name—considered to be a very distinct biogeographic unit transitional between the Antarctic and Sub-Antarctic. In this scheme, the McDonald-Heard islands were included in the Antarctic Province.

The most recent arrangement, based on the distribution of the fauna in general, is that proposed by Hedgpeth (1969:fig. 10) for the *Antarctic Map Folio Series*. Within an Antarctic Region a Continental (or High Antarctic) Subregion (or Province) is recognized and within the Subregion three special areas were identified: (1) an "Extension" to include the McDonald-Heard Islands and Bouvet Island, (2) a Scotia Subregion for the Antarctic Peninsula, and (3) a South Georgia District for the island of that name.

On the other hand, Dell (1965:148, 1968:115) in his general discussions about the benthic fauna of the Antarctic continent, emphasized that there was little evidence to warrant biogeographic subdivision and that as our knowledge becomes more precise it becomes obvious that a high percentage of each group of animals has a circum-Antarctic distribution.

Fishes

Ichthyological investigation of the South Polar Province began with the voyage of H.M.S. *Erebus* and *Terror* in 1839 to 1843 under the command of Sir J. C. Ross. Three fishes were briefly described by Richardson (Richardson and Gray 1844–1875) but the specimens were not saved.[1] Dollo (1904), in reporting on the results of the voyage of the S.Y. *Belgica*, 1897 to 1899, defined the Antarctic fauna as that which occurred within the Antarctic Polar Circle. The species described were listed according to the quadrant in which they were found (American, African, Australian, Pacific). Lönnberg (1905:4) disagreed with Dollo's definition and stated that the fish fauna of the South Shetland Islands and all areas to the south should also be considered truly Antarctic.

[1] Richardson related that one of them was thrown up by spray against the bows of the *Terror* and frozen there. The specimen was rescued, but unfortunately, before a detailed drawing or description could be made, it was stolen by the ship's cat. Norman (1938:89) considered it probable that it belonged to the species later described by Dollo (1904) as *Cryodraco antarcticus*.

Regan (1914) was of the opinion that the mean annual isotherm of 6°C approximated the northern boundary of an Antarctic zone. This zone was considered to be comprised of two districts, a "Kerguelen" and a "Glacial." Antarctica proper, the Scotian Arc islands, and "probably" Bouvet Island were included in the latter district. Norman (1938) essentially followed Regan's arrangement but definitely included Bouvet Island in the Glacial District. Norman also called attention to the Antarctic Convergence as a boundary for Antarctic coast fishes, but Nybelin (1947) considered it to be a poor boundary because it cut through the middle of the Kerguelen area. Instead, Nybelin suggested that the northern borderline be placed at about the 4°C isotherm for the warmest month of the year and that the extreme limit of pack ice could also be taken as a line of delimitation.

According to Nybelin (1947), the Antarctic should be divided as follows: (1) a High-Antarctic Region with an East Antarctic Subregion and a West Antarctic Subregion (the subregions being divided at the Ross and Weddell Seas); (2) a Low-Antarctic Region to include South Georgia, Shag Rocks, South Sandwich, and Bouvet Islands. Later, Nybelin (1952) decided, on the basis of material collected on the east side of the Weddell Sea by the Norwegian, British, Swedish Antarctic Expedition, 1949 to 1952, that the whole of the east coast of the continent may be considered as an ichthyogeographic unit.

Andriashev (1959), in discussing the ichthyological investigations of the Soviet Antarctic Expedition, 1955 to 1958, recognized an Antarctic Region divided into two subregions, one "Continental" for Antarctica proper and the other "Insular" for certain sub-Antarctic areas, including Bouvet Island. The Continental Subregion was further divided into East and West Antarctic Provinces. Later, Andriashev (1965) altered his arrangement for the Antarctic Region, recognizing a Glacial Subregion and a Kerguelen Subregion, the former being divided into a Continental Province and a South Georgian Province. The Continental Province was subdivided into an East Antarctic District and a West Antarctic District.

Rofen and DeWitt (1960) provided a complete list of the shore fishes of the Antarctic exclusive of South Georgia and Bouvet Islands. Finally, DeWitt (1968), having taken into account a number of new species that were described since the 1960 checklist, observed that about 74 percent of the Antarctic fish genera and about 90 percent of the species were endemic. He also recognized two coastal zoogeographic districts, the East Antarctic District for the Antarctic Peninsula to the South Orkney Islands and the West Antarctic District for the rest of the continental coast.

Other Groups   In regard to recently published works on the invertebrate fauna, Millar (1960), who studied the ascidians, found that 8 of 27 (29.6 percent) of the South Polar species were endemic. Also, he noted that the great majority of species occurring on the coast of Antarctica proper (20 of 23) also had been taken at Graham Land. Kott (1969) published a monograph on this group that included some general distributional data. In regard to Antarctic echinoderms, Fell (1961), after working on the ophiuroid starfishes, concluded that the Antarctic benthial faunal region extended to all the shallow-water areas colder than 2°C and included Antarctica, the Palmer Archipelago, and all of the islands of the Scotian Arc. According to his data, 14 of 33 Ross Sea species, or 42.4 percent, are apparently confined to Antarctic waters. Clark (1963:80) observed that the asteroid fauna of the Ross Sea was mainly comprised of circumpolar species, and Arnaud (1964:68) found that 77 percent of the littoral echinoderms of Adélie Land were endemic to the Antarctic. Pawson (1969a:37) concluded that 22 (58 percent) of the 38 holothurians and (1969b:39) 34 (77 percent) of the 44 species of sea urchins were endemics.

For the Mollusca, Powell (1965) published a general discussion of origin and distribution in Antarctic and sub-Antarctic seas, Nicol (1966:10) studied 36 species of pelecypods and found that only 7, or 19 percent, appeared to be endemics, but Akimushkin (1963) reported that the Antarctic cephalopod fauna consisted of 25 species and that 19 of them, or 76 percent, were endemics. Foster (1969:21) determined that at least 11 of the 16 (69 percent) of the Antarctic species of brachiopods were endemics.

The sponges of the Antarctic have been studied by Koltun (1964:335, 1969:13), who observed that over 300 species are now known and that more than half of them are limited to these waters. In regard to the bryozoans, Bullivant (1969:23) recorded 310 species and subspecies from south of the Antarctic Convergence, and it appeared to him that 180, or 58 percent, were endemics. Finally, Squires (1969:17), who worked on the scleractinian corals, concluded that 4 of the 7 species known from the Antarctic were endemics.

Because of the relative ease of observation, the local distributional patterns of seals and penguins offer dependable information. Four of the five true seals (Ross, Crabeater, Leopard, and Weddell) breed only in the Antarctic and are circumpolar (Davies 1958:489, Mackintosh 1960). Three species of penguins (Emperor, Adelie, and Ringed) have breeding areas only around the edge of the continent and nearby islands; three others are shared with the sub-Antarctic ranging from South Shetland and South Georgia Islands to the Kerguelen group and other cold-temperate areas (Mackintosh 1960).

Conclusions

Although most authors have preferred to subdivide the Antarctic continent into faunal provinces or subprovinces, the better the area becomes known, the more homogeneous the fauna appears. Species that were once thought to be confined to relatively well-known localities such as the Ross Sea or Graham Land have turned up elsewhere. This seems to be true for the fishes, at least some of the invertebrate groups, the seals, and the penguins. The continental shelf of the Antarctic including the South Shetland, South Orkney, and South Sandwich Islands is, therefore, treated as a single zoogeographic province.

In most groups, the degree of species endemism for the South Polar Province is very high. It is about 90 percent for the fishes, 80 percent for the seals, 50 percent for the penguins, and 19 to 77 percent for some of the better-known groups of marine invertebrates. Furthermore, endemic genera are very common.

South Georgia

South Georgia is a mountainous island about 100 miles long and 20 miles wide. It lies about 300 miles west of the South Sandwich Islands and about 800 miles east of the Falkland Islands. It is surrounded by water more than 3,000 meters deep and lies within the mean limit of drift ice. Although the surface temperatures (1.3 to 3.7°C) are distinctly warmer than those of the South Polar Province, they average 5 to 6°C colder than those of the Falkland Islands. Many Falkland species are undoubtedly carried to South Georgia by the West Wind Drift, but in most cases, the temperature barrier is probably successful in preventing colonization.

Apparently, the first natural history collections from South Georgia were taken by the German "Deutschen Polarkommission," 1882 to 1883, the fishes being described by Fischer (1885). Next, the Swedish South Polar Expedition of 1901 to 1903 made extensive collections. The fishes were treated by Lönnberg (1905), who noted a high degree of endemism and interpreted this to mean a long and complete isolation from other shores or shallow waters. The fishes from the British *Discovery* expeditions were reported on by Norman (1938). Finally, Nybelin (1947) examined material from a series of collections made between 1904 and

1926 by Norwegian whalers and from the *Norvegia* Expedition of 1927 to 1928.

At this date, the shore fishes of South Georgia seem to be fairly well known. A few years ago, it was noted (Briggs 1966:156) that, according to the available literature, about 57 percent of the species were endemics. However, additional work has now been done on the fishes of this island and the adjacent areas so that a more up-to-date calculation (Dr. H. H. DeWitt, personal communication) reduces this figure to about 34 percent. In regard to the mollusks, Powell (1951:57) indicated that about 40 of a total of 69 species of pelecypods and gastropods (58 percent) were endemic. A more recent estimate (Dell 1968:115) has shown that 98 of 164 species, or about 60 percent of the total molluscan fauna, consists of endemics.

So far, it appears that the rate of endemism in some of the other invertebrate groups is considerably less than that of the mollusks. Millar (1960) reported that 32 ascidian species were known from South Georgia and Shag Rocks. Of these, 3 were considered to be endemic and 7 were shared only with the South Polar Province. One of the 15 holothurian species (Pawson 1969a:37) and 2 of the 10 sea urchins (Pawson 1969b:39) were identified as endemics.

It may be recalled that Fell (1961) included South Georgia with the other Scotian Arc islands on the basis of the echinoderm (ophiuroid) relationships and that Hedgpeth (1969:fig. 10) considered South Georgia to be a district within his Continental Subregion. All together, the evidence seems quite strong in favor of an Antarctic affinity for South Georgia, but the distributional patterns of the ascidian species are admittedly puzzling. Considering its moderate degree of spatial isolation, the extent of faunal independence in the mollusks and fishes is remarkable.

Bouvet   Bouvet Island is extremely isolated, lying about 1,000 miles from the Antarctic continent opposite Princess Astrid Land. It is situated well within the mean position of the Antarctic Convergence and the mean limit of drift ice.

Ekman (1953:221) included Bouvet Island within the Antarctic Region because of its cold temperature regime (-1 to $+1°C$) and what little was known about its fauna. He referred to the presence of 12 echinoderms, 1 of them endemic and the great majority of the others shared with Antarctica. So far, only 2 shore fishes are definitely known (Nybelin 1947), 1 endemic and the other ranging to South Georgia and the South

Polar Province. Powell (1951) reported 5 gastropod mollusks, 2 endemic, 1 shared only with the Antarctic, and 2 extending to both Antarctic and cold-temperate localities. The one ascidian species is shared with Antarctica (Millar 1960).

While the fauna of Bouvet is probably sparse and admittedly poorly known, there may be considerable endemism in both mollusks and fishes. The recognition of a Bouvet Province at this time is based more on a prediction of future findings than on present fact.

Bipolar Distribution

In his discussion on geographic distribution in *On the Origin of Species*, Darwin (1859:374–382) compiled an impressive body of evidence, indicating a very close relationship between certain elements of the temperate biota of the Southern Hemisphere to that of the Northern Hemisphere. To explain these relationships, he assumed that northern plants and animals had, during the glacial period when the climate was much colder, been able to migrate southward across the equator.

As Ekman (1953:238) stated in his detailed discussion, the concept of bipolarity, involving at first the relationship between the Arctic and Antarctic faunas, probably stemmed from the statements of the famous polar explorer Sir James Ross. Ross, who took the H.M.S. *Erebus* and *Terror* to the Antarctic in 1839 to 1843, observed that some of the animals taken on that voyage appeared to be the same as those he had seen in the Arctic. He assumed that such animals had been able to migrate between the two polar seas by means of the cold, abyssal regions.

Bipolar distribution, or bipolarity, became a frequently discussed and controversial subject for about 100 years, from the 1850s to the 1950s. The concept soon became broadened to cover almost all facets of relationships between the nontropical parts of the Northern and Southern Hemispheres. As the individual species comprising the faunas of the higher latitudes became better known, the apparent relationship at this level became less. Now, few examples of authentic bipolar species (in the original sense of the term) can be found. Those that still remain in this category, such as the four species of sponges mentioned by Koltun (1969:14), belong mainly to groups that are beset with systematic difficulties. In the shelf faunas at least, even species that have only antitropical or bitemperate distributions are exceedingly scarce.

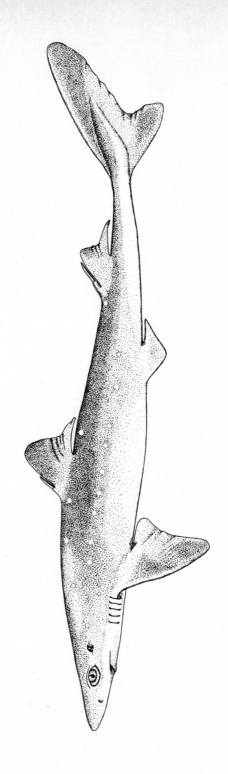

Figure 7-6. The spotted spiny dogfish (*Squalus acanthias*). A shelf species with a broad, antitropical distribution in both hemispheres. After Garman (1913).

At the generic level, one can find considerable evidence of both bipolar and antitropical distributions. In many cases, this is clearly attributable to the ability of certain groups to gradually migrate between the hemispheres by means of isothermic submergence. However, some genera are restricted to the surface waters to the extent that another explanation appears to be necessary. For them, the most popular theory has been that the deterioration of the climate during the glacial periods was such that temperate zone animals could cross the equatorial regions (Darwin 1859:376, Regan 1916b, Berg 1933, Hubbs 1952).

Since we now have good evidence that the surface waters of the oceans did not cool down to the extent necessary to permit the crossing of the tropics by cool-water animals (p. 416), the climatic theory is no longer tenable. Theél (1885), in his work on the holothurians of the *Challenger* Expedition, proposed a "relict theory" to account for bipolar distributions. Although Theél's theory was abandoned by later biologists, most of whom were in favor of the climatic change theory, it is basically the logical explanation. Considering that some marine animal groups are very widespread in a latitudinal sense, one can visualize the occurrence of a competitive displacement in the tropics, leaving relict populations in the cooler waters to the north and south.

Although the long controversy over bipolar distribution and its causes died down when the scarcity of truly bipolar or antitropical species became apparent, it remains an important concept for a few species (especially in the pelagic environment) and for many genera. Almost all such north-south relationships are explainable by means of isothermic submergence or the relict theory.

Literature Cited

Akimushkin, I. I. 1963. *Cephalopods of the seas of the U.S.S.R.* Academy of Science, U.S.S.R., Institute of Oceanology, viii + 223 pp., 60 figs. (English translation, Jerusalem 1965).

Andriashev, A. P. 1959. Ichthyological investigations of Soviet Antarctic Expedition (1955–1958) and zoogeography of the Antarctic waters. *Int. Oceanog. Congr. Preprints, Amer. Ass. Adv. Sci., Wash., D. C.*, pp. 129–130.

———. 1965. A general review of the Antarctic fish fauna. *in* P. van Oye and J. van Mieghem (editors), *Biogeography and ecology in Antarctica*. W. Junk, The Hague, pp. 491–550, 18 figs.

Arnaud, P. 1964. Echinodermes littoraux de Terre Adelie et pelecypodes commensaux d'echinides Antarctiques. *Sta. Marine d'Endoume, Rec. Trav.*, 54:1–69, 4 figs.

Balech, E. 1954. Division zoogeografica del litoral sudamericano. *Rev. Biol. Marina*, 4(1–3):184–195, 2 figs.

Bennett, I., and E. C. Pope. 1953. Intertidal zonation of the exposed rocky shores of Victoria, together with a rearrangement of the biogeographical provinces of temperate Australian shores. *Australian J. Marine Freshwater Res.*, 4(1):105–159, 5 figs., 6 plates.

———, and ———. 1960. Intertidal zonation of the exposed rocky shores of Tasmania and its relationship with the rest of Australia. *Australian J. Marine Freshwater Res.*, 11(2):182–221, 7 plates.

Berg, L. S. 1933. Die bipolare Verbreitung der Organismen and die Eiszeit. *Zoogeographica*, 1:449–484.

Bergquist, P. R. 1961. Demospongiae (Porifera) of the Chatham Islands and Chatham Rise, collected by the Chatham Islands 1954 Expedition. *Bull. New Zealand Dept. Sci. Ind. Res.*, 139(5):169–206, 20 figs.

Blanc, M. 1951. Poissons recueillis aux Iles Kerguelen par le Docteur Aretas. *Bull. Museum Natl. Hist. Nat., Paris*, 23(5):493–496.

———. 1954. Poissons recueillis aux Iles Kerguelen par P. Paulian (1951) et M. Angot (1952). *Bull. Museum Natl. Hist. Nat., Paris*, 26(2):190–193.

———. 1958. Sur quelques poissons des Iles Kerguelen rapportes par le Dr. Bourland. *Bull. Museum Natl. Hist. Nat., Paris*, 30(2):134–138.

Briggs, J. C. 1955. A monograph of the clingfishes (Order Xenopterygii). *Stanford Ichthyol. Bull.*, 6:1–224, 114 figs.

———. 1966. Oceanic islands, endemism, and marine paleotemperatures. *Syst. Zool.*, 15(2):153–163, 4 figs.

Brodie, J. W. 1960. Coastal surface currents around New Zealand. *New Zealand J. Geol. Geophysics*, 3(2):235–252, 7 figs.

Bullivant, J. S. 1969. Bryozoa. Distribution of selected groups of marine invertebrates in waters south of 35°S. latitude. *Antarctic Map Folio Ser.*, (11):22–23.

Carmichael, D. 1817. Description of four species of fish found on the coast of Tristan da Cunha. *Trans. Linn. Soc. London*, 12:500–513, 4 plates (not seen).

Clark, H. E. S. 1963. The fauna of the Ross Sea. Part 3. Asteroidea. *Bull. New Zealand Dept. Sci. Ind. Res.*, (151):1–84, 20 figs., 17 plates.

Darwin, C. 1859. *On the origin of species.* John Murray, London, ix + 1–490 pp., 1 chart.

Davies, J. L. 1958. The Pinnipedia: an essay in zoogeography. *Geograph. Rev.*, 48(4):474–493, 10 figs.

Day, J. H. 1954. The Polychaeta of Tristan da Cunha. *Res. Norwegian Sci. Expedition Tristan da Cunha 1937–38*, (29):1–35, 4 figs.

Dell, R. K. 1952. *in* F. A. Simpson, A. H. and A. W. Reed (editors), *The Antarctica today.* Wellington, New Zealand, pp. 129–150 (not seen).

———. 1960a. Crabs (Decapoda, Brachyura) of the Chatham Islands 1954 expedition. *Bull. New Zealand Dept. Sci. Ind. Res.*, 139(1):1–7, 2 plates.

———. 1960b. Chatham Island marine Mollusca based upon the collections of the Chatham Islands expedition, 1954. *Bull. New Zealand Dept. Sci. Ind. Res.*, 139(4):141–157, 1 fig.

———. 1963. The littoral marine Mollusca of the Snares Islands. *Records Dominion Museum, Wellington*, 4(15):221–229, 1 fig.

————. 1964. Marine Mollusca from Macquarie and Heard Islands. *Records Dominion Museum, Wellington*, 4(20):267–301, 36 figs.

————. 1965. Marine biology. *in* T. Hatherton (editor), *Antarctica*. F. A. Praeger, New York, pp. 129–152, 4 figs.

————. 1968. Benthic faunas of the Antarctic. *Symp. Antarctic Oceanog., Scott Polar Res. Inst., Cambridge, England*, pp. 110–118.

DeWitt, H. H. 1966. *A revision of the Antarctic and southern genus Notothenia (Pisces, Nototheniidae)*. Ph.D. dissertation, Stanford University, Stanford, xiii + 1–469 pp., 42 figs.

————. 1968. Coastal fishes. *Australian Nat. Hist.*, Dec., pp. 119–123, 3 figs.

————, and J. C. Tyler. 1960. Fishes of the Stanford Antarctic biological research program, 1958–1959. *Stanford Ichthyol. Bull.*, 7(4):162–199, 6 figs.

Dieffenbach, E. 1843. *Travels in New Zealand; with contributions to the geography, geology, botany, and natural history of that country*. J. Murray, London (not seen).

Dietrich, G. 1963. *General oceanography, an introduction*. John Wiley, New York, xv + 588 pp., 228 figs.

Dollo, L. 1904. *Résultats du voyage du S.Y. Belgica en 1897–1898–1899*. Rapports Scientifiques, Zoologie, Poissons, J.-E. Buschmann, Anvers., pp. 1–239, 6 figs., 12 plates.

Ekman, S. 1935. *Tiergeographie des Meeres*. Akademische Verlagsgesellschaft, Leipzig, xii + 512 pp.

————. 1953. *Zoogeography of the sea*. Sidgwick & Jackson, London, xiv + 417 pp., 121 figs.

Fell, H. B. 1960. Archibenthal and littoral echinoderms of the Chatham Islands. *Bull. New Zealand Dept. Sci. Ind. Res.*, 139(2):55–75, 10 plates.

————. 1961. The fauna of the Ross Sea. Part I. Ophiuroidea. *Bull. New Zealand Dept. Sci. Ind. Res.*, (142):1–75, 9 figs., 19 plates.

————. 1962. West-wind-drift dispersal of echinoderms in the southern hemisphere. *Nature*, 193(4817):759–761, 2 figs.

Finlay, H. J. 1925. Some modern concepts applied to the study of Cainozoic Mollusca from New Zealand. *Verbeek Mem. Birthday Vol.*, pp. 161–172 (not seen).

Fischer, J. G. 1885. Ichthyologische und herpetologische Bemerkungen. *Jahrb. Hamburg Wiss. Anstalten*, 2:49–119, 4 plates (not seen).

Fleming, C. A. 1962. New Zealand biogeography. A paleontologist's approach. *Tuatara*, 10:53–108, 15 figs.

Forbes, E. 1856. Map of the distribution of marine life. *in* Alexander K. Johnston, *The physical atlas of natural phenomena* (new edition). W. and A. K. Johnston, Edinburgh and London, plate no. 31.

Foster, M. W. 1969. Brachiopoda. Distribution of selected groups of marine invertebrates in waters south of 35°S. latitude. *Antarctic Map Folio Ser.*, (11):21–22.

Fraser-Brunner, A. 1941. Noted on the plectognath fishes. VII. the Aracanidae, a distinct family of ostraciontoid fishes, with descriptions of two new species. *Ann. Mag. Nat. Hist., Ser. 11*, 8:306–313, 3 figs.

Günther, A. 1879. Observations on the collection of fishes of Kerguelen land made during the transit of Venus expeditions in the years 1874–75. *Phil. Trans. Roy. Soc. London*, 168(extra vol.):166 (not seen).

Haig, J. 1955. The Crustacea Anomura of Chile. Rep. Lund Univ. Chile Exped. 1948–49. *Lunds Univ. Årsskrift, N.F. Avd. 2*, 51(12):1–68, 13 figs.

Hedgpeth, J. W., 1957. Marine biogeography. *in Treatise on marine ecology and paleoecology.* vol. 1. *Geol. Soc. Amer., Mem.* (67):359–382, 16 figs., 1 plate.

———. 1969. Introduction to Antarctic zoogeography. *Antarctic Map Folio Ser.*, (11):1–9, 15 figs.

Holdgate, M. W. 1960. The fauna of the mid-Atlantic islands. *Proc. Roy. Soc. London, Ser. B*, 152(949):550–567, 5 figs.

Hubbs, C. L. 1952. Antitropical distribution of fishes and other organisms. Symposium on problems of bipolarity and of pan-temperate faunas. *Proc. 7th Pacific Sci. Congr.*, 3:324–329.

Kidder, J. H. 1875. Contributions to the natural history of Kerguelen Island. *Bull. U. S. Nat. Museum*, (2):ix + 1–122.

Knox, G. A. 1957. General account of the Chatham Islands 1954 expedition. *Bull. New Zealand Dept. Sci. Ind. Res.*, (122):1–37, 24 figs.

———. 1960. Littoral ecology and biogeography of the southern oceans. *Proc. Roy. Soc., Ser. B*, 152(949):577–624, 73 figs.

———. 1963a. Problems of speciation in intertidal animals with special reference to New Zealand shores. *Syst. Ass. Publ.*, (5):7–29, 10 figs.

———. 1963b. The biogeography and intertidal ecology of the Australasian coast. *Oceanog. Marine Biol. Ann. Rev.*, 1:341–404, 5 figs.

Koltun, V. M. 1964. Sponge collection assembled by the Soviet Antarctic Expedition from 1955–58. *Soviet Antarctic Expedition*, Elsevier, New York, 1:333–336, 1 fig.

———. 1969. Porifera. Distribution of selected groups of marine invertebrates in waters south of 35°S latitude. *Antarctic Map Folio Ser.*, (11):13–14.

Kott, P. 1969. Antarctic Ascidiacea. *Antarctic Res. Ser., Amer. Geophys. Union.*, 13:xv + 1–239 pp., 243 figs., 2 plates.

Lönnberg, E. 1905. The fishes of the Swedish South Polar expedition. *Wiss. Ergeb. Schwed. Südpolar Expedition, 1902–03*, 5(6):1–69, 4 figs., 5 plates.

Mackintosh, N. A. 1960. The pattern of distribution of the Antarctic fauna. *Proc. Roy. Soc. London, Ser. B*, 152(949):624–631, 3 figs.

MacPherson, J. H., and C. J. Gabriel. 1962. *Marine molluscs of Victoria.* Melbourne University Press, Australia, xv + 475 pp., 486 figs.

Madsen, F. J. 1956. Asteroidea. Rept. Lund Univ. Chile Exped. 1948–49. *Lunds Univ. Årsskrift., NF Avd. 2*, 52(2):1–53, 1 fig., 6 plates.

Mann, G. 1954. *La vida de los peces en aguas Chilenas.* Ministerio de Agricultura y Universidad de Chile, Santiago. 342 pp., illus.

McCulloch, A. R. 1929–1930. A checklist of the fishes recorded from Australia. *Australian Mus., Sydney, Mem. 5*, pts. 1–4:1–534.

Menzies, R. J. 1962. The zoogeography, ecology, and systematics of the Chilean marine isopods. Rep. Lund. Univ. Chile Exped. 1948–49. *Lunds. Univ. Årsskrift., Avd. 2*, 57(11):1–162, 51 figs.

Millar, R. H. 1960. Ascidiacea. *Discovery Rep.*, 30:1–160, 72 figs., 6 plates.

Moore, L. B. 1949. The marine algal provinces of New Zealand. *Trans. Roy. Soc. New Zealand*, 77:187–189 (not seen).

———. 1961. Distribution patterns in New Zealand seaweeds. *Tuatara*, 9(1):18–23, 3 figs.

Moreland, J. M. 1957. Report on the fishes. *in* G. A. Knox, General account of the Chatham Islands 1954 expedition. *Bull. New Zealand Dept. Sci. Ind. Res.*, (122):34.

———. 1959. The composition, distribution and origin of the New Zealand fish fauna. *Proc. New Zealand Ecol. Soc.*, (6):28–30.

Munro, I. S. R. 1956—. *Handbook of Australian fishes.* Vol. 15. Fisheries Newsletter, Sydney (series still in course of publication).

Nicol, D. 1966. Descriptions, ecology, and geographic distributions of some Antarctic pelecypods. *Bull. Amer. Paleont.*, 51(231):1–394, 10 plates.

Norman, J. R. 1937a. Coast fishes. Part II. The Patagonian region. *Discovery Rep.*, 16:1–150, 76 figs., 5 plates.

———. 1937b. Fishes. Reports—Series B (Zoology and Botany), B.A.N.Z. Antarctic Research Expedition 1929–1931, 1(2):49–88, 11 figs.

———. 1938. Coast fishes. Part III. The Antarctic zone. *Discovery Rep.*, 18:104, 62 figs., 1 plate.

Nybelin, O. 1947. Antarctic fishes. Scientific Results of the Norwegian Antarctic Expeditions 1927–1928 et Seq. No. 26, Det Norske Videnskaps—Akademi i Oslo, 76 pp., 6 plates.

———. 1951. Subantarctic and Antarctic fishes. Scientific Results of the "Brategg" Expedition 1947–48 No. 2, *Kommandor Chr. Christensens Hvalfangstmuseum I Sandefjord* Pub. No. 18, 32 pp., 3 figs.

———. 1952. Fishes collected during the Norwegian-British-Swedish Antarctic Expedition 1949–52. *Medd. Goteborgs Musei Zoologiska Avdelning* 124, 13 pp.

Parrott, A. W. 1958. Fishes from the Auckland and Campbell Islands. *Records Dominion Museum*, Wellington, 3(2):109–119.

Pawson, D. L. 1969a. Holothuroidea. Distribution of selected groups of marine invertebrates south of 35°S. latitude. *Antarctic Map Folio Ser.*, (11):36–38.

———. 1969b. Echinoidea. Distribution of selected groups of marine invertebrates south of 35°S. latitude. *Antarctic Map Folio Ser.*, (11):38–41.

Penrith, M. J. 1967. The fishes of Tristan da Cunha, Gough Island and the Vema Seamount. *Ann. S. African Museum*, 48(22):523–548, 2 figs., 1 plate.

Powell, A. W. B. 1937. New species of marine Mollusca from New Zealand. *Discovery Rep.*, 15:153–222.

———. 1951. Antarctic and subantarctic Mollusca: pelecypoda and gastropoda. *Discovery Rep.*, 26:47–196, 14 figs., 6 plates.

———. 1962. *Shells of New Zealand.* 4th edition. Whitcomb and Tombs, Christchurch, 203 pp., 36 plates.

————. 1965. Mollusca of antarctic and subantarctic seas. *in* J. Van Mieghem and P. Van Oye (editors), *Biogeography and ecology in Antarctica.* W. Junk, The Hague, pp. 333–380, 6 figs.

Regan, C. T. 1914. Fishes. British Antarctic ("Terra Nova") expedition, 1910, *Zoology*, 1(1):1–54, 8 figs., 13 plates.

————. 1916a. Antarctic and Sub-antarctic fishes. *Ann. Mag. Nat. Hist. Ser. 8,* 18:377–379 (not seen).

————. 1916b. The British fishes of the subfamily Clupeinae and related species in other seas. *Ann. Mag. Nat. Hist., Ser. 8,* 18(103):1–18, 2 plates.

Richardson, J., and J. E. Gray. 1844–75. *The zoology of the voyage of H.M.S. Erebus and Terror, under the command of Capt. Sir J. C. Ross, 1839–43.* 2 vols. Jamson and Sons, London, illus.

Rofen, R. R., and H. H. DeWitt. 1960. Antarctic fishes. *in Science in Antarctica.* Part I: *The life sciences in Antarctica.* Pub. 839, National Academy of Science, National Research Council, pp. 94–112.

Rowan, M. K., and A. N. Rowan. 1955. Fishes of Tristan da Cunha. *S. African J. Sci.,* 52:129.

Schilder, F. A. 1956. *Lehrbuch der Allgemeinen Zoogeographie.* Gustav Fischer, Jena, viii + 150 pp., 134 figs.

Scott, T. D. 1962. *The marine and fresh water fishes of South Australia.* Govt. Printer, Adelaide, pp. 1–338, illus.

Sivertsen, E. 1945. Fishes of Tristan da Cunha. Results of the Norwegian Scientific Expedition to Tristan da Cunha 1937–1938. No. 12. Det Norske Videnskaps—Akademi, Oslo, 44 pp., 8 plates.

Soot-Ryen, T. 1959. Pelecypoda. Rept. Lund Univ. Chile Exped. 1948–49. *Lunds Univ. Årsskrift N.F. Avd. 2,* 55(6):1–86, 6 figs., 4 plates.

Squires, D. F. 1969. Scleractinia. Distribution of selected groups of marine invertebrates in waters south of 35°S. latitude. *Antarctic Map Folio Ser.,* (11):15–18.

Stephenson, T. A. 1947. The constitution of the intertidal fauna and flora of South Africa, part III. *Ann. Natal Museum,* 11(2):207–324, 11 figs., 2 plates.

Studer, T. 1879. Die fauna von Kerguelensland. *Arch. Naturg.,* 45(1):114–141 (not seen).

Theél, H. 1885. Report on the Holothuroidea. Part II. *in* Voyage of H.M.S. Challenger. *Zoology,* 14(part 39):1–290, 16 plates.

Waite, E. R. 1916. Fishes. Australasian Antarctic Expedition, *Sci. Rep. Ser. C,* 3(1):1–92, 5 plates (not seen).

Whitley, G. P. 1932. Marine zoogeographical regions of Australasia. *Australian Nat.,* 8(8):166–167, map.

Wilson, J. T. 1963. Evidence from islands on the spreading of ocean floors. *Nature,* 197(4867):536–538, 4 figs.

Womersley, H. B. S., and S. J. Edmonds. 1958. A general account of the intertidal ecology of South Australian coasts. *Australian J. Marine Freshwater Res.,* 9(2):217–260, 2 figs., 12 plates.

The Northern
Ocean

Northern
Hemisphere
Warm-Temperate
Regions

We see Aristotle, the first and in many ways the greatest of all naturalists, actually watching the creatures he loves. He is leaning out of a boat in the great gulf that indents the island of Lesbos, intent on what is going on at the bottom of the shallow water. In the bright sun and in the still, clear water of the Mediterranean every detail, every movement, can be discerned. Hour after hour, he lies there, motionless, watching, absorbed, and he has left for us his imperishable account of some of the things he has seen with his own eyes.

Charles Singer *in* A History of Biology, *1950*

Of the four major marine faunal zones, the warm-temperate has been the most poorly defined and the least understood. Ekman (1953) did not recognize the presence of a warm-temperate fauna in either southern Japan or along the California coast. In the Western Atlantic, he (p. 139) described the fauna between Cape Cod and Cape Hatteras as warm-temperate rather than cold-temperate. But it is now apparent that warm-temperate associations with their characteristic species (and often genera or even families) border on the tropical faunas not only in the Northern Hemisphere but everywhere in the world.

Wherever one takes the trouble to examine closely the ranges of marine species that occur along the periphery of the tropics, many warm-temperate forms can be identified. In some places such as southern Japan or the southern Atlantic coast of the United States, the warm-temperate group may be overshadowed by numerous eurythermic tropical species and by strictly tropical species that happen to be carried to higher latitudes by strong currents. However, in all such places the warm-temperate species are present and comprise an appreciable portion of the fauna. Some experimental work (Graham 1970) appears to indicate that the endemic warm-temperate fauna is comprised of species of tropical ancestry that have become adapted to lower winter temperatures and species of cold-temperate ancestry that have adapted to higher summer temperatures.

The amount of endemism in the warm-temperate areas varies, on a regional basis, from about 28 to 32 percent for Japan to about 40 to 50 percent for the Mediterranean-Atlantic. Stephenson and Stephenson (1952) were of the opinion that the populations of warm-temperate areas were made up primarily of eurythermic tropical species. This is true of most of them but not all. A notable exception is the San Diego Province of the California Region where about two-thirds of the nonendemics are eurythermic temperate instead of tropical species.

The relationships of faunas across the North Pacific has been of interest to zoogeographers for many years. Although the warm-temperate areas of each side are now widely separated, they may have been, at one time, situated somewhat closer together. In fishes, the serranid genus *Stereolepis* contains only two species, one in California waters and the other in Japan; the blenniid genus *Neoclinus* is represented by three species in California and one in Japan; and the surfperch family Embiotocidae has 20 species in California (some extending north to

Relationships Among the Regions

southern Alaska) and 2 in Japan. In the latter case, the species on each side belong to different genera.

Another example of amphi-Pacific distribution is the pichard genus *Sardinops*; one subspecies is found off the United States–Canadian coast and the other off the Japan-Russian coast. In this instance, however, the species are eurythermic temperate rather than warm-temperate. A similar degree of relationship can be seen in the sea lion genus *Zalophus*; one subspecies is found off the California coast and another in the Sea of Japan (a third is found on the Galápagos Islands).

Rather than indicating earlier continuities in distribution (Ekman 1953:156), such patterns may mean that the cold-water barrier was, at one time, less extensive so that a few species made their way across. Another and perhaps better explanation, in view of the relatively stable climatic history of the North Pacific (p. 427), is that the genera concerned were once widespread in the tropical Pacific but, because of competitive displacements, now reside as relicts in the warm-temperate zone.

In his article on the biogeographic features of the Mediterranean fish fauna, Tortonese (1964:100) called attention to the presence of a number of amphi-Atlantic species. It must be noted, however, that all of these are eurythermic tropical species that could easily have migrated to the Mediterranean from the more southern waters of the eastern Atlantic. There is no indication of any special relationship between the warm-temperate areas on each side of the North Atlantic. The fauna of each region has evolved independently and was derived mainly from the adjacent tropical fauna. As far as the warm-temperate faunas are concerned, there are also no good indications of any special amphi-American or amphi-Eurasian relationships.

The warm-temperate waters of the Northern Hemisphere may be divided into four distinct zoogeographic regions: the Mediterranean-Atlantic, the Carolina, the California, and the Japan. The first is by far the most extensive, and it is subdivided into four provinces. The California Region consists of two provinces, while each of the others contains but a single faunal complex.

The Mediterranean-Atlantic Region

The North Atlantic Current, sometimes called the North Atlantic Drift, is a continuation of the Gulf Stream that flows east-northeast. Along its route, filaments of warm, blue water extend toward the European coastline. Some of the current takes a northerly course and runs past the British Isles, along the Norwegian coast, and to some extent, as far as the Barents Sea.

The major portion of the North Atlantic Current apparently stays within the main subtropical gyre of the North Atlantic. Opposite the Iberian Peninsula, it turns southward and forms the Canary Current which parallels the northwestern coast of Africa. It is the relatively warm water provided by the North Atlantic Current that makes it possible for a warm-temperate fauna to exist as far north as the English Channel. Even during the coldest month of the year, the sea surface temperature of southern England is about the same as that of North Carolina just above Cape Hatteras. Yet, the difference in latitude is 14°!

From the open Atlantic, a surface current passes through the Straits of Gibraltar, extends along the coast of northern Africa, turns northward at the eastern end of the Mediterranean Basin, and flows in a westerly direction along the southern coast of Europe. The average salinity of the Mediterranean is rather high (38 $^o/oo$), and it is even higher (39.5 $^o/oo$) in the eastern end of the sea where the greatest evaporation occurs. Equilibrium with the Atlantic is maintained by a deep current of highly saline water that passes out of the Mediterranean via the Straits of Gilbraltar.

Black Sea and Mediterranean water is exchanged through the narrow Bosporus. An upper, low salinity current flows southward from the Black Sea, and a deeper, more saline current from the Mediterranean flows northward. Within the Black Sea, there tend to be two counterclockwise circulations, one for the eastern basin and one for the western (Zenkevitch 1963:384). Although at one time it was possible for some marine animals to migrate from the Black Sea to the Caspian and Aral Seas, the latter two bodies of water are essentially isolated.

The extensive coastline between the English Channel and Cape Verde, including the whole of the Mediterranean, forms the largest warm-temperate area in the world. The waters of the Black, Caspian, and Aral Seas are markedly colder, but they received their marine faunas mainly from the Mediterranean and, therefore, are included within the region.

History

Our written knowledge of marine biology began with the works of Aristotle who observed Mediterranean life some 2,300 years ago. Following the Renaissance, men who once again became interested in observing marine animals turned first to the Mediterranean. Pierre Belon published his *De Aquatilibus Libri Duo* (in Paris, 1553), the first book devoted entirely to marine animals. Soon after, Hyppolyto Salviani brought out his beautifully illustrated *Aquatilium Animalium Historiae Liber Primus* (in Rome, 1554–1558), Gulielmus Rondelet published his *Libri de Piscibus Marinis* (in Lyons, 1554), and Conrad Gesner brought out his encyclopaedic *Historia Animalium*, of which parts three and four were devoted to aquatic animals (in Zurich, 1556–1558).

Following the early burst of enthusiasm in the mid-sixteenth century, the Mediterranean was relatively neglected until quite recent times. The modern era in Mediterranean-Atlantic biology probably began with the opening of the famous zoological station at Naples in 1874 (although important work had been done in France and in Italy in the early part of the nineteenth century). There are now many additional marine stations in the area, the most productive being the Institute Océanographique at Monaco; the Sea Fisheries Research Station at Haifa, Israel; the Station d'Aquiculture et de Peche at Castiglione, Algeria; the Station Marine d'Endoume, Marseille, France; and the Institut Scientifique Chérifien, Rabat, Morocco (the well-known laboratories at Plymouth, England and Roscoff, France are located close to the northern boundary of the region). In addition, significant contributions have been made as the result of the Danish Oceanographical Expeditions from 1908 to 1910, by the work of the Commission Internationale pour l'Exploration Scientifique de la Méditerranée in Paris, and from research sponsored by the Instituto Espanol de Oceanografia in Madrid.

Although the modern era in Mediterranean-Atlantic biology has resulted in a vast increase in knowledge about local marine organisms, particularly in regard to ecology and life history, there have been few attempts to study the various groups over the region as a whole or even throughout the Mediterranean. Consequently, very little is known about distributional patterns within the area, and the systematics of many of the families and genera is still in a state of confusion.

Distributional Schemes

The earliest zoogeographers recognized the distinctive nature of the Mediterranean-Atlantic fauna and were well informed about its extent. Forbes (1856:map) depicted a "Lusitanian Province," occupying the area from the northern edge of the Bay of Biscay to the African coast as far south as the Canary Islands and west to the Azores. Forbes also recognized a "Mediterranean Province" for the whole of the Mediterranean and the greater part of the Black Sea, and the "Aralo-Caspian" as an outlying and exceptional province. Woodward (1856:361) showed close agreement with Forbes in regard to geographic limits but with the exception of the Aralo-Caspian Province, considered the entire region to lie within the Lusitanian Province. In a later work, Forbes (1859:240) referred to the existence of "a good physical boundary in the breadth of the English Channel." Ortmann (1896:map), on the basis of decapod crustacean distribution, placed the Mediterranean and Black Seas in a tropical "Mediterrane Subregion" and the Atlantic coast from Gibraltar north in an "Atlantisch-boreale Subregion."

More recently, Ekman (1953:80) emphasized that the Straits of Gibraltar do not form an important boundary, but at the same time, he recognized a Lusitanian Region extending from the Straits north to the western entrance to the English Channel and a Mauretanian Region reaching south to the African tropics and including the island groups of Madeira, Azores, Canaries, and Cape Verdes. Schilder (1956:75) outlined a "Mittelmeerisch" Subregion with a northern boundary at about Cape Finisterre, Spain. This was subdivided into two provinces—a "Sarmatisch" for the Aral-Caspian area and a "Südeuropäisch" for the rest. The latter was split into two subprovinces—a "Mediterran" for the Mediterranean and a "Makaronesisch" for the Atlantic coast.

The distributional map published by Hedgpeth (1957) showed warm-temperate waters extending north to southern Ireland, the north entrance to the English Channel, and eastward through the Mediterranean and Black Seas. Mars (1963:72–75), on the basis of Pleistocene stratigraphy, recognized a "Province mediterraneo-atlantic" extending from the western entrance to the English Channel south to Cape Bojador, Spanish Sahara. This province was divided into the following regions: (1) a "Franco-ibérique" from the northern boundary to Cape St. Vincent, (2) a "Mediterraneene" to include the Mediterranean plus the Gulf of Cadiz, and (3) a "Marocaine" from the Straits of Gibraltar to the southern boundary. Finally, Zenkevitch (1963:379) referred to a "Mediterranean-Lusitanian subregion" of the Boreal Region and recognized a Black Sea–Azov Province and a Caspian Province as parts of the subregion.

For the invertebrates, Hyman (1955:673–674) called attention to a more or less uniform echinoderm fauna extending from Iceland and the northern limit of Norway south as far as the Cape Verde Islands and including the Mediterranean. However, she noted that the littoral region in question is roughly divisible into the Boreal (from the northern boundary to the western entrance of the English Channel), the Lusitanian (south to the Straits of Gibraltar), and the Mauretanian (to the Cape Verde Islands) Provinces. Coomans (1962:map), from evidence provided by the Mollusca, depicted a single "subtropical" Lusitanian (or

Mediterranean) Province from Cape Finisterre to Cape Verde and including the Mediterranean, Black, and Caspian Seas. Hall (1964:231), also a malacologist, proposed a Moroccan "outer tropical" Province from Cape St. Vincent to Cape Blanco and extending into the southern Mediterranean and a Lusitanian "warm temperate" Province from about Lands' End, England, to Cape St. Vincent but also extending into the northern Mediterranean.

Fishes   Although a comprehensive faunal work on Mediterranean-Atlantic fishes is lacking, certain areas and their relationships have been quite well investigated. In the Mediterranean proper, Tortonese (1938, 1954, 1960a, 1960b, 1963, 1964) published a series of important contributions on faunal affinities and local distributions; Kosswig (1955, 1956) wrote about faunal history; Ben-Tuvia (1953, 1962, 1966), Steinitz (1949a, 1949b, 1950a, 1950b, 1952), and Fowler and Steinitz (1956) were concerned about the fishes of the eastern basin and their origin. Also, the fishes of the Adriatic (Sôljan 1948), Algeria (Dieuzeide, Novella, and Roland 1953–1955, Dieuzeide and Roland 1957), Italy (Griffini 1903, Tortonese 1956) and France (Moreau 1881) are now reasonably well known.

Thanks to several recent works, the fish fauna of the Atlantic coast is now much better understood. Of particular importance are the publications of Navarro (1943) and Lozano Cabo (1950) on Spanish West Africa; Dollfus (1955) and Boutière (1958) on Morocco; Lozano Rey (1928, 1947, 1952a, 1952b, 1960) on Spain; Albuquerque (1954–1956) on Portugal; and Bauchot, Bauchot, and Lubet (1957) on France. The general ichthyological treatise edited by Blegvad (1929–1938) is also most useful for this area. The pertinent literature on the fauna of the offshore islands (Madeira and the Azores) is reviewed elsewhere (pp. 207–209). The marine fauna of the Canary and Salvaje Islands is apparently somewhat reduced but continuous with that of the mainland.

In discussing the general distribution of Mediterranean fishes, Tortonese (1964) recognized, among others, a group of "boreal" species that extends chiefly north of Gibraltar and a "tropical Atlantic" group that ranges mainly south of Gibraltar. However, some of the species that he listed as boreal have been taken on the Moroccan coast (Dollfus 1955) and several of the tropical Atlantic group have been recorded from Portugal (Albuquerque 1954–1956).

When the fish fauna of the Atlantic coasts, particularly Morocco and Portugal, are closely compared to that of the Mediterranean some interesting relationships become apparent. The fauna of the Mediterranean is comparatively rich, consisting of about 362 shore species (Tortonese 1963), and that of mainland Portugal totals about 248 species (Albuquerque, op. cit.). Only about 30 of the Portuguese species are *not* found in the Mediterranean; of these 30, 17 seem to extend northward well into the cold-temperate waters; 9 are eurythermic tropical species reaching their northern limits in Portugal; and only 3 could be interpreted as warm-temperate Atlantic coast species. The shore fish fauna of Portugal can, therefore, be said to be composed of about 88 percent Mediterranean species with almost all the rest being cool-water species shared with the cold-temperate area to the north or tropical species shared with the western coast of Africa south of Cape Verde.

The same type of comparison can be made for Morocco. Here the total number of shore species is about 216 (Dollfus, op. cit.) with only 17 *not* also occurring in the Mediterranean; of these 17, 7 extend into colder waters north of the English Channel, 2 are shared with the African tropics, 4 range broadly both to the north and south, and the remaining 4 are apparently endemic to the warm-temperate Atlantic coast. Therefore, about 92 percent of the total for Morocco are Mediterranean species with the majority of the rest being shared with adjoining zoogeographic regions to the north or south.

In the case of both Portugal and Morocco, only a very few species can possibly be considered as Atlantic coast, warm-temperate endemics. Some of these are rare and others are not systematically well defined so that the actual amount of endemism may even be lower than indicated by the above data.

The distributional patterns of the shore fishes lend strong support to the early scheme of Woodward (1856:361) wherein the Atlantic coasts and the Mediterranean were incorporated in a single Lusitanian Province. This is in contrast to many of the modern workers who have tended to subdivide the area. There is no firm evidence in either fishes or invertebrates to indicate that separate Atlantic coast faunas need to be recognized.

The Provinces

Lusitania

The existence of the southern boundary for the Lusitania Province at Cape Verde on the western coast of Africa is well documented and has been discussed (p. 87). To the north, it seems likely that the area with the highest rate of species change is the western entrance to the English Channel. Ekman (1953:82) noted that 28 percent of the Mediterranean decapod crustaceans stop at this point, Hyman (1955:673) described a change in the echinoderm fauna here, and the work of Holme (1961) on the bottom fauna of the English Channel indicated a boundary effect. Also, the investigations of Crisp and Southward (1958), Crisp and Fischer-Piette (1959), and Cabioch (1968) on the benthic organisms of the English and French coasts of the Channel revealed the presence of many range terminations.

In regard to fishes, species belonging to the comparatively rich Lusitanian fauna are well represented along the Atlantic coasts of Spain and Portugal, and although many seem to reach the northern limits of their ranges in the Bay of Biscay, a good number extend as far as the western end of the English Channel. For example, more of the sea breams (family Sparidae), a well-represented family, appear to reach their range limits in the latter area than at any other place (Wheeler 1969).

Data published (Southward 1963) on the demersal fishes of the Plymouth area indicate that a faunal change has taken place since the 1919–1922 period. Many of the northern species have become scarce or have disappeared, while many southern forms have increased in abundance. The once-thriving herring (*Clupea harengus*) fishery has suffered a dramatic decline, but the niche of this species has been filled by its warm-temperate equivalent, the pilchard (*Sardina pilchardus*). Southward concluded that the best explanation for this phenomenon was a gradual rise in sea temperature and a possible accompanying change in water movement.

The Mediterranean The Mediterranean is an interesting and special part of the Lusitania Province. Its fauna is a good deal richer than that of the Atlantic coasts with many endemic species. Estimates of endemism for some of the invertebrate groups have been made by Pérés and Cuard (1964) and Pérés (1967): of the 129 species of decapod crustaceans, 13.2 percent were considered to be endemic; 26.1 percent of the 107 species of echinoderms; 27.1 percent of the 192 hydroids; and 50.4 percent of the 132 ascidians.

For the fishes, the comprehensive list of Tortonese (1963) included a total of 540 species, of which 362 were shore forms. Of the latter, 62 were listed as endemics, but it should be noted here that 7 of them have

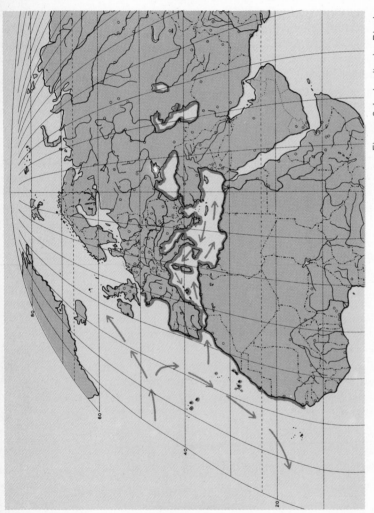

Figure 8-1 Lusitania, Black Sea, Caspian, and Aral Provinces.

been reported from the Atlantic coast of Morocco (Dollfus 1955), 2 from the coast of Portugal (Albuquerque 1954–1956), and 1 from the Azores (Albuquerque, op. cit.). Considering then that a total of 52 probably do not occur outside the Mediterranean, the endemism is about 14.4 percent.

The greater richness of the Mediterranean fauna with its many endemic species compared to the Atlantic coast with its predominance of species shared with the Mediterranean and extremely low endemism leads to an important conclusion: the Mediterranean basin has probably functioned as *the primary center* for the evolution and radiation of the eastern Atlantic warm-temperate fauna. A good many of the Atlantic coast species, possibly the majority, must have come from the Mediterranean.

The principal features of the history of the Mediterranean and its fauna have been reviewed by Ekman (1953), Kosswig (1955, 1956), Tortonese (1960, 1964), and Pérés and Picard (1964). For a long time, the Mediterranean basin and much of the surrounding area that is now land was occupied by the Tethys Sea. During the Mesozoic and early Tertiary periods, this sea was very large and served to link the eastern Atlantic to the Indian Ocean.

In the early Tertiary, the coral reefs of the Tethys Sea were well developed, and the associated tropical fauna was very rich. Material from the fossil beds of Lebanon and from Monte Bolca in Italy have yielded abundant remains of genera that are now confined to the tropical regions of the world. During the course of the Tertiary, a gradual cooling was superimposed on the climatic fluctuations so that, by the Pliocene, conditions were almost as they are now (Schwarzbach 1963:180). Fossil fish evidence (Arambourg 1965) has shown that the Mediterranean began to be affected by cooler temperatures in the Oligocene and that by Miocene times it had lost much of its tropical character.

Although Ekman (1953) placed considerable emphasis on the effect of Tertiary climatic changes, Kosswig (1955:71) decided it was more probable that the Tethys fauna died out as the result of the further lowerings of temperature or changes in salinity in the Mediterranean during the glacial periods. It now seems likely that *both* the climatic deterioration of the Tertiary and the Pleistocene glaciations had important effects. The existence of the ancient Tethys Sea was apparently terminated in the Lower Miocene by the establishment of a land barrier separating the Mediterranean basin from the Indian Ocean. This may have been followed by a temporary closure of the passage to the Atlantic Ocean, isolating the Mediterranean Sea and allowing the development of physical changes that were detrimental to the original marine fauna (Ruggieri 1967:284).

Even though the major portion of the faunal change probably took place before the end of the Pliocene, one could still expect to find, remaining in the warmer parts of the Mediterranean, a number of eurythermic species of tropical origin. It was this group that was probably eliminated during the glacial periods, and since the Red Sea was then cut off by a land barrier, reinvasion was prevented. In recent years, a good many such forms have successfully established themselves in the Mediterranean by migrating through the Suez Canal (p. 111).

Certain species, such as the fishes *Myoxocephalus bubalis* and *Anarhichas lupas*, now found only in the northern Mediterranean (and in the cold-water areas of the North Atlantic), probably owe their presence to the lower temperatures during the glacial periods. In the late Pleistocene, the mean summer temperature of the surface sea water in the Mediterranean is believed to have dropped to about 12°C (Emiliani 1961). This would make it possible for many northern species to move southward and enter the Mediterranean. Fossil remains of a number of northern mollusk species have been found. Today, similar summer temperatures are found along the Norwegian coast, well within the cold-temperate zone.

In the southern parts of the Mediterranean, many wide-ranging tropical species are found that are absent in the northern reaches (Postel 1956, Tortonese 1964:101). There is also apparent a notable faunal change correlated with depth. In fact, it was from studies in the Aegean Sea that this phenomenon was first described by Forbes (1844). He compared the depth changes to the altitudinal changes of fauna and flora found on the sides of mountains. Such changes may be attributed primarily to isothermic submergence and happen to be particularly noticeable along the shelf in warm-temperate areas.

For the province in its entirety, including both the Mediterranean and the warm-temperate Atlantic coast, the degree of endemism in most of the animal groups is certainly very high. Only one modern estimate has been made by Ekman (1953:85) in regard to three echinoderm groups (asteroids, ophiuroids, and echinoids). He considered that about 40 percent of the species were endemic. In fishes, it can be said that the endemism is probably even higher, possibly about 50 percent.

The Azores   The Azores, a group of nine islands at 36°50′ to 39°44′N and 25° to 31°16′W, are of unusual zoogeographic interest mainly because, aside from Iceland, they comprise the only island group to be found near the northern portion of the Mid-Atlantic Ridge. They are isolated from the mainland on both the east and the west by extensive deep-water regions as well as a considerable surface distance; the nearest mainland coast is that of Portugal about 900 miles away.

The literature on the marine fishes of the Azores or Western Islands is quite voluminous, beginning, apparently, with the account of Drouet (1861). A modern list was compiled by Collins (1954) with the assistance of G. E. Maul. Some of the fishes of this archipelago are also included by Albuquerque (1954–1956) in her extensive faunal work, and a new species of scorpaenid fish was described by Eschmeyer (1969). Considering the number of collections that have been reported during the past 100 years, it can be said that the shore fishes of the Azores are at least reasonably well known.

There are 99 species of shore fishes now known from the Azores of which 21 are widespread, transatlantic species. Of the nontransatlantic species that are shared with the mainland (or their adjacent island areas), 77 are found in the eastern Atlantic and *none* in the western Atlantic! This is, without doubt, overwhelming evidence of a very close faunal relationship to the east. It is also interesting to note that 80 of the total 99 shore fishes also occur in the Mediterranean. In fact, there are more such species shared with the Mediterranean than with Madeira (about 70 species in common) which lies about 400 miles closer.

The surface current chart of Sverdrup, Johnson, and Fleming (1946), and those of other authors as well, shows the Azores lying within the influence of the eastward flow of the North Atlantic Current—the continuation of the Gulf Stream. If this be accepted as the usual state of affairs, then it becomes difficult to explain the very close faunal relationship to the eastern Atlantic and the Mediterranean. Certainly, it is reasonable to suppose that a good many of the 21 transatlantic species migrated to the islands by means of the North Atlantic Current, but the relationship to the west still must be considered to be relatively weak.

Purely on the basis of the affinity of the shore fish fauna, it can be predicted that the Azores are influenced most by surface currents from the east instead of the west. Since this kind of current development is not shown on the modern current charts, either the actual pattern of water movement must be quite different, or it has undergone a change since the archipelago received its present quota of species.

The most striking aspect of the Azorian shore fish fauna is the almost complete absence of endemism. Only 1 of the 99 recorded species, a recently described scorpaenid, is apparently confined to the island. Here is a well-isolated group of islands, apparently of Miocene age (Wilson 1963:536), that should possess a large percentage of peculiar species, yet it does not. As in the case of the Cape Verde Islands, the cause is probably attributable to the severe drops in Pleistocene sea surface temperature (Emiliani 1958) that may have wiped out the older fauna. If the Azores have been repopulated within the past 12,000 years, very little evolutionary change could be expected.

It can be said that the Azores possess an offshore extension of the Lusitania fauna that shows a close relationship to the Mediterranean area. Several of its transatlantic species do not occur along the mainland shores at this latitude and probably immigrated by means of the North Atlantic Current.

The marine invertebrates are apparently quite poorly known and no reliable data could be found.

Madeira

Of the oceanic islands and island groups which occur in the tropical and warm-temperate waters of the Atlantic, the best known from the standpoint of its fish fauna is Madeira. Fortunately, this isolated area, consisting of two islands at 32°3′ to 33°7′N and 16°13′ to 16°37′W, has appealed to several competent ichthyologists over the years, three of whom have been outstanding: Richard Thomas Lowe published on Madeiran fishes from 1833 to 1854, James Yate Johnson wrote a number of contributions between 1862 and 1890, and the present curator of the Municipal Museum at Funchal, Dr. G. E. Maul, began his series in 1945.

Although Madeira lies nearly 500 statute miles from the nearest mainland (the coast of Morocco), Maul in Noronha e Sarmento (1948:134–181) and Albuquerque (1954–1956) have shown that the island shore fish fauna is quite rich with some 123 species represented. As might be expected from the geographic location of the islands, the great majority of the shore fishes are also found along the eastern Atlantic mainland or at other island localities on that side of the ocean. The relationship to the Mediterranean fauna seems especially close since reference to the recent list of Tortonese (1963) indicates that about 94 species (76.4 percent of the Madeiran shore fauna) are shared by the two areas.

Twenty-nine of the shore fishes recorded from Madeira may be considered transatlantic species, since they have also been taken somewhere in the western Atlantic. It is most interesting to see that, for two of these (*Gymnura hirundo* Lowe and *Antennarius radiosus* Garman), Madeira is the only known eastern Atlantic locality. Maul (1959:14) considered it likely that the latter colonization took place as the result of individuals being carried from Bermuda or the Caribbean by the Gulf Stream (via the North Atlantic Current), since Madeira is occasionally influenced by the South East Drift from that current.

Despite its isolation, Madeira shows the same interesting lack of endemism that characterized the other oceanic islands of the North Atlantic. Although a relatively large number of fishes were originally described from specimens taken there, almost all have subsequently been taken elsewhere so that only four species can now be considered autochthonous: *Gobius ephippiatus* Lowe, *Blennius bufo* (Lowe), *Lepadogaster zebrina* Lowe, and *Paraconger macrops* (Günther).

Due to its strong faunal ties to the east and very low rate of endemism (3.2 percent), Madeira is considered to be an isolated component of the Lusitania Province. In comparison to the Azores, it has the richer fauna, mainly because of the presence of a number of eurythermic tropical species. The complement of four endemic species, as compared to one for the Azores, probably indicates that Madeira did not suffer quite as great a temperature drop during the Pleistocene. Its age (Miocene, says Wilson 1963:536) seems about the same as that of the Azores.

As in the case of the Azores, no data on the invertebrate fauna could be found that was comparable to the information about the fishes.

Black Sea   The whole Black Sea–Caspian Sea region has an interesting history. As long ago as 1859, Forbes (p. 283) recognized, from evidence provided by fossil materials, that the region had at one time possessed an ancient and peculiar freshwater biota and that this assemblage had been superseded by an influx of marine species from the Mediterranean. The entire series of post-Tethyian changes that took place have been reviewed by Caspers (1957:802–809) and by Zenkevitch (1963:354–367).

Briefly, it can be said that the Black Sea–Caspian Sea region was occupied by a portion of the vanishing Tethys Sea until the mid-Miocene when the connection to the ocean was probably lost. The basin thus isolated has been given the name "Sarmatian" and a recognizable Sarmatian fauna developed. Later, the surface water became quite fresh, while the saline water stagnated at greater depths and was permeated by hydrogen sulfide. The result was a loss of most, if not all, of the Sarmatian fauna.

Still later, the oceanic connection was once again established, the salinity of the water consequently increased, and a host of Mediterranean genera entered the area. At this time, the basin began to separate into eastern and western parts and was given the name "Maeotic." Then, the connection to the Mediterranean again became restricted, and the salinity dropped, initiating a third major faunal change.

By the beginning of the Pliocene, the basin was once again completely isolated and the drop in salinity continued—this stage being called the "Pontic." An elevation of land then severed the connection between the eastern and western portions, and separate Black Sea and Caspian Sea faunas began to develop. By the end of the Tertiary, the Black Sea had attained essentially its present shape.

During the Pleistocene, the salinity of the Black Sea fluctuated as the Bosporus became opened and closed. Generally, the glacial advances corresponded to periods of freshening and the interglacial periods with incursions of salty water. The immediate cause of the fluctuations was the eustatic changes of sea level transmitted from the Mediterranean.

The Modern Fauna

Our knowledge of the Black Sea (and Caspian Sea) fauna began with the explorations of Peter Simon Pallas in 1793 to 1794. Many of his findings were published in his famous *Zoographia Rosso-Asiatica* (1811). From 1856 to 1859, Karl Kessler made several important contributions to the ichthyology and geology of the area. The first biological station was established at Odessa (1871 to 1872) and was later moved to Sevastopol. There are now five such stations in operation on the Russian shores (Zenkevitch 1963:382), and since the turn of the century, an imposing body of information has accumulated.

Three hydrologic properties of the Black Sea have important effects upon the composition and distribution of its fauna: (1) the salinity is low, being usually 17 or 18 ‰ in the upper layers of the open sea and becoming lower near the mouths of the large rivers and in the Sea of

Azov; (2) the winter temperatures are cold with means of about 0°C for the extreme northern parts and about 10°C for the southernmost part; (3) below about 150 meters, the oxygen content is very low and hydrogen sulfide is produced so that almost all of the deeper waters can be inhabited only by a few anaerobic organisms.

Compared to the Mediterranean, the fauna of the Black Sea is sparse with only 20 to 25 percent as many species. Furthermore, many characteristic marine groups are either absent or very poorly represented. There are no scaphopods, cephalopods, pteropods, or brachiopods; and the gastropods, echinoderms, tunicates, and cirripeds are represented by just a few species.

Sometime ago, Knipovitch (1933) classified the Black Sea fauna into several different categories based on origin and present habitat. Those of zoogeographic interest dealt with the distribution of the Pontic relicts, the species of Mediterranean origin that had become permanent residents, and the species that undertook occasional migrations from the Mediterranean. Yakubova (1935) studied the bottom fauna of the shelf above the hydrogen sulfide zone and divided the area into three coastal zones: the eastern half of the sea with the most typical fauna, a more saline southwestern zone with many Mediterranean migrants, and a northwestern area adjacent to the large river mouths with a euryhaline and freshwater fauna.

Perhaps for economic reasons, the fishes are better known than any other segment of the Black Sea fauna. The monograph of Slastenenko (1955–1956) was subsequently summarized by two articles in which the same author (Slastenenko 1955, 1959) sought to clarify the distributional relationships. The most recent work is that of Svetovidov (1964) and, as Collette and Bănărescu (1965) noted in their review, the origins of the marine fishes may be summarized as follows: 99 species are shared with the Mediterranean, 18 endemics are derived from Mediterranean populations, 8 species are shared with the Caspian, and 15 endemics are derived from Caspian forms. The total number of species is 140, and of these, 33, or about 24 percent, are endemic at the specific or subspecific level.

Strictly speaking, the Black Sea (and the Caspian) should not be listed under a warm-temperate heading, since the annual temperature range is equivalent to that of cold-temperate waters in other parts of the world. On the other hand, the major portion of its fauna is shared with the Mediterranean, and many of its autochthonous species were obviously derived from Mediterranean stocks. The rather high level of endemism can be attributed to several periods of spatial isolation, a temperature barrier, and a salinity barrier. Also, it is clear that the Black Sea is not merely a boundary or transitional region (Ekman 1953:94) but forms an interesting and distinct zoogeographic province.

The Sea of Azov is a constriction of the northernmost portion of the Black Sea. Rather than another sea, it actually forms a broad, shallow estuary for the Don River. The mean temperature for February is about 0°C, and the average salinity is close to 11 ⁰/₀₀. Unlike the Black Sea proper, a vertical circulation takes place at times so that nutrients are brought to the surface. This, plus additional nutrients that are brought in by the river, help to make the Sea of Azov a very productive area; Zenkevitch (1963:478) claims it is the most productive in the world.

The Sea of Azov

Due to the extreme cold in winter and relatively low salinity, the fauna is rather poor in number of species. Including freshwater forms, there are 79 fishes and about 226 invertebrates; of the total of 305 species, 75 are shared with the Caspian Sea and have been called "Caspian Relicts" (Zenkevitch, op cit., 479).

Apparently, only a few species are confined to the Sea of Azov, and since the great majority of the fauna is found also in the Black Sea, it is considered to be zoogeographically a part of that province.

The Caspian Sea is the largest enclosed body of water in the world, over 700 miles in length and about 200, miles in width. Although its level fluctuates considerably, it averages about 25 meters below sea level. It acts as a collecting basin for water from several rivers, but the largest by far is the Volga, which contributes more than 75 percent of the annual inflow.

Caspian

Because of its north-south alignment, the surface temperature for February (the coldest month) is 0°C in the northern end but 9°C in the southern part. The salinity of the surface waters is between 12 and 13 ⁰/₀₀, but Caspian water differs from normal sea water in that it is rather low in sodium and chlorine but rich in calcium and sulphates.

Zenkevitch (1963:564) has reviewed the recent Russian literature on the fauna of the Caspian Sea and provided some interesting information about its origin and affinity: the total number of free living species is 476; of these, 315 are endemic to the Black Sea–Azov–Caspian area, while 222 of the latter are found only in the Caspian. This means that the

Caspian endemics comprise 46.6 percent of the total and that most of the rest do not range beyond the Black Sea. The animal groups best represented in terms of species are fishes, 78 (25 endemic); amphipods, 72 (38 endemic); mollusks, 58 (50 endemic); and copepods, 50 (23 endemic). Such typical marine groups as the siphonophores, anthozoans, nemerteans, sipunculids, chaetognaths, chitons, cephalopods, and echinoderms are entirely absent.

As was noted earlier (p. 202), the Caspian became separated from the Black Sea probably in the early Pliocene. The brackish-water fauna of the Pontic Sea was then apparently able to continue its development in the former basin, while in the Black Sea, it became decimated as a result of invasions from the Mediterranean. There is some evidence that, in the late Pliocene, Caspian Sea waters became more saline and that a Black Sea connection may have become reestablished. Even so, many of the Pontic forms were able to survive, perhaps in areas of low salinity near the river mouths.

Although the various Russian authorities are not in complete agreement (Zenkevitch 1963), it seems likely that some kind of restricted passage between the Caspian and the Black Seas became available during one or more of the Pleistocene glacial periods. This would help explain the presence of such an impressive number of Black Sea–Azov–Caspian endemics (some 75 species) and also the fact that certain Mediterranean species were able to reach the Caspian.

During the 1920s, additional Mediterranean species were accidentally or purposefully introduced by man, and further immigrations occurred with the establishment of the Volga-Don Canal, a freshwater route between the Caspian and Azov Seas. The present Mediterranean group in the Caspian Sea is now estimated (Zenkevitch, op. cit., 574) at 28 species, including 1 diatom, 10 benthic algae, 1 angiosperm, 1 medusa, 1 bryozoan, 2 barnacles, 2 shrimps, 1 crab, 2 polychaetes, 3 mollusks, 1 cladoceran, and 3 fishes.

In addition to its original Pontic (sometimes called Sarmatic) fauna plus the admixture of Mediterranean forms, the Caspian Sea contains a small but interesting group that are obviously of Arctic derivation. These are euryhaline species that probably migrated south via a freshwater route provided by Pleistocene glacial activity. Included are the fishes *Stenodus leucichthys* and *Salmo trutta*, the seal *Pusa caspica*, and the polychaete *Manayunkia caspia*. It has been stated, however, that the Caspian seal may be a relict of an earlier (late Miocene) connection to the Arctic Basin (McLaren 1960).

In summary, it can be said that the Caspian Sea possesses a highly distinct, brackish and freshwater fauna composed mainly of late Tertiary relicts, but with Pleistocene and modern immigrants from the Mediterranean via the Black Sea, and a few glacial relicts from the Arctic Ocean. It is one of the most distinct of all zoogeographic provinces.

The Aral Sea is the most easterly of the large bodies of water located along the South Russian Geosyncline. It is of comparatively recent origin, being formed in the middle or late Pleistocene. The average salinity is about 10 %o, but as in the case of the Caspian Sea, the proportions of the various ions are quite different from that of sea water. Here there are more sulphates, calcium, and magnesium, while the sodium and chlorine content is low. In February, the mean temperature at the surface is 1°C and in August it reaches 24°C.

Aral

Shortly after its formation, the Aral Sea was apparently connected, at least to some extent, to the Caspian. The majority of Aral species are also found in the Caspian, and it is interesting to note that two Mediterranean marine species (the benthic eelgrass *Zostera nana* and the clam *Cardium edule*) have succeeded in migrating the entire distance via the Black and Caspian Seas. Logvinenko and Starbogatov (1962) concluded that the mollusks of the Aral Sea were derived from those of the shallow part of the Caspian. Nikolsky (1940), in his treatise on the fishes of the Aral Sea, stated that 14 of the 24 species were shared with the Caspian and that almost all the others were only subspecifically distinct. Compared to the Caspian, the Aral fauna is quite depauperate; furthermore, the great majority of the species are of freshwater rather than marine origin. Nikolsky (op. cit.) considered that nine, or 37.5 percent, of the fish species were endemics, but equivalent information on the other animal groups is lacking. Tentatively, the Aral Sea is considered to be a separate province but is very closely related to the Caspian Province.

The Carolina Region

Surface currents continually enter the Gulf of Mexico from the Caribbean Sea through the Yucatán Channel, a relatively narrow passage about 130 miles across. In the summer, this tropical water is widely diffused into the northern Gulf, but in the winter, it tends to circulate somewhat farther to the south (Leipper 1954:figs. 34, 35). The cause of this seasonal difference is undoubtedly the frequent occurrence of northerly winds during the winter. Such winds also lower the surface temperature of the onshore water until it is below the tolerance of most tropical organisms.

With marked seasonal current and water temperature changes, it is perhaps not surprising that considerable faunal changes should also take place. Along the Texas coast in the latter part of April, when the average water temperature climbs to about 20°C there is a significant change in the fish fauna. Certain cold-loving species migrate offshore into deep water, others that are apparently attracted by the higher temperature will move inshore, and still others will migrate into the area from the south, most probably from the Gulf of Campeche. This change has been documented by Miller (1965).

The Florida Current carries the outflow from the surface of the Gulf of Mexico. It is a swift stream that extends eastward through the Straits of Florida and then turns north toward Cape Hatteras. As it begins to flow northward, the Florida Current becomes reinforced by the Antilles Current—a stream that comes from the east side of the Greater Antilles, having originated from the North Equatorial Current. The Florida Current is considered a part of the extensive Gulf Stream system.

There is no doubt that the Florida Current has a profound effect upon the biota of the Atlantic coast from northern Florida to North Carolina. So many tropical organisms are carried into the area that their numbers have tended to obscure those that are permanent residents. Also present are many eurythermic tropical species that are able to stand the winter temperature of the onshore waters. For these reasons, many zoogeographers have completely overlooked the presence of a native, warm-temperate fauna.

Opinions and Patterns   In the northwestern part of the Atlantic, the distribution of mollusks has been studied more intensively than that of any other group. In his classic work of 1856, Woodward (p. 379) proposed the name "Trans-Atlantic Province" for the coastline between Cape Cod and the Florida east coast. The northern coast of the Gulf of Mexico was considered to belong to the tropics. Woodward's scheme was followed by Fischer (1881, 1882) and P. H. Fischer (1950). In 1934, however, Johnson recognized two subdivisions (proposed earlier by Forbes) for the Transatlantic Province with a dividing line at Cape Hatteras; these were the "Virginian" to the north and the "Carolinian" to the south.

More recently, there has been a tendency, among malacologists, to be

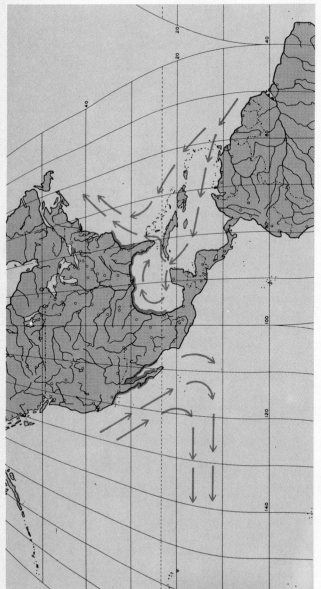

Figure 8-2 Carolina Region and the Cortez and San Diego Province of the California Region.

aware of the presence of a warm-temperate fauna in the northern Gulf of Mexico. Pulley (1952), on the basis of bivalve distributions, recognized the usual Virginian and Carolinian Provinces for the Atlantic coast, but for the Gulf coast, he proposed four different provinces—far too many, at least for the fauna in general. Rehder (1954:472) mentioned the presence of a Carolinian Province from Cape Hatteras to Cape Kennedy and across the northern Gulf. Abbott (1957) suggested the name "Appalachian Province" to replace Woodward's term and also recognized Virginian, Carolinian, and Texan subregions, the latter pertaining to the northern Gulf. Abbott further suggested that the peculiar area of local endemism around Tampa Bay might be designated the "Conradian Sub-region."

Bousfield (1960) utilized a concept of overlapping zones and thought that the area from Labrador to Cape Hatteras could be called "Boreal" and the region from the Gulf of St. Lawrence to northern Florida could be called "Virginian." Coomans (1962:98) delineated a Virginian and a Carolinian Province with the northern Gulf included in the latter. The latest work on general mollusk distribution was that of Hall (1964:226). He attempted to describe Northern Hemisphere provinces on the theoretical basis of temperatures supposedly required for reproduction and early growth. Unfortunately, the result is a scheme that does not fit observed distributional patterns. For this area, Hall considered the northern Gulf to be tropical and included the Atlantic coast Virginian and Carolinian areas respectively in "mild temperate" and "outer tropical" zones. He restricted the warm-temperate zone to the eastern Atlantic.

In regard to other marine groups, Ortman (1896:map), who worked on the decapod crustaceans, did not recognize a warm-temperate fauna and considered the dividing line between the boreal and tropical regions to lie at Cape Hatteras. Van Name (1945) mentioned an "Eastern U. S. Region" for ascidians extending from Cape Cod to northern Florida but noted that an admixture of West Indian species began at North Carolina. Humm (1969:45) observed that the inshore algal flora of the northern and northwestern Gulf of Mexico is related to that of the Atlantic coast between Cape Kennedy and Cape Hatteras.

Several workers have discussed distributional patterns from the standpoint of the marine fauna in general. Forbes (1856:map) was the first to name and depict a Virginian Province (New York to Cape Hatteras) and a Carolinian Province (Cape Hatteras to about the Florida-Georgia border). Ekman (1953:46) was of the opinion that the tropical boundary belonged at Cape Hatteras or a little to the south and (p. 139) that the fauna between Cape Hatteras and Cape Cod could be described as warm-temperate. Stephenson and Stephenson (1952:41–42) found a distinctly marked warm-temperate region between Cape Kennedy and Cape Hatteras and noted further that this fauna existed in two distinct sections, one for the Gulf and one for the Atlantic coast.

Hedgpeth (1953:204) followed Woodward and the other malacologists in recognizing a transatlantic fauna and considered it to extend from Cape Cod to Cape Kennedy and across the northern Gulf. This was subdivided at Cape Hatteras into Virginian and Carolinian components. Schilder (1956:75) depicted only a tropical-boreal boundary at Cape Hatteras with the northern Gulf being considered tropical. For his 1957 map, Hedgpeth outlined a warm-temperate area in the northern Gulf and, on the Atlantic coast, running from about Cape Kennedy all the way north to Cape Cod.

Northern Gulf of Mexico

Although the fauna of the northern Gulf of Mexico is comparatively rich and possesses an interesting endemic component, less is known about it than that of any other coast of the continental United States (excluding Alaska). A few modern biologists have still considered the northern Gulf to be inhabited by a tropical or Caribbean biota (Van Name 1954:495, Taylor 1955:261, Schilder 1956:75, Hall 1964:231), but the majority, including almost all of those who have personally worked in the area, have called attention to the presence of warm-temperate or nontropical species. Evidence (p. 67) has already been presented for the location of the two tropical versus warm-temperate boundaries in the Gulf of Mexico. These are most probably located at Cape Rojo, just below Tampico, Mexico, and at Cape Romano on the Florida west coast.

Most of the invertebrate groups are very poorly known, but the following information is available: Phleger and Parker (1951) found a 15 percent endemism in the Foraminifera; Pulley (1952) recognized four different provinces on the basis of the bivalve distributions; Hedgpeth (1953) noted that 3 out of 24 echinoderm species were endemic and that about a 10 percent endemism occurred in most invertebrate groups; Hartman (1954:413) referred to "a large endemic population" and also "unique genera" of polychaetes; and Behre (1954:453) mentioned several genera of endemic decapods.

The shoreline of Texas lies well within the warm-temperate part of the Gulf of Mexico and occupies a large percentage of that area. Since its fish fauna is fairly well known (Hoese 1958, Briggs et al. 1964), it may provide a reasonably good indication of the relationship of the northern Gulf as a whole. The total known shore species now number 300, and of these, no less than 214 (more than two-thirds) are found on the Atlantic coast of the United States from Florida to North Carolina. Thirty-nine of the total, or 13.0 percent, are endemic to the northern Gulf. Purely southern species, those that extend only southward from the Gulf, are surprisingly few—only 47.

Considering the rate at which tropical species are being discovered along the deeper portions of the Texas shelf, the actual number of shore fish species is probably between 375 and 400. As such species are discovered, the proportion of northern Gulf endemics will drop, possibly to about 10 percent. So, it can be estimated at this time that the overall levels of endemism for the fishes and invertebrates of the northern Gulf of Mexico are about the same—in the vicinity of 10 percent.

Except for the presence of low salinity water at the surface near the Mississippi Delta, the chemical and physical qualities of the shelf waters are similar on both sides of the northern Gulf of Mexico. However, when one looks at the distribution of the fauna some interesting differences become apparent. The fish fauna of the northeastern Gulf is the richer, mainly because of the presence of a number of eurythermic tropical species. It has been suggested (Briggs 1958:244) that this contrast may be the result of a difference in the type of bottom community.

The work of Hedgpeth (1954:206) on the bottom communities of the Gulf showed a virtually continuous association of coral patches and sponges covering the broad continental shelf west of Florida from the Keys almost to the western boundary of the state. West of this point, the coral-sponge association is abruptly replaced by the shrimp ground community. There are a few scattered coral reefs on the deeper parts of the shelf off the Texas coast, but so far, their fauna is not well known.

Baughman (1950:118) believed that the western Gulf had an entirely different faunal complex, cut off from Florida by the "vast and silt-laden flood of the Mississippi." Ginsburg (1952:101) indicated agreement with this view and, in addition, suggested that in some past geological epoch a peninsular barrier existed between Cape San Blas, Florida and Mobile Bay, Alabama. Since tropical shore fishes in particular tend to become highly specialized and dependent upon certain types of bottom fauna for food and shelter, there seems to be a good ecological

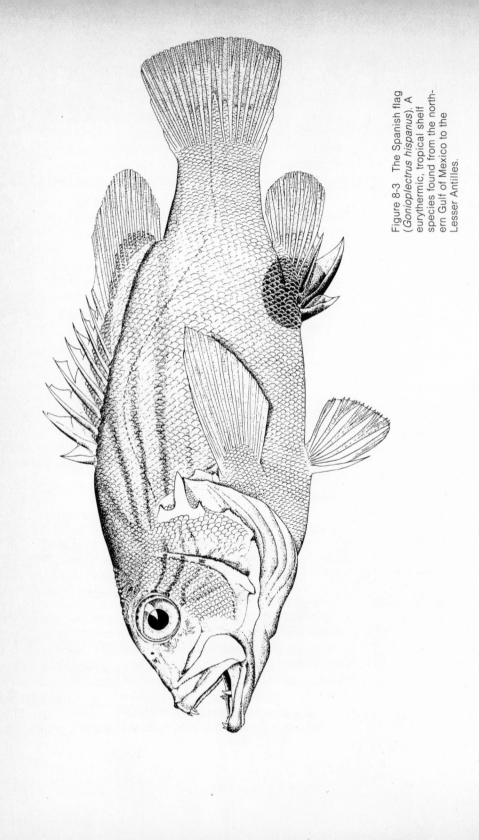

Figure 8-3 The Spanish flag (*Gonioplectrus hispanus*). A eurythermic, tropical shelf species found from the northern Gulf of Mexico to the Lesser Antilles.

basis for the differences apparent in the fish faunas of the two sides of the northern Gulf and thus no need to look for a physical barrier.

In the northern Gulf, there is also some indication of speciation within certain warm-temperate genera. For example, one species of menhaden, *Brevoortia gunteri*, is found in the western Gulf, while another, *B. smithi*, exists in the eastern part (Dahlberg 1970). Similar differences have been found in the blenniid genus *Chasmodes* (Springer 1959:332) and in the flatfish genus *Gymnachirus* (Dawson 1964). In each of these cases, small areas of overlap have been found, but these overlaps do not occur in the same section of the northern Gulf.

Atlantic Coast For many years, since the publication of Forbes' (1856) treatise, Cape Hatteras has been regarded as a significant zoogeographic boundary. But the important question is, does it represent the northern edge of the tropics (Ortman 1896, Ekman 1953, Schilder 1956), the northern border of a warm-temperate region (Coomans, 1962), or simply a dividing line between two warm-temperate provinces (Forbes 1856, Johnson 1934, Pulley 1952, Hedgpeth 1953, 1957; Abbott, 1957)? The first supposition is easily disposed of since it was previously decided (p. 64) that the northern boundary of the tropics lies on the Florida Peninsula at Cape Kennedy.

If one were to be swayed by the weight of opinion, Cape Hatteras would have to be considered a dividing line between two portions of a warm-temperate region that extends from Florida all the way to Cape Cod. However, upon close examination, the evidence seems rather weak. There is no doubt that, during the summer and early fall, many warm-temperate species, especially fishes, are found occupying the area north of Cape Hatteras.

The fish family Sciaenidae has, in the southeastern United States area, 23 shore species. A number of these (9) have quite well-defined, warm-temperate distributions in the Gulf of Mexico, yet on the Atlantic coast they have been taken as far north as Virginia, Maryland, and Massachusetts. But in instances where dependable catch records were found, it could be seen that such occurrences took place only during times of relatively high, sea surface temperatures. Even in the case of the Atlantic croaker (*Micropogon undulatus*), a species that has been taken in Massachusetts and supports an important commerical fishery in Chesapeake Bay, a migration to the south of Cape Hatteras apparently takes place when temperatures drop in the fall (Haven 1959).

In 1933, Parr published a geographic-ecological analysis of the seasonal changes in temperature conditions along the Atlantic coast. In this work, he emphasized that, in the summer, southern forms could move as far north as Cape Cod and only encounter a slow and gradual temperature decline on the way. It is also true, however, that for many of the more sedentary warm-temperate fish species, Cape Hatteras seems to be a good, continuous barrier.

Cape Hatteras also serves as an important, virtually continuous, barrier for many invertebrate species. Wells and Gray (1960) have written about the inability of the mussel (*Mytilus edulis*) to form permanent colonies on the Carolina coast below the Cape. In his thorough analysis of mollusk distribution, Coomans (1962:101) emphasized that the faunas north and south of the Cape are very different and that the Transatlantic Province of the earlier malacologists is an unnatural creation. He concluded that only the Carolinian part (below Cape Hatteras and including the northern Gulf of Mexico) might be considered a province of its own, comparable to the Lusitanian area in Europe and northern Africa. In his work on the decapod crustaceans of the Carolinas, Williams (1965:3) noted that Cape Hatteras formed a natural barrier to the northward distribution of 27.7 percent of the shallow-water forms.

On the basis of recent evidence about the distribution of the fishes, mollusks, and decapod crustaceans, it is suggested that Cape Hatteras does form a northern barrier for the warm-temperate fauna but that it is subject to frequent transgressions by the more motile species during periods of high water temperature.

Relationship to the Gulf Coast

In comparing the shore fish fauna of the warm-temperate Atlantic coast with that of the northern Gulf of Mexico, it can be seen that the latter is a good deal richer. For example, it was noted earlier (p. 219), that 300 species had so far been found on the Texas coast and that the actual number was probably between 375 and 400. In contrast, Bearden (1961, plus an addendum in 1962) has recorded a total of 244 for the South Carolina coast. This same relationship is also evident along the Florida peninsula where the Gulf coast fauna has been found to be appreciably richer than the fauna of the Atlantic coast (Briggs 1958:241).

A most interesting characteristic of the Atlantic coast segment is its very high degree of relationship to the Gulf coast. About 85 percent of the Atlantic shore fishes are also found in the Gulf, the majority of the remainder being eurythermic temperate species that extend into that area from the north. Furthermore, there is very little endemism along the warm-temperate part of the Atlantic coast where, at most, only about six species could be placed in this category.

Since the Florida peninsula, with its southern tip projecting into tropical waters, effectively separates the two components of the Carolina fauna, one would expect to find some evolutionary divergence. But, in addition to the little endemism on the Atlantic coast, only a few geminate or twin species may be found. In fishes, the total is about six including both species and subspecies. Examples are (1) *Alosa sapidissima*, *A. alabamae*; (2) *A. mediocris*, *A. chrysochloris*; (3) *Brevoortia tyrannus*, *B. patronus*; (4) *Cynoscion regalis*, *C. arenarius*; (5) *Prionotus tribulus tribulus*, *P. t. crassiceps*; and (6) *Centropristis striatus striatus*, *C. s. melanus* (the first three are best considered eurythermic temperate species rather than warm-temperate).

From a zoogeographic standpoint, the significant attributes of the Atlantic coast fauna are the very high degree of relationship to the Gulf coast, the almost negligible amount of endemism, and the occurrence of only a few twin species. These are good indications that the two warm-temperate faunas must have been in communication a relatively short time ago. Because of its richer fauna and appreciable amount of endemism, it can be surmised that the northern Gulf of Mexico, rather than the Atlantic coast, has functioned as the main evolutionary center.

The relative paucity of the Atlantic coast fauna suggests that, wherever the route of communication was located, the passage from the Gulf to the Atlantic was somewhat restricted—that is, it may have functioned more as a filter bridge than a wide-open thoroughfare. For many years, biologists and geologists working in Florida assumed that the peninsula had undergone extensive inundations during the interglacial periods of the Pleistocene. The distributional patterns of group after group of animals and plants, both terrestrial and freshwater, appeared to offer good evidence for this view (Neill 1957).

In 1950, Deevey found that a significant number of hydroid species had disjunct patterns broken by the Florida peninsula, but he supposed that their ranges had been continuous around the tip of the peninsula during the glacial periods. His conclusion has been supported by recent geological evidence. It is now apparent that the high marine terraces are considerably older than originally suspected and that the peninsula was not inundated during the Pleistocene (Tanner 1968). It seems clear that the Florida peninsula has not been able to function as an effective zoogeographic barrier because the lowering of sea surface temperature during the glacial stages permitted circumvention of the peninsula by warm-temperate species (p. 412).

In summary, it may be noted that the Carolina Region is a distinctive warm-temperate area consisting of two closely related but spatially separated parts. The northern Gulf of Mexico section includes those shelf waters north of Cape Romano in the east and Cape Rojo in the west. The Atlantic coast section extends from Cape Kennedy to Cape Hatteras. Of the two, the northern Gulf is the richer in species and also demonstrates a fairly high degree of endemism (about 10 percent) in both invertebrates and fishes. This area has probably functioned as the primary evolutionary center and has contributed species to the Atlantic coast section by means of a glacial stage migratory route around the tip of the Florida peninsula. The relative paucity of the Atlantic coast fauna suggests that the migratory route functioned in the manner of a filter bridge rather than a wide-open thoroughfare.

Since the Atlantic coast section demonstrates a very close relationship to the northern Gulf and shows only a negligible degree of endemism, the two areas are included within the same zoogeographic region with no provincial distinction. Considering the region in its entirety, the amount of endemism is quite high, probably about 30 to 40 percent.

The California Region

South of Point Conception, the central California shoreline takes an abrupt curve toward the east. This, and probably the presence of the Channel Islands, gives the inshore waters of the area some protection from the influence of the cool, southbound California Current. These sheltering effects, plus the help of some onshore current movement from the south during the winter, permit the existence of a relatively warm-water area along the southern California coast. Water of similar temperature extends, aside from some local interruptions caused by upwelling, south all the way to the temperate-tropical boundary at Magdalena Bay, Baja California.

The geographic position of the Gulf of California or Sea of Cortez is most interesting, for although its mouth is in the tropics, the remainder is situated far enough north to be warm-temperate. This condition has apparently prevailed at least from the beginning of the Pleistocene since, by then, the peninsula of Baja California had attained essentially its present outline (Durham and Allison 1960:63).

Until Garth (1955) cited the presence of certain outer coast species in the northern Gulf of California as evidence for recognition of a warm-temperate fauna, the area had traditionally been considered to support a tropical or "Panamic" fauna (Ekman 1953:38). It needs to be emphasized that there is an important difference between a tropical fauna *per se* and a warm-temperate fauna of tropical derivation.

During the Pleistocene, apparently only tropical organisms had open access to the Gulf of California. The entrance to this body of water is relatively wide and deep, and for the most part, the distribution of salinity and oxygen are such that there is no restriction to the dispersal of most marine organisms. Walker (1960:126) considered the long stretch of sandy shore between Mazatlán and Guaymas to act as a barrier for fish groups that are confined to the rocky shore habitat, but this explanation does not seem to work well for all such groups. Certainly, the principal reason for the development of a highly endemic fauna and flora is the presence of a temperature barrier.

The evidence indicates that, given about 3 million years (the approximate duration of the Pleistocene) a distinct warm-temperate fauna will develop from a tropical one and that the presence of a temperature barrier alone will produce sufficient isolation to permit this process to take place. The work of Roden and Groves (1959:15) showed that, at least in February 1957, the 20°C isotherm extended across the mouth of the Gulf of California from just north of La Paz to just south of Topolobampo. This corresponds almost exactly with the boundaries that were previously (p. 45) suggested on the basis of animal distribution.

The two provinces of the California Warm-Temperate Region are quite different from one another. The endemic fauna of the Gulf of California has been almost entirely derived from the Eastern Pacific Tropical Region to the south, while that of the San Diego Province demonstrates a dual origin, about half being related to northern families and genera and about half to tropical groups. In the Gulf, with the exception of a small group of species that are shared with the San Diego fauna, virtually all the nonendemics may be classified as eurythermic tropical species. In contrast, about two-thirds of the nonendemics along the outer coast section are eurythermic temperate species that range into the area from north of Point Conception, and only about one-third have tropical affinities.

The Provinces
Cortez

Ekman (1935:71) was apparently the first to recognize the Gulf of California as a distinct zoogeographic entity (as a tropical subregion).

He reiterated this observation in his 1953 book and referred to the work of Glassell (1934), who found that 40 percent of the brachyuran crab. species were endemic. In a new survey of this group (canceroid, grapsoid, and spider crabs), Garth (1960) showed that 35 percent of the total were indigenous. It is interesting to note that the marine flora has been found to demonstrate exactly the same degree of endemism (Dawson 1960:97).

As far as fishes are concerned, the above observations are supported by the findings of those who have done revisions of groups that are well represented in the Gulf of California (C. Hubbs 1952, Briggs 1955, Springer 1958, Stephens 1963). In 1960, Walker presented a thoughtful analysis of the distribution and affinities of the fish fauna as a whole. He showed that 526 shore and pelagic species have been recorded and that 17 percent of these were endemic. In regard to the shore fishes alone, it can be said, on the basis of Walker's evidence, that the level of endemism is about 19 percent.

The general faunal relationships of the Gulf of California are compli-cated by the occurrence, in the northernmost end of this area, of a group of species that are common on the outer Baja California and southern California coasts but are absent from the southern Gulf. However, it seems likely that these species represent a minor invasion that has taken place around the tip of the Baja California peninsula by means of isothermic submergence. The drops in sea surface temperature that took place during the Pleistocene glaciations were apparently not sufficient to permit a general circumvention of the peninsula by warm-temperate species (p. 415).

Walker (1960:131) considered it most probable that the northern Gulf "disjuncts" accomplished their invasion through a late Pleistocene seaway in the La Paz region at a time when the water temperatures in the Gulf were slightly cooler than at present. This seems unlikely for a time of cooler water temperature would almost necessarily coincide with a glacial advance. This means that the ocean level would be lower, tending to preclude the existence of a seaway across the peninsula. Also, Durham and Allison (1960:63) stated that by the beginning of the Pleistocene, Baja California had attained essentially its present outline, and that the Vizcaino Peninsula and the Cape San Lucas area finally had become attached to the peninsula before the end of the Pliocene.

The Gulf of California is considered to be a distinct warm-temperate province. It is wholly isolated from and has relatively few species in common with the San Diego Province. Its many endemic species were independently derived from tropical forms to the south, and the rest of the fauna (with the exception of some warm-temperate invaders from the north) is comprised of eurythermic tropical species.

In the Spanish colonial period, the Gulf of California was known as the Sea of Cortés. This, because it was first discovered by Hernando Cortés in 1536. The name "Cortez Province" seems, therefore, appropriate.

San Diego
Although marine zoologists have realized for many years—at least since the study of Bartsch (1912) on the pyramidellid mollusks—that Point Conception is an important area of species change, it was not recognized as the northern limit of a distinct warm-temperate fauna until this was pointed out by Garth (1955). Previously, Ekman (1953:144) had believed that the whole of the Pacific American coast above northern Baja California corresponded to the Atlantic boreal (cold-temperate) region.

In general, the fish fauna of the San Diego Province reflects the dominant influence of the rich, cold-temperate area of the North Pacific. Hubbs (1960), in concentrating on the coast of northwestern Baja California, found that 130 (44.7 percent) out of 219 species ranged into that area from north of Point Conception. In comparison, only 65 were tropical species of southern origin and 96 were provincial endemics.

In southern California the northern faunal element becomes even more pronounced due to the appearance of additional wide-ranging, cold-temperate species and a dropping out of the tropical forms. It should be emphasized, however, that this change is not a gradual one throughout the province. Along the open, western Baja California coast as far south as Magdalena Bay, there are local zones of upwelling that commonly have surface temperatures 3 to 9°C lower than the surrounding waters (Hubbs, op. cit.). In such places, northern California fishes, invertebrates, and algae (Dawson 1960:94) are apt to reappear having partially or completely bypassed the inshore waters of southern California.

There is little doubt that the designation of a northern provincial boundary at Point Conception fits quite well the facts of fish distribution although some warm-temperate species range north to Monterey Bay. As Hubbs (1948) has shown, a series of unusually warm years or even a single warm season will be accompanied by the excursion of some warm-temperate species north of Point Conception and by an influx of tropical forms into southern California. For example, between 1853 and 1860 many San Diego Province fishes were taken at Monterey and a few at San Francisco. For the genus *Sebastes* (family Scorpaenidae), the most pronounced barrier effect appears to occur somewhat farther north at 36 to 38°N (Chen 1971:81).

Although Hubbs' (1960:141) work on the northwestern Baja California fishes showed that only 96 of the 291 species, or 32.9 percent, were endemic to the San Diego Province, the actual level of endemism for the entire province is probably somewhat higher, since Hubbs included in his calculations some of the "deep-bottom fish" fauna and at least a few pelagic species. A rough approximation of about 40 percent can be given at this time. So far, there seem to be 21 or 22 endemic genera. Although Valentine (1967:158) placed the warm-temperate molluscan fauna of the outer coast in two provinces, his data indicate that, if a single province were recognized, the endemism would be about 21 percent.

Garth (1955) based his case for the existence of a warm-temperate fauna upon examples from the brachyuran Crustacea, but other invertebrate groups also have many provincial endemics and indicated that Point Conception is an important boundary. In discussing the Pleistocene invertebrate faunas in general, Emerson (1956:329) referred to a "Transitional Province" occurring between Point Conception and Cape San Lucas, and Ross (1962) found this concept fitted the distribution of the littoral balanomorph Cirripedia. In regard to the Mollusca, the works of Bartsch (1912), Newell (1948), Cox (1962), and Valentine (1967) also indicate the importance of Point Conception. The same can be said for the work of Osburn (1950–1953) on the Bryozoa. Hedgpeth (1957:plate 1) has shown, on his map of the littoral provinces of the world, a warm-temperate region in the appropriate position.

It can be said that the San Diego Warm-Temperate Province is, in a sense, an interesting zone of mixing where eurythermic species of both southern and northern origin are brought into contact. On the other hand, an impressive portion of the fauna (possibly about 30 percent) is made up of endemic species. These also demonstrate a dual origin, about half belonging to tropical genera or families and half with northern affinities. The deeper waters of the shelf tend to hold more northern species due to the distributional principle of "isothermic submergence" where, in general, animals of wide latitudinal distribution occur in deeper water to the southward. Hubbs and Barnhart (1944) and Hubbs (1948) have given examples of this with California fish species.

Guadalupe Island
: Guadalupe Island is small, about 22 miles long and 4 to 6 miles wide, and is located about 162 miles off the coast of northern Baja California. It is surrounded by deep water so that its littoral fauna is well isolated from that of the mainland. Despite repeated collections made since 1945, the fish fauna is still poorly known because very little work has been done on the material. Briggs (1955) described two new cling-fishes, *Rimicola sila* and *Gobiesox eugrammus*, from specimens collected by Dr. Carl Hubbs. Hubbs and Rechnitzer (1958), in describing a new butterfly fish, *Chaetodon falcifer*, from the island, did give a list of 33 other species that occurred in the same collection.

Although Hubbs and Rechnitzer (op. cit., 274) spoke of a "high incidence of endemism" in the littoral marine fauna, only about four Guadalupe shore fishes can still be considered well-defined endemics. Several of those described as such, including *Chaetodon falcifer* (Hubbs 1960b) and *Gobiesox eugrammus* (Briggs 1965), have since been taken at other localities.

Because of the work of Strong and Hanna (1930), Strong (1954), and Chace (1958), the mollusks are better known than any other component of the marine fauna of Guadalupe Island. Chace listed a total of 193 species with 149, or 77.2 percent, belonging to the southern California fauna, 34 (17.6 percent) belonging to the Panamic fauna, and 10 (5.2 percent) endemic.

Considering that so few fish species have, thus far, proved to be undoubted endemics and that the autochthonous mollusks make up a relatively small part of their group, Guadalupe Island must still be considered a closely related part of the San Diego Province. If, someday, it is decided that 10 percent or more of the shallow-water fauna is endemic, then the island would qualify as a separate province.

Alijos Rocks
: This is a very small group of rocks that barely project out of the water. They lie about 185 miles westward of Cape Lazardo, Baja California. Few fishes have been collected, and no endemics have been described. Dawson (1957, 1960) found a lack of endemism in the marine flora and noted that, despite the high surface temperatures, the flora in

the vicinity of the surface showed a distinctly northern facies. Mainly for this reason, the Alijos Rocks are included in the San Diego Province.

The Japan Region

As the northern part of the North Equatorial Current enters the Philippine Sea and begins to turn toward Japan, it becomes called the Kuroshio. Its main stream is a strong current that cuts through the Ryukyu chain of islands just east of Taiwan (Fig. 8-4). It then continues turning until it flows in a northeasterly direction. Below the island of Kyushu, a split occurs whereby a lesser, western branch, called the Tsushima Current, runs through the Korean Strait and into the Sea of Japan. The main branch of the Kuroshio passes to the east of Kyushu Island and remains close to the coast until it reaches Cape Inubo where it meets with the cold Oyashio Current from the north.

The sharp temperature change at Cape Inubo is accompanied by an equally sharp faunal change. However, there are occasional captures of warm-water organisms to the north as far as Hokkaido Island. These occurrences are probably due to surface intrusions of warm water from the Kuroshio system since it tends to overlap the relatively cold, dense water of the Oyashio. To the west, the Tsushima Current is not ordinarily strong enough to bring the warm-water fauna more than a short distance into the Sea of Japan.

As is indicated in the modern works of Ekman (1953) and Schilder (1956:75), the shelf fauna of southern Japan, below the vicinity of Tokyo, has been referred to almost invariably as a tropical or subtropical assemblage. Apparently, the only indications to the contrary are the map of Hedgpeth (1957) which shows a warm-temperate area and the recognition of such fauna by Briggs (1966:7).

Beginning with the founding of the Usa Marine Biological Station in 1954, Kamohara (1964) maintained a catalog of the fish fauna of Kochi Prefecture on the east coast of Shikoku Island. The total number of shore fish species recorded is 924 with 269, or 29.1 percent, endemic to the area from Cape Inubo or Ibaraki Prefecture south to Taiwan and the east China coast. Of the nonendemics, 562 are shared with the tropical Western Pacific south of Taiwan and 93 extend northward past the vicinity of Tokyo.

It has already been noted (p. 24) that the fauna of the Amami Islands just north of Okinawa is mainly tropical, and this fits in well with the fact

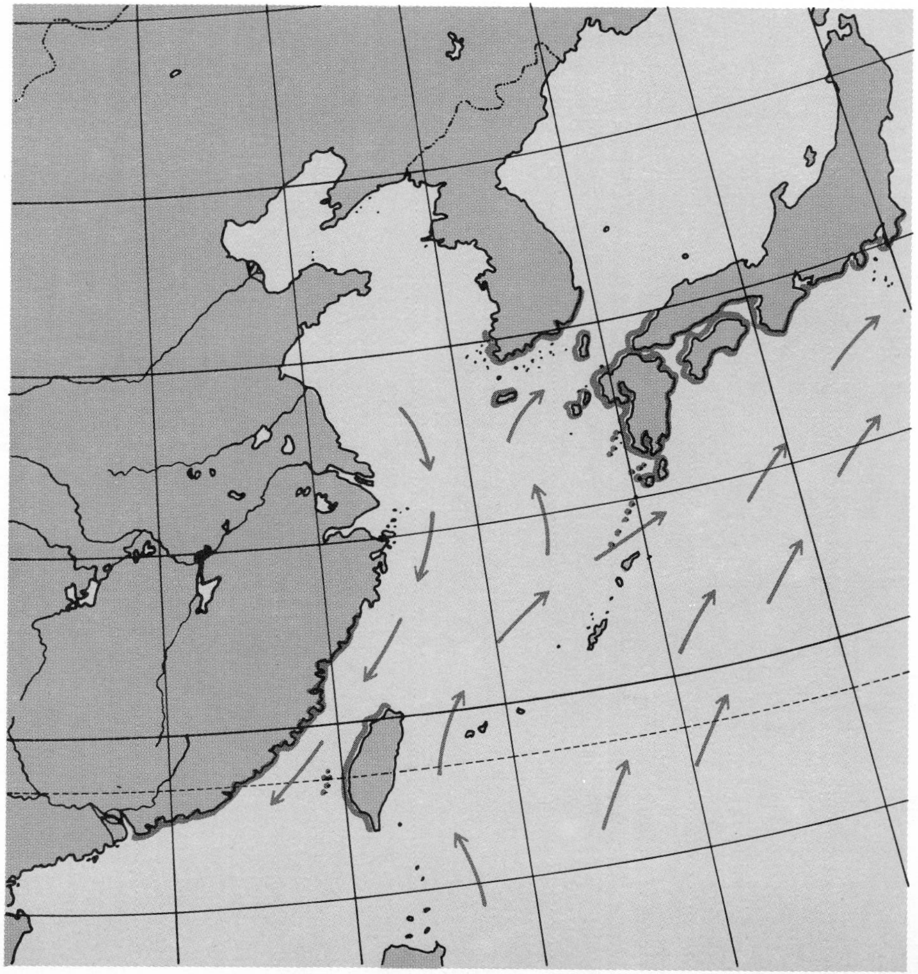

Figure 8-4   Japan Region.

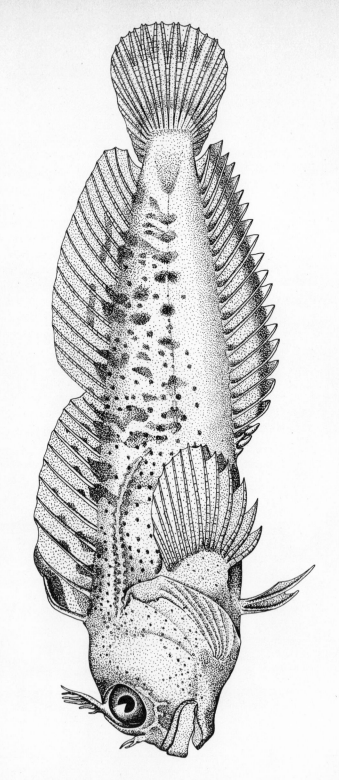

Figure 8-5 The blenny (*Blennius yatabei*). A warm-temperate shelf species confined to the Japan Region. After Jordan, Tanaka, and Snyder (1913).

that they lie about on the February 20°C surface isotherm. However, a number of warm-temperate species have been recorded from Taiwan. This apparent inconsistency becomes resolved when the detailed, surface temperature charts of Tsukuda (1937) are examined. After passing through the southern Amami Islands, the 20°C isocryme dips southward and bisects Taiwan. Most of the west coast of the latter island should, therefore, be able to support a warm-temperate fauna.

The flatfishes of Japan have been the object of distributional studies by Kuronuma (1942) and Ochiai (1957, 1959). These were essentially based on the "midpoints" method of Schenk and Keen (1936) whereby great emphasis was given to the latitudinal average of each species under consideration. This type of analysis makes the faunal change, in the area between the Kurile Islands and Taiwan, seem to be gradual. In reality, the changes at the warm-temperate versus cold-temperate boundary are quite dramatic, particularly on the Pacific coast. For example, 30 shore species belonging to the warm-water flatfish family Bothidae have been taken at Kochi Prefecture (Kamohara 1964), but only 2 of them range north of Ibaraki Prefecture (just northwest of Tokyo) and 16 are apparently endemic to the area between Cape Inubo and Taiwan. A similar but perhaps not so sharp a change takes place on the west coast, for Honma (1952:224) reported only two bothid flatfishes from Niigata Prefecture on the Sea of Japan, a cold-temperate locality.

The most probable reason why a Japanese warm-temperate fauna has not received general recognition is that the warm Kuroshio Current has been so effective in transporting tropical species into the area. The result has been a swamping of the resident fauna to the extent that if one depended on distributional records alone, a tropical fauna would appear to extend clear to Tokyo and, in some cases, even to Hokkaido. This situation is similar to that found along the east coast of the United States where the Gulf Stream is responsible for the temporary occurrence of many tropical organisms as far north as Massachusetts.

As Kamohara's (1964) catalog shows, the warm-temperate endemic proportion in certain fish families is quite large: 21 out of 45 in the Scorpaenidae, 8 out of 15 in the Triglidae, 8 out of 11 in the Cottidae, and all 6 species in the Atherinidae. Kanazawa (1958), in revising the eels of the genus *Conger*, showed that one species, *C. jordani*, has a quite well-defined warm-temperate distribution extending from Tokyo to Okinawa and that it recurs as the same or a closely related form at Port Alfred which is within the Southern Africa Warm-Temperate Region.

Only about 50 or *one-fifth* of the endemic fishes belong to typical cold-water fish families or genera, the rest being tropical derivatives. Even in the case of the 93 species that are shared with northern Japan, the great majority do not belong to northern groups but are of tropical origin probably having been dropped off by surface intrusions of the Kuroshio Current. The rather sparse northern elements in the warm-temperate fauna are apt to occur, due to isothermic submergence, in the deeper waters. From the data given by Watanabe (1960), it can be seen that 6 of the 11 marine cottids recorded from Kochi Prefecture have been taken only below 60 fathoms. Ishiyama (1958) notes that two of the wide-ranging Japanese rays, *Raja tengu* and *R. kenojei*, inhabit relatively deeper waters in the southern parts of their ranges.

It is interesting to see that the extent of endemism in the southern Japanese echinoderms (Asteroidea, Ophiuroidea, Echinoidea) is almost the same as for the fishes, i.e., 28 percent (Ekman 1953:23). The beautiful monograph (Sakai 1965) on the crabs collected from Sagami Bay by the Emperor of Japan described 240 species; of these, 108, or about 32 percent, are apparently warm-temperate endemics; 180, or 53 percent, are shared with the Indo-Pacific tropics; and the majority of the remainder are eurythermic temperate species. Despite the fact that Ekman preferred to call the southern Japanese fauna subtropical, these high rates of endemism strongly indicate that a distinct warm-temperate region should be recognized.

The Barriers

Since Japan was not generally accessible to Western naturalists until late in the nineteenth century, the earlier writers had very little data upon which to base their zoogeographic conclusions. Forbes (1856) placed the entire island of Honshu in a temperate "Japonian Province" leaving the area from Kyushu south in the tropics, and Woodward (1856:371) did the same, saying that Honshu represented the Lusitanian Province. Ortmann (1896:map) and Jordan (1901:546) finally recognized the important faunal break at Cape Inubo near Tokyo but thought that it marked the northern boundary of the tropical fauna.

In 1931, Tanaka (p. 60) published a study of the distribution of fishes in Japanese waters and also called attention to the importance of the change in fauna at Cape Inubo. Since then, this boundary has been so well documented that there is little doubt of its existence. There is considerably less agreement about the location of a warm-temperate boundary on the west coast of the Japanese islands. In regard to fishes, Tanaka (op. cit.) favored Hamada (or Yoshida a little to the west) in Shimane Prefecture. Yoshida and Itō (1957) remarked that "southern fishes have superiority south of Shimane," Honma (1952) found that the fauna of the Sado Island–Niigata area was far enough north to resemble that of the southern part of Hokkaido, the list of the fishes of the Oki Islands by Mori (1956) included relatively few tropical species, and Ishiyama (1958) noticed that certain northern rays (Rajidae) extended south to Shimane Prefecture.

For the invertebrates, Ortmann (1896:map), working on the decapod crustaceans, depicted a tropical fauna extending northward all the way to the west coast of Hokkaido. Nomura and Hatai (1936:plate 13), mainly on the basis of mollusk and brachiopod distribution, published a map of the zoological provinces in the Japanese seas. In the Sea of Japan, their major boundaries were located at about 34°N and 37°30′N; on the Pacific coast, they were placed at about 33°30′N, 35°45′N, and 41°N. Yoshida, Amio, and Nojima (1957) were able to divide the shellfish fauna of the Sea of Japan into three parts with boundaries at 35°N, 41°N, and 46°N. These writers also mentioned a "remarkable line of distinction" between Shimane Prefecture and the Noto Peninsula. The malacologists Kawakami and Habe (1961), working on cowries and cones, published a table that shows a significant drop in number of species between Yamaguchi Prefecture and Fukui Prefecture.

In regard to the marine fauna in general, Ekman (1953:22) identified a species break at about the northern part of the Korea Strait and Hedgpeth (1957:map) illustrated a warm-temperate zone extending only about as far north as the entrance to the Korea Strait. All in all, the foregoing evidence, particularly that which deals with the distribution of specific animal groups, is indicative of a faunal change somewhere in the vicinity of Shimane Prefecture, and there seems to be no reason why the locality of Hamada, as originally suggested by Tanaka, should not still be considered the boundary.

There is no doubt that the southern tip of the Korean Peninsula supports a warm-temperate fauna. No less than 88 shore fish species, otherwise confined to the area between Cape Inubo, Taiwan, and the east China coast, are also found here. Most of the warm-temperate endemics plus a fair number of tropical visitors are recorded only from Pusan or Quelpart Island (Mori 1952).

As can be seen by the checklist of J. T. F. Chen (1951–1953), many warm-temperate species are present at Taiwan, but very little is known about their local distribution. Works on the fishes of the Pescadore Islands (Chu 1957) and Quemoy Island (T. Chen 1960) also include many warm-temperate species offering some indication that the fauna of the west side of Taiwan and extending to the Chinese mainland may be linked with that of southern Japan. This fits in well with evidence from surface temperature records (Tsukuda 1937) which show February means of less than 20°C.

Very little has been published dealing with modern collections of fishes or invertebrates along the Chinese coast. An extensive work on the flatfishes of China by Wu (1932) and the synopsis of Chinese fishes by Fowler (1930–1962) show that changes from a tropical to a warm-temperate to a cold-temperate fauna take place. Previously (p. 24) it was suggested that the tropics probably ended at Hong Kong, but it is most difficult to tell where the next boundary occurs. Wu (1931) did report on a small fish collection from the coast of Fuchou, just opposite northern Taiwan, and warm-temperate relationships are indicated here by the presence of such typical species as *Plecoglossus altivelis*, *Atherina bleekeri*, *Thysanophrys meerdervoortii*, *Neopercis sexfasciata*, and *Blennius yatabei* (Fig. 8-5).

The February mean surface temperature at Shanghai is about 8°C (Tsukuda 1937), making it difficult to imagine that a warm-temperate fauna could extend that far north. Solely on the basis of temperature, it can be surmised that a zoogeographic boundary is situated somewhere near Wenchou on the coast of Chekiang Province. It would be interesting to know if this is an accurate estimate.

Literature Cited

Abbott, R. T. 1957. The tropical western Atlantic province. *Proc. Phila. Shell Club*, 1(2):7–11.

Albuquerque, R. M. 1954–1956. Peixes de Portugal e ilhas adjacentes chaves para a sua determinacao. *Portug. Acta Biol. (B)*, 5:i–xvi + 1–1164, 445 figs.

Arambourg, C. 1965. Considérations nouvelles au sujet de la faune ichthyo-logique paleomediterranéenne. *Senckenbergiana Lethara Frankfurt am Main*, (46a):13–17.

Bartsch, P. 1912. A zoogeographic study based on the pyramidellid mollusks of the west coast of America. *Proc. U. S. Nat. Museum*, 42(1906):297–349, plate 40 (map).

Bauchot, M. L., R. Bauchot, and P. Lubet. 1957. Etude de la Faune ichthyo-logique de Bassin d'Arcachon (Gironde). *Bull. Museum Hist. Nat. Paris, Ser. 2*, 29(5):385–406, 2 figs.

Baughman, J. L. 1950. Random notes on Texas fishes. Part I. *Texas J. Sci.*, 2(1):117–138.

Bearden, C. M. 1961. *List of marine fishes recorded from South Carolina*. Bears Bluff Laboratories, Wadmalow Island, South Carolina, pp. 1–12.

Behre, E. H. 1954. Decapoda of the Gulf of Mexico. *Fish. Bull. U. S. Fish Wildlife Serv.*, 55(89):451–455.

Ben-Tuvia, A. 1953. Mediterranean fishes of Israel. *Bull. Sea Fish. Res. Sta. Israel*, (8):1–40, 20 figs.

——. 1962. Collection of fishes from Cyprus. *Bull. Res. Council, Israel*, 11(3):132–145.

——. 1966. Red Sea fishes recently found in the Mediterranean. *Copeia*, (2):254–275, 2 figs.

Blegvad, H. (editor). 1929–1939. *La faune ichthyologique de l'Atlantique Nord.* Conseil Permanent International pour L'Exploration de la Mer, A. F. Host, Copenhagen, 418 plates.

Bousfield, E. L. 1960. Canadian Atlantic sea shells. National Museums of Canada, Ottawa, 72 pp. (not seen).

Boutière, H. 1958. Les scorpaenidés des eaux marocaines. *Trav. Inst. Sci Cherifien, Sér. Zool.*, (15):1–83, 12 figs., 5 plates.

Briggs, J. C. 1955. A monograph of the clingfishes (Order Xenopterygii). *Stanford Ichthyol. Bull.*, 6:1–224, 114 figs.

——. 1958. A list of Florida fishes and their distribution. *Bull. Florida State Museum, Biol. Sci.*, 2(8):223–318, 3 figs.

——. 1965. The clingfishes (Gobiesocidae) of Guadalupe island, Mexico. *Calif. Fish Game.*, 51(2):123–125.

——. 1966. The warm-temperate marine fauna of Japan, Taiwan, and the east China coast. *Proc. 11th Pacific Sci. Congr.*, 7(7):7.

——, H. D. Hoese, W. F. Hadley, and R. S. Jones. 1964. Twenty-two new marine fish records for the northwestern Gulf of Mexico. *Texas J. Sci.*, 16(1):113–116.

Cabioch, L. 1968. Contribution à la conaissance des peuplements benthiques de la Manche Occidentale. *Cahiers Biol. Mar.*, 9(5)(supple.):493–720, 44 figs., 6 plates.

Caspers, H. 1957. Black Sea and Sea of Azov. *in* J. W. Hedgpeth, *Treatise on marine ecology and paleoecology.* Geol. Soc: Amer., Mem., Geol. Soc. Amer. (publisher) 1(67):801–890, 37 figs.

Chace, E. P. 1958. The marine molluscan fauna of Guadalupe Island, Mexico. *Trans. San Diego Soc. Nat. Hist.*, 12(19):321–322, 1 fig.

Chen, J. T. F. 1951–1953. Checklist of the species of fishes known from Taiwan (Formosa). *Quart. J. Taiwan Museum*, 4(3–4):181–210; 5(4):305–341; 6(2):102–140.

Chen, L. 1971. Systematics, variation, distribution, and biology of rockfishes of the subgenus *Sebastomus* (Pisces, Scorpaenidae, *Sebastes*). *Bull Scripps Inst. Oceanog.*, 18:1–107, 16 figs., 6 plates.

Chen, T. 1960. Contributions to the fishes from Quemoy (Kinmen). *Quart. J. Taiwan Museum*, 13(3–4):191–213.

Chu, K. 1957. A list of fishes from Pescadore Islands. *Rep. Inst. Fishery Biol.,* *Taiwan*, 1(2):14–22.

Coe, W. R. 1940. Revision of the nemertean fauna of the Pacific Coasts of North, Central and northern South America. *Allan Hancock Pacific Expedition*, 2(13):247–323, 8 plates.

Collins, B. L. 1954. Lista de Peixes dos Mares dos Acores. *Acoreana*, 5(2):102–142.

Collette, B. B., and P. Bănărescu, 1965. Review of A. N. Svetovidov, "Fishes of the Black Sea," 1964, *Copeia*, (4):523–525.

Coomans, H. E. 1962. The marine mollusk fauna of the Virginia area as a basis for defining zoogeographical provinces. *Beaufortia*, 9(98):83–104.

Cox, K. W. 1962. California abalones, family Haliotidae, *Calif. Dept. Fish Game, Fish Bull.*, (118):1–133, 61 figs.

Crisp, D. J., and E. Fischer-Piette. 1959. Répartition des principales espéces intercotidales de la côte Atlantique Française en 1954–1955. *Ann. Inst. Oceanog. Paris*, 36(2):275–388, 21 figs.

———, and A. J. Southward. 1958, The distribution of intertidal organisms along the coasts of the English Channel. *J. Marine Biol. Ass. U. K.*, (37):157–208.

Dahlberg, M. D. 1970. Atlantic and Gulf of Mexico menhadens, genus *Brevoortia* (Pisces: Clupeidae). *Bull. Florida State Museum, Biol. Sci.*, 15(3):91–162, 8 figs.

Dawson, E. 1957. Notes on the eastern Pacific insular marine algae. *Contrib. Sci. Los Angeles Co. Museum*, (8):1–8, 4 figs.

———. 1960. A review of the ecology, distribution, and affinities of the benthic flora. Symposium: The biogeography of Baja California and adjacent seas. *Syst. Zool.*, 9(3–4):93–100.

Dawson, C. E. 1964. A revision of the Western Atlantic flatfish genus *Gymnachirus* (the naked soles). *Copeia*, (4):646–655, 12 figs.

Deevey, E. S. 1950. Hydroids from Louisiana and Texas, with remarks on the Pleistocene biogeography of the western Gulf of Mexico. *Ecology*, 3:334–367, 11 figs.

Dieuzeide, R., M. Novella, and J. Roland. 1953–1955. Catalogue des poissons des côtes Algeriennes. *Station d'Aquiculture Pêche Castiglione, Nouv. Ser.*, (4):1–274, (5):1–258, (6):1–384, illus.

———, and J. Roland. 1957. Complement au catalogue des poissons des côtes Algeriennes. *Bull. Trav. Publ. Stat. d'Aquicult. Pêche Castiglione (NS)*, 8:83–106, 8 figs.

Dollfus, R. P. 1955. Première Contribution à l'Establissment d'Un Fichier Ichthyologique de Maroc Atlantique. *Trav. Inst. Sci. Cherifien, Ser. Zool. No. 6*, 226 pp.

Drouet, H. 1861. Eléments de la Faune açoréene. Mem. Soc. Acad. Aube, 25: 1–245.

Durham, J. W., and E. C. Allison. 1960. The geologic history of Baja California and its marine faunas. Symposium: the biogeography of Baja California and adjacent seas. *Syst. Zool.*, 9(2):47–91, 7 figs.

Ekman, S. 1935. *Tiergeographie des Meeres.* Akademische Verlagsgesellschaft, Leipzig, xii + 512 pp.

———. 1953. *Zoogeography of the sea.* Sidgwick & Jackson, London, xiv + 417 pp., 121 figs.

Emerson, W. K. 1956. Pleistocene invertebrates from Punta China, Baja California, Mexico. With remarks on the composition of the Pacific Coast Quaternary faunas. *Bull. Amer. Museum Nat. Hist.*, 111(4):313–342, 1 fig. 2 plates.

Emiliani, C. 1958. Paleotemperature analysis of core 280 and Pleistocene correlations. *J. Geol.*, 66:264–275, 5 figs.

————. 1961. Cenozoic climatic changes as indicated by the stratigraphy and chronology of deep-sea cores of globigerina-ooz facies. *Ann. N. Y. Acad. Sci.*, 95(1):521–536, 10 figs.

Eschmeyer, W. N. 1969. A systematic review of the scorpionfishes of the Atlantic Ocean (Pisces: Scorpaenidae). *Occas. Papers Calif. Acad. Sci.*, (79):iv + 1–143, 13 figs.

Fischer, P. H. 1880–1887. *Manuel de Conchyliologie et de Paleontologie Conchyliologique*, Librairie F. Savy, Paris, xxiv + 1369 pp., 1,138 figs., 24 plates, 1 map.

Fischer, P. 1950. *Vie et moeurs des Mollusques*. Payot, Paris, pp. 1–312, 180 figs., 1 map.

Forbes, E. 1844. Report on the mollusca and Radiata of the Aegean Sea, and on their distribution, considered as bearing on geology. *Rep. Brit. Ass.*, 1843(1844):130–193, 1 plate.

————. 1856. Map of the distribution of marine life. *in* Alexander K. Johnston, *The physical atlas of natural phenomena* (new edition). W. and A. K. Johnston, Edinburgh and London, plate no. 31.

————. 1859. *The natural history of European seas* (edited and continued by Robert Godwin-Austen). John Van Voorst, London, viii + 306 pp., 1 map.

Fowler, H. W., 1930–1962. A synopsis of the fishes of China. A series of more than 60 separate papers appeared under this title. They were published over a period of 32 years in *Hong Kong Naturalist, J. Hong Kong Fisheries Res. Sta.*, and *Quart. J. Taiwan Museum.*

Garth, J. S. 1955. A case for a warm-temperate marine fauna on the west coast of North America. *in Essays in the Natural Sciences in honor of Captain Allen Hancock on the occasion of his birthday, July 26, 1955.* University of Southern California Press, Los Angeles, pp. 19–27.

————. 1960. Distribution and affinities of the brachyuran Crustacea. Symposium: the biogeography of Baja California and adjacent seas. *Syst. Zool.*, 9(3–4):105–123, 3 figs.

Ginsburg, I. 1952. Eight new fishes from the Gulf Coast of the United States, with two new genera and notes on geographic distribution. *J. Wash. Acad. Sci.*, 42(3):84–101, 9 figs.

Glassell, S. A. 1934. Affinities of the brachyuran fauna of the Gulf of California. *J. Wash. Acad. Sci.*, 24:296–302 (not seen).

Graham, J. B. 1970. Temperature sensitivity of two species of intertidal fishes. *Copeia*, (1):49–56, 9 figs.

Griffini, A. 1903. *Ittiologia Italiana*. Ulrico Hoepli, Milan, xii + 475 pp., 244 figs.

Hall, C. A. 1964. Shallow-water marine climates and molluscan provinces. *Ecology*, 45(2):226–234, 6 figs.

Hartman, O. 1954. Polychaetous annelids of the Gulf of Mexico. *Fish. Bull. U. S. Fish Wildlife Serv.*, 55(89):413–417.

Haven, D. S. 1959. Migration of the croaker, *Micropogon undulatus. Copeia* (1):25–30, 4 figs.

Hedgpeth, J. W. editor, 1953. An introduction to the zoogeography of the northwestern Gulf of Mexico with reference to the invertebrate fauna. *Publ. Inst. Marine Sci. Univ. Texas*, Geol. Soc. Amer., 3:107–224, 46 figs.

———. 1954. Bottom communities of the Gulf of Mexico. *In* Gulf of Mexico, its origin, waters, and marine life. *Fish. Bull. U.S. Fish Wildlife Serv.*, 55(89):203–214, 4 figs.

———. 1957. Marine biogeography. *in Treatise on marine ecology and pale-oecology*. vol. 1. *Geol. Soc. Amer., Mem.*, (67):359–382, 16 figs.; 1 plate.

Hoese, H. D. 1958. A partially annotated checklist of the marine fishes of Texas. *Pub. Inst. Marine Sci.*, Texas 5:312–352.

Holme, N. A. 1961. The bottom fauna of the English Channel. *J. Marine Biol. Ass. U. K.*, 41(2):397–461.

Honma, Y. 1952. A list of fishes collected in the Province of Echigo, including Sado Island. *Jap. J. Ichthyol.*, 2(4–5):220–229.

Hubbs, C. L. 1948. Changes in the fish fauna of Western North America correlated with changes in ocean temperature. *J. Sears Found. Marine Res.*, 7(3):459–482, 6 figs.

———. 1960. The marine vertebrates of the outer coast. Symposium: the biogeography of Baja California and adjacent seas. *Syst. Zool.*, 9(3–4):134–147.

———, and P. S. Barnhart. 1944. Extension of range for blennioid fishes in southern California. *Calif. Fish Game*, 30(1):49–51.

———, and A. B. Rechnitzer. 1958. A new fish, *Chaetodon falcifer*, from Guadalupe Island, Baja California, with notes on related species. *Proc. Calif. Acad. Sci.*, 29(8):273–313, 3 plates.

Hubbs, C. 1952. A contribution to the classification of the blennioid fishes of the family Clinidae, with a partial revision of the Eastern Pacific forms. *Stanford Ichthyol. Bull.*, 4(2):41–165, 64 figs.

Humm, H. J. 1969. Distribution of marine algae along the Atlantic coast of North America. *Phycologia*, 7(1):43–53.

Hyman, L. H. 1955. *The invertebrates*. vol. 4. *Echinodermata*. McGraw-Hill, New York, 763 pp., 280 figs.

Ishiyama, R. 1958. Studies on the rajid fishes (Rajidae) found in the waters around Japan. *J. Shimonoseki Coll. Fish.*, 7(2–3):193–394, 86 figs., 3 plates.

Johnson, C. W. 1934. List of the marine mollusca of the Atlantic coast from Laborador to Texas. *Proc. Boston Soc. Nat. Hist.*, 40(1):1–204.

Jordan, D. S. 1901. The fish fauna of Japan, with observations on the geographical distribution of fishes. *Science*, new series 14(354):545–567.

Kamohara, T. 1964. Revised catalogue of fishes of Kochi Prefecture, Japan. *Rep. Usa Marine Biol. Sta.*, 11(1):1–99.

Kanazawa, R. H. 1958. A revision of the eels of the genus *Conger* with descriptions of four new species. *Proc. U. S. Nat. Museum*, 108(3400):219–267, 7 figs. 4 plates.

Kawakami, I., and T. Habe. 1961. The characteristic aspects of the molluscan fauna in the west coast of Kyushu, Japan. *Records Oceanog. Works Japan*, special no. 5:195–197.

Knipovitch, N. 1933. Hydrological investigations in the Black Sea. *Trans. Azov-Black Seas Sci. Fisheries Expedition*, 10 (not seen).

Kosswig, C. 1955. Zoogeography of the Near East. *Syst. Zool.*, 4(2):49–73 + 96, 14 figs.

———. 1956. Beitrag zur Faunengeschichte des Mittelmeeres. *Pubbl. Staz. Zool. Napoli*, 28:78–88.

Kuronuma, K. 1942. "Latitudinal value" a quantitative indication of fish distribution and its application to the flatfishes of the Japanese Islands. *Bull. Biogeogr. Soc. Japan*, 12(4):85–91.

Leipper. D. F. 1954. Physical oceanography of the Gulf of Mexico. *in* Gulf of Mexico its origin, waters, and marine life. *Fish. Bull. U. S. Fish Wildlife Serv.*, 55(89):119–137, figs. 34–43.

Logvinenko, B. M., and Y. L. Starobogatov. 1962. The molluscs of the Caspian Sea and their geographical relations (in Russian). *Byul. Mosk. Obshichestva Ispytatelei Prirody Otd. Biol.*, 67(1):153–154.

Lozano Cabo, F. 1950. Datos sobre repartición geográfica de especies de peces de la cuesta del N. W. de Africa. *Bol. Real Soc. Espan. Hist. Nat. Secc. Biol.*, 48(1):5–14.

Lozano Rey, L. 1928. *Fauna Ibérica*: Peces. vol. I. Museum Nacional Ciencias Natural, Madrid, xi + 692 pp., 197 figs., 20 plates.

———. 1947. Peces ganoideos y fisóstomos. *Mem. R. Acad. Cien. Madrid, Ser. Cien. Nat.*, 11:xv + 839, 190 figs., 20 plates.

———. 1952a. Peces fisoclistos. Primera parte. *Mem. R. Acad. Cien. Madrid, Ser. Cien. Nat.*, 14:xv + 378, 20 figs., 30 plates.

———. 1952b. Peces fisoclistos. Segunda parte. *Mem. R. Acad. Cien. Madrid, Ser. Cien. Nat.*, 14:387–703, figs. 21–31, plates 31–51.

———. 1960. Peces fisoclistos. Tercera parte. *Mem. R. Acad. Cien., Madrid, Ser. Cien. Nat.*, 14:xvi + 613, 173 figs., 7 plates.

Mars, P. 1963. Les faunes et la stratigraphie du Quaternaire Mediterraneen. *Rec. Trav. St. Mar. End., Bull. 28*, (43):61–97, 6 figs.

Maul, G. E. 1948. Peixes. *in* A. C. de Noronha and A. A. Sarmento (editors), *Vertebrados da Madeira*, 2:1–181, Junta Geral, Distrito Autónomo do Funchal.

———. 1959. *Aulostomus*, a recent spontaneous settler in Madeiran waters. *Bocagiana*, (1):1–18, 1 fig.

McLaren, I. A. 1960. On the origin of the Caspian and Baikal seals and the paleoclimatological implication. *Amer. J. Sci.*, 258:47–65, 5 figs.

Miller, J. M. 1965. A trawl survey of the shallow Gulf fishes near Port Aransas, Texas. *Publ. Inst. Marine Sci. Texas*, 10:80–107, 1 fig.

Moreau, E. 1881. *Histoire naturelle des poissons de la France*. 3 vols. G. Masson, Paris, vii + 480, 572, 697 pp., 220 figs.

Mori, T. 1952. Check list of the fishes of Korea. *Mem. Hyogo Univ. Agri.*, 1(3):, *Biol. Ser.* (1):1–228, 1 map.

———. 1956. Fishes of San-in District including Oki Islands and its adjacent waters (Southern Japan Sea). *Mem. Hyogo Univ. Agri.*, 2(3):1–62, 6 figs.

Navarro, F. de P. 1943. La pesca de arrastre en los fondos del Cabo Blanco y del

Banco Arguin (Africa Sahariana). *Trab. Inst. Espan. Oceanog.*, No. 18, 1–225 pp., 38 plates.

Neill, W. T. 1957. Historical biogeography of present-day Florida. *Bull. Florida State Museum, Biol. Sci.*, 2(7):175–220.

Newell, I. M. 1948. Marine molluscan provinces of Western North America: a critique and a new analysis. *Proc. Amer. Phil. Soc.*, 92:155–166, 7 figs.

Nikolsky, G. 1940. The fishes of the Aral Sea. Contr. Connaiss. Faune Flore U. R. S. S. Moscow N. S. Sect. Zool., 1:1–216 (in Russian with English summary).

Nomura, S., and K. Hatai. 1936. A note on the zoological provinces in the Japanese seas. *Bull. Biogeogr. Soc., Japan*, 6(21):207–214, 1 plate.

Ochiai, A. 1957. Zoogeographical studies on the Soleoid fishes found in Japan and its neighboring regions. Parts 1–3. *Bull. Japan. Soc. Fish.*, 22(9):522–535, 3 figs.

———. 1959. Morphology, taxonomy and ecology of the soleoid fishes found in Japan. (in Japanese) publisher? iv + 236 pp., 70 figs., 2 plates.

Ortmann, A. E. 1896. *Grundzüge der marinen Tiergeographie.* Gustav Fischer, Jena, iv + 96 pp., 1 map.

Osburn, R. C. 1950–1953. Bryozoa of the Pacific Coast of America. *Allan Hancock Pacific Expeditions*, 14(1–3):1–841, 82 plates.

Parr, A. E. 1933. A geographic-ecological analysis of the seasonal changes in temperature conditions in shallow water along the Atlantic coast of the United States. *Bull. Bingham Oceanog. Lab.*, 4(3):1–90, 28 figs.

Pérés, J. M. 1967. The Mediterranean benthos. *Oceanog. Marine Biol. Ann. Rev.*, 5:449–533, 11 figs.

———, and J. Picard. 1964. Nouveau manuel de bionomie benthique de la Mediterranee. *Rec. Trav. Sta. Mar. d'Endoume*, 31(47):1–137, 7 figs.

Phleger, F. B., and F. L. Parker. 1951. Ecology of foraminifera, northwest Gulf of Mexico. Part II. Foraminifera species. *Geol. Soc. Amer., Mem.*, 46:1–64, 20 plates (not seen).

Postel, E. 1956. Les Affinités tropicales de la Faune Ichthyologique du golfe du Gabés. *Bull. Sta. Oceanog. Salammbô*, (53):64–68, 1 fig.

Pulley, T. E. 1952. A zoogeographic study based on the bivalves of the Gulf of Mexico. Ph.D. thesis, Harvard University, 1–215 pp., 8 figs., 19 plates.

Rehder, H. A. 1954. Mollusks. Gulf of Mexico its origin, waters, and marine life. *Fish. Bull., U. S. Fish Wildlife Serv.*, 55(89):469–474.

Roden, G. I., and G. W. Groves. 1959. Recent oceanographic investigations in the Gulf of California. *J. Marine Res.*, 18(1–3):10–35, 16 figs.

Ross, A. 1962. Results of the Puritan-American Museum of Natural History Expedition to Western Mexico 15. The littoral balanomorph Cirripedia. *Amer. Museum Novitates*, (2084):1–44, 24 figs.

Ruggieri, G. 1967. The Miocene and later evolution of the Mediterranean Sea. *Syst. Ass. Publ.*, (7):283–290, 2 figs.

Sakai, T. 1965. *The crabs of Sagami Bay.* Maruzen Co., Tokyo, xvi + 206 + 92 + 32 pp., 100 plates.

Schenk, H. G., and A. M. Keen. 1936. Marine molluscan provinces of Western North America. *Proc. Amer. Phil. Soc.*, 76:921–938, 6 figs.

Schilder, F. A. 1956. *Lehrbuch der Allgemeinen Zoogeographie.* Gustav Fischer, Jena, viii + 150 pp., 134 figs.

Schwarzbach, M. 1963. *Climates of the past.* Van Nostrand, London, xii + 328 pp., 134 figs.

Slastenenko, E. P. 1955–1956. *Karadeniz Havzasi Baliklari. The fishes of the Black Sea Basin.* Hanif Altan, Istanbul, 711 + xlix pp., 142 figs. (in Turkish).

———. 1955. A review of the Black Sea fish fauna and general marine life conditions. *Copeia,* (3):230–235.

———. 1959. Zoogeographical review of the Black Sea fish fauna. *Hydrobiologia,* 14(2):177–188.

Šoljan, T. 1948. *Faune et Flora Adriatica.* vol. 1. *Pisces.* Institut za Oceanografia i. Ribartvo, Split, Yugoslavia, 437 pp., 1,350 figs.

Southward, A. J. 1963. The distribution of some plankton animals in the English Channel and approaches III. Theories about long-term biological changes including fish. *J. Marine Biol. Ass. U. K.,* 43(1):1–29, 6 figs.

Springer, V. G. 1958. Systematics and zoogeography of the clinid fishes of the subtribe Labrisomini Hubbs. *Publ. Inst. Marine Sci., Texas,* 5:417–492, 4 figs., 7 plates.

———. 1959. Blenniid fishes of the genus *Chasmodes. Texas J. Sci.,* 11(3):321–334, 5 figs.

Steinitz, H. 1949a. Contributions to the knowledge of Blenniidae of the eastern Mediterranean I. *Rev. Fac. Sci. Univ. Istanbul, Ser. B,* 14(2):129–152, 4 figs.

———. 1949b. Contribution to the knowledge of the Blenniidae of the eastern Mediterranean II. *Rev. Fac. Sci. Univ. Istanbul, Ser. B,* 14(3):170–197, 12 figs.

———. 1950a. Contribution to the knowledge of the Blenniidae of the Eastern Mediterranean III. *Rev. Fac. Sci. Univ. Istanbul, Ser. B,* 15(1):60–87, 19 figs.

———. 1950b. On the zoogeography of the teleostean genera *Salarias, Ophioblennius* and *Labrisomus. Arch. Zool. Italiano,* 35:325–348, 5 figs.

———. 1952. Notes on fishes from Cyprus. *Bull. Inst. Oceanog. Monaco,* 49(1004):1–12.

Stephens, J. T. 1963. A revised classification of the blennioid fishes of the American family Chaenopsidae. *Univ. Calif. Publ. Zool.,* 68:1–133, 11 figs., 15 plates.

Stephenson, T. A., and Anne Stephenson. 1952. Life between tide-marks in North America. II. Northern Florida and the Carolinas. *J. Ecol.,* 40(1):1–49, 9 figs., 6 plates.

Strong, A. M. 1954. The marine molluscan fauna of Guadalupe Island, Mexico. *Min. Conch. Club S. Calif.,* 142:6–10 (not seen).

———, and G. D. Hanna. 1930. Marine Mollusca of Guadalupe Island, Mexico. *Proc. Calif. Acad. Sci., Ser. 4,* 19(1):1–6.

Sverdrup, H. V., M. W. Johnson, and R. H. Fleming. 1946. *The oceans.* Prentice-Hall, Englewood Cliffs, N. J., pp. 1–1087, 265 figs., 7 charts.

Svetovidov, A. N. 1964. Fishes of the Black Sea. *Akad. Nauk., SSSR, Fauna of USSR,* (86):1–552, 191 figs. (in Russian).

Tanaka, S. 1931. On the distribution of fishes in Japanese waters. *J. Fac. Sci. Imp. Univ. Tokyo. Sect. 4, Zool.,* 3(1):1–90, 3 plates.

Tanner, W. F. 1968. Multiple influences on sea-level changes in the Tertiary. *Palaeogeogr., Palaeoclimatol., Palaeoecol.,* 5:165–171.

Taylor, W. R. 1955. Marine algal flora of the Caribbean and its extension into neighboring seas. *in Essays in the natural sciences in honor of Captain Allan Hancock*. University of Southern California Press, Los Angeles, pp. 259–270, 8 figs.

Tortonese, E. 1938. L'ittiofauna mediterranea in rapporto alla zoogeografia. *Boll. Mus. Zool. Anat. Comp. Univ. Torino, 46, Ser. 3*, (84):1–35, 6 figs.

———. 1954. "Zoogeography of the Mediterranean sea perches" (Pisces Serranidae) Comm. Internat. Explor. Sci. Mer Méditerranée, Rapp. et Proc.— Verb., vol. 12, (N. S.), pp. 93–103.

———. 1956. *Fauna d'Italia.* vol. 2. *Leptocardia, Ciclostomata, Selachii.* Edizioni Calderini, Bologna, viii + 332 pp., 163 figs.

———. 1960a. The relations between the Mediterranean and Atlantic fauna. *Publ. Hydrobiol. Inst., Fac. Sci., Univ. Istanbul*, 5(1–2):30–34, 1 fig.

———. 1960b. General characters of the Mediterranean fish fauna. *Publ. Hydrobiol. Res. Inst., Fac. Sci., Univ. Istanbul*, 5(1–2):43–50, 3 figs.

———. 1963. Elenco riveduto dei Leptocardi, Ciclostomi, pesci cartilaginei e ossei del mare Mediterraneo. *Ann. Mus. Civico Stor. Nat., Genova*, 74:156–185.

———. 1964. The main biogeographical features and problems of the Mediterranean fish fauna. *Copeia*, (1):98–107.

Tyron, G. W. 1882–1884. *Structural and systematic conchology.* 3 vols. Tyron, Philadelphia, viii + 312, 430, and 453 pp., 140 plates, map.

Tsukuda, K. 1937. On the surface temperature of the neighboring seas of Japan. *Mem. Imp. Marine Observatory, Kobe*, 6(3):239–257, 13 plates.

Valentine, J. W. 1967. The influence of climatic fluctuation on species diversity within the tethyan provincial system. *Syst. Ass. Publ.*, (7):153–166, 3 figs.

Van Name, W. G. 1945. The North and South American ascidians. *Bull. Amer. Museum Nat. Hist.*, 84:1–476, 31 plates.

Walker, B. W. 1960. The distribution and affinities of the marine fish fauna of the Gulf of California. Symposium: the biogeography of Baja California and adjacent seas. *Syst. Zool.*, 9(3–4):123–133, 1 fig.

Watanabe, M. 1960. *Fauna Japonica. Cottidae (Pisces).* Biogeographical Society of Japan, Tokyo, vii + 1–218 pp., 75 figs., 40 plates.

Wells, H. W., and I. E. Gray. 1960. The seasonal occurrence of *Mytilus edulis* on the Carolina coast as a result of transport around Cape Hatteras. *Biol. Bull*, 119(3):550–559.

Wheeler, A. 1969. *The fishes of the British Isles and north-west Europe.* Macmillan, London, xviii + 613 pp., 16 plates, figs.

Williams, A. B. 1965. Marine decapod crustaceans of the Carolinas. *Fish. Bull. U.S. Fish Wildlife Serv.*, 65(1):1–298, 252 figs.

Wilson, J. 1963. Evidence from islands on the spreading of ocean floors. *Nature*, 197(4867):536–538, 4 figs.

Woodward, S. P. 1851–1856. *A manual of the mollusca.* John Weale, London, viii + 1–486 pp., 24 plates + 24 pp., 1 map.

Wu, H. 1931. Notes on the fishes from the coast of Foochow Region and Ming River. *Contrib. Biol. Lab. Sci. Soc. China, Nanking*, 7(1):1–64, 10 figs.

————. 1932. *Contribution a l'étude morphologique, biologique et systématique des poissons hétérosomes de la Chine.* Jouve et Cie, Paris, pp. 1–179, 25 figs.

Yakubova, L. I. 1935. Zur regionalen Gliederung des Schwarzen Meeres auf Grund der Benthos-Fauna und ihre Verbreitung an den Küsten des Schwarzen Meeres. *Akad. Nauk. S.S.S.R., Leningrad*, 1(4) (not seen).

Yoshida, H., M. Amio, and S. Nojima. 1957. Shell fish fauna in the Japan Sea off Yoshimi in Yamaguchi Prefecture. *J. Shimonoseki Coll. Fisheries*, 7(1):117–120.

————, and T. Itō. 1957. Fish fauna of the Japan Sea. *J. Shimonoseki Coll. Fisheries*, 6(2):261–270.

Zenkevitch, L. 1963. *Biology of the seas of the U.S.S.R.* George Allen and Unwin, London, pp. 1–955, 427 figs.

Chapter
9

Northern
Hemisphere
Old-Temperate
and Arctic Regions

*And just as the final aim of taxonomic research is not the graduated scale per se
but the unravelling of the historical (phylogenetic) relationships between the
taxonomic categories and thus the history of the animal kingdom, in the same
way the final aim of zoogeography is not the graduated regional system in itself
but the history which this system reflects, that is, the history of the faunas.*

Sven Ekman in Zoogeography of the Sea, *1953*

When one considers the diversity and general distribution of the marine faunas that occupy the cold waters of the world, it becomes clear that the North Pacific has functioned as a center of evolutionary radiation. It has supplied species to, and to a large extent has controlled, the faunal complexion of the Arctic and the North Atlantic (p. 437). Furthermore, North Pacific species have, by means of isothermic submergence, bypassed the equatorial region to invade the cold waters of the Southern Hemisphere.

In the North Pacific, the rich boreal fauna of the shelf may be divided into two regions, one on each side of the ocean. These regions may, in turn, be subdivided into five distinct provinces, three on the western side and two on the east. The relatively sparse boreal fauna of the Atlantic also belongs in two regions but they are not subdivisable. The arctic fauna, being essentially homogeneous, is also placed in a single region without subdivisions.

The Western Atlantic Boreal Region

As the Florida Current passes Cape Hatteras on its northeasterly course, it becomes the Gulf Stream. Previously, it had been flowing near or over the continental shelf, but beyond the Cape, it extends over deep water. It is composed of a series of meandering filaments, some of which may flow as fast as 5 knots, an extreme speed for an ocean current. For most of its course, the Gulf Stream stays well out to sea, but it approaches shallow water once more over the Grand Banks, which extend to the southeast of Newfoundland.

Zoogeographically, the most important function of the Gulf Stream is that it forms a very effective barrier between a cold-water area to the northwest and the warm water of the Sargasso Sea (the western North Atlantic) on the southeast. Between the Gulf Stream and the shore of North America, there is a southwest-flowing, coastal current that tends to form an elongated, counterclockwise eddy (Pickard 1964:128). The coastal current is supplied with cold, low salinity water from the Labrador Current, which flows south out of the Labrador Sea and around the tip of Newfoundland (Fig. 9-1). It is this coastal current that has in large part determined the nature of the fauna that occupies the shelf from the latter island south to Cape Hatteras.

In his thorough study of temperature conditions and their effects on the ecology of the Atlantic coast of the United States, Parr (1933:62) called attention to the extreme seasonal fluctuation in temperature that takes

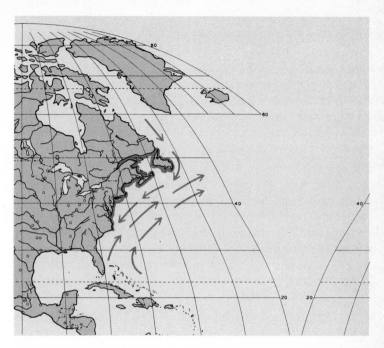

Figure 9-1 Western Atlantic Boreal Region.

place in the shallow waters of the area between Cape Cod and Cape Hatteras. He concluded that, under such conditions, it would be difficult for either the northern or southern faunistic elements to contribute any permanent residents to the area.

The Middle Atlantic Seaboard

In considering the geographic extent and distinctiveness of the cold-temperate, western North Atlantic, the zoologist is confronted with a good deal of conflicting opinion. Most of the disagreement has been concerned with the relationship of the fauna that occupies the region between Cape Hatteras and Cape Cod—often called the Middle Atlantic Seaboard. The area is penetrated, during the summer months for the most part, by a huge number of tropical and warm-temperate organisms. This has often resulted in its being allied with the Carolina Region to the south of Cape Hatteras.

It was concluded previously (p. 222), on the basis of considerable evidence, that Cape Hatteras represented the boundary between the warm-temperate and cold-temperate faunas. The question that now needs to be considered is, does the Middle Atlantic Seaboard comprise a separate cold-temperate (boreal) province or is its fauna mainly a continuation of that which is present just north of Cape Cod?

Coomans (1962:101) made a detailed study of the molluscan fauna, referring to the area as the "Virginian" after Forbes (1856), and found that 10.5 percent of the species were endemic. Most of the nonendemics were considered to be species of northern rather than southern origin. Coomans concluded that there was no reason to consider the Virginian area as an autonomous zoogeographic province and that the percentage of boreal mollusks was large enough to include it in the boreal part of the Western Atlantic.

According to the recent ichthyological literature, about 250 shore fish species have been recorded from the Middle Atlantic Seaboard. However, the rate of endemism is extremely low. Only one species seems to definitely belong in this category. This is the northern stargazer (*Astroscopus guttatus*). Of the rest, the majority (about 190 species) are eurythermic tropical or warm-temperate forms that range into the area from the south, usually during the warmer months of the year.

Lest the 250 species mentioned above give the impression of a rather rich fish fauna, it should be emphasized that many of the records represent warm season migrations or infrequent occurrences due to Gulf Stream transport, so the number of resident species is actually quite small. For example, a study of the winter fish fauna of the seaboard area was carried out by Edwards, Livingstone, and Harmer (1962) in which a total of 53 trawling operations were completed by the vessel *Albatross III* in January and February of 1959. Both the shallow and deep waters of the shelf were covered in an attempt to secure a representative sample of the fauna. But this operation succeeded in capturing a total of only 36 shore species. About half were typically boreal, and most of the rest were species from the warm-water areas to the south.

Although the degree of endemism in the mollusks (10.5 percent) seems sufficient to consider the seaboard area as a distinct province, the almost negligible amount in the fishes does not reinforce this concept. The situation in other marine groups seems to be similar to that noted for the fishes. After having examined the fauna in general, Stephenson and Stephenson (1954:66) noted that the area did not appear to be a province with a distinct population of its own. Taylor (1962:3) observed that the marine flora from southern New England to New Jersey was boreal rather than tropical. Therefore, the Middle Atlantic Seaboard is recognized as simply a southern portion of a cold-temperate faunal area that extends to the north beyond Cape Cod.

North of Cape Cod

The cold-temperate nature of the fauna that lies immediately to the north of Cape Cod has been recognized for many years. Forbes (1856) referred to the coastline between Newfoundland and New York as the "Bostonian sub-region" of a Boreal Province that extended to both sides of the Atlantic. Ekman (1953:139) called the fauna north of Cape Cod "boreal" but gave no northern boundary for the region. Schilder (1956:75) depicted an "Ostamerikanisch Provinz" between about southern Newfoundland and Cape Cod, and Hedgpeth (1957:map) delineated a cold-temperate fauna for the latter area. Stephenson and Stephenson (1954) described a "subarctic province" that included Labrador, much of Hudson Bay, and extending to the northern Gulf of St. Lawrence; from the latter place to Cape Cod, they recognized a "Nova Scotian Province."

Specifically for the molluscan fauna, Woodward (1856:map) showed a Boreal Province (extending to both sides of the Atlantic) from the southern Gulf of St. Lawrence to Cape Cod. Approximately the same area was recognized and called either Boreal or Nova Scotian by Johnson (1934), Pulley (1952), Coomans (1962), and Hall (1964). Bousefield (1960), utilizing a concept of overlapping zones, considered the boreal fauna to extend from Labrador to Cape Hatteras and the Virginian fauna from the Gulf of St. Lawrence to northern Florida. A detailed, geographic study of the clam *Spisula polynyma* (Chamberlin and Stearns 1963) showed the western North Atlantic population to extend from the Strait of Belle Isle to a little south of Cape Cod.

Some pertinent information is available for other marine groups: On the basis of the distribution of the little arctic crustacean, *Mysis oculata*, Madsen (1936:69) considered the sub-Arctic (here included in the Arctic Region) to extend southward as far as Cape Charles, Labrador, a locality at the northern border of the Strait of Belle Isle. Pettibone (1956:542) showed that almost all of the 68 species of Labrador polychaetes ranged northward into the Arctic, and Grainger (1964:32) found that all of the 13 species of Labrador sea stars (Asteroidea) were shared with the Arctic. Three species of arctic seals (hood, sharp, and bearded) extend as far south as the Gulf of St. Lawrence, the common seal ranges from the Arctic to about New York, and the gray seal seems to be limited (in the Western Atlantic) to the area between the Strait of

Belle Isle and northern Nova Scotia (King 1964). An endemic, boreal hermit crab (*Pagurus arcuatus*) is found from the Strait of Belle Isle to Cape Hatteras (Squires 1964), and the American lobster, *Homarus americanus*, occupies an almost identical range (Squires 1966:plate 3).

Thanks to the thorough work on the fishes of the Gulf of Maine by Bigelow and Schroeder (1953), the ichthyology of this area is better known than that of any other part of the eastern coast of North America. A total of 129 species of shore fishes have been recorded; of these, about 19 percent are endemic to the area between Labrador and Cape Hatteras, about 23 percent are shared with the Arctic, and some 53 percent are warm-water species of only occasional occurrence.

In his study on Labrador fishes, Backus (1957) found that most of the shore species had, primarily, arctic distributions, and the majority of species that were restricted to the western North Atlantic found their northern range limits in southern Labrador (in the vicinity of the Strait of Belle Isle). In his list of the marine fishes of Canada, McAllister (1960) considered the boundary between the arctic and Atlantic faunas to be located at Battle Harbour, Labrador. However, the latter lies near the southern tip of Labrador and only a few miles north of the Strait of Belle Isle.

Traditionally, Cape Cod was considered to mark the dividing line between the western North Atlantic warm-temperate and cold-temperate faunas (Ekman 1953, Hedgpeth 1957, many others). But since the latter fauna is now known to be more or less continuous around the Cape, it can no longer be regarded as a major zooge-ographic boundary; yet, some of the reasons for considering it a boundary still hold. That is, a remarkable number of eurythermic tropical and warm-temperate organisms do occasionally migrate or are carried as far north as Cape Cod, and a good many wide-ranging arctic-boreal species manage to penetrate this far south. In a sense, then, the Cape does function as a boundary but this involves primarily only the most motile, eurythermic species.

On the basis of available evidence, particularly for the fishes, it can be said that the island of Greenland does not seem to provide a suitable habitat for boreal species. Jensen's (1942, 1944, 1948) contributions on the fishes of Greenland reveal quite well the basic geographic relation-ships. In these works, a total of 34 shore species were reported. Aside from 2 species that are extremely rare, the rest are distributed as follows: 29 are shared with arctic waters, either on the Canadian side or

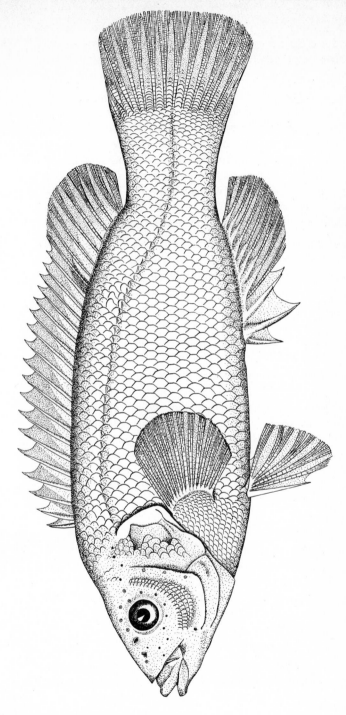

Figure 9-2 The cunner (*Tautogolabrus adspersus*). A shelf species confined to the Western Atlantic Boreal Region. After Bigelow and Schroeder (1953).

(in two instances) to Spitsbergen on the east side, 2 (*Molva molva* and *Brosmius brosome*) seem otherwise to have strictly boreal distributions, and 1 is known only from northwestern Greenland to the Barents Sea.

It has been said (Marty 1962:57) that the herring (*Clupea harengus harengus*), a common Greenland species, has a distribution that coincides with the boreal zone, but in reality, this species is quite eurythermic. It is found along both the east and west coasts of Greenland, north to Spitsbergen (Jensen 1948), and in northern Labrador (McAllister 1960:9). Its close relative in the Pacific (*C. h. pallasii*) ranges through the Bering Strait and well into the Arctic Ocean. For the fauna in general, Hedgpeth's (1957) zoogeographic map indicates an arctic rather than a boreal relationship for the entire coast of Greenland.

In summary, it can be said that the evidence from mollusks, polychaetes, sea stars, fishes, and marine mammals indicates quite clearly that the fauna of Labrador, except for possibly the southernmost tip, is primarily arctic. The distribution of these groups seems to indicate also that the arctic-boreal boundary is located at about the Strait of Belle Isle (the northern entrance to the Gulf of St. Lawrence). According to the evidence provided by fish distribution and by studies of the shore fauna in general, all of Greenland is surrounded by an arctic fauna.

The extent of endemism for the entire Western Atlantic Boreal Region is particularly difficult to assess. The difficulty is due mainly to the fact that the composition of the fauna is greatly influenced by relatively large numbers of southern species that are either seasonal migrants or sporadic invaders. If this group is left out of consideration, the endemic level would appear to be quite high—about 25 percent for the fishes and perhaps 30 to 40 percent for the mollusks.

The Eastern Atlantic Boreal Region

Beyond the Grand Banks of Newfoundland, the eastward continuation of the Gulf Stream becomes the North Atlantic Current. The latter forms the upper portion of the great, clockwise North Atlantic Gyre and its main flow is to the northeast toward the British Isles. In the middle of the North Atlantic, a major division takes place whereby one branch turns southward to remain within the main gyre while the other extends northward to flow into the Norwegian Sea. Upon separation, the northern branch becomes the Norwegian Current.

By means of the Norwegian Current, relatively warm, saline water is carried into much higher latitudes than anywhere else in the world. The

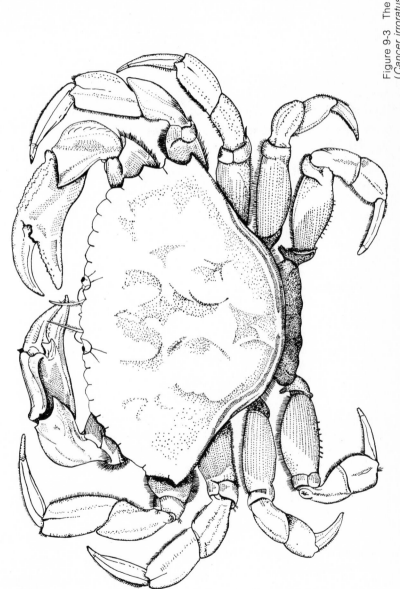

Figure 9-3 The rock crab (*Cancer irroratus*). A shelf species confined to the Western Atlantic Boreal Region. After Williams (1965).

various minor branches of this current have a profound effect on the shore fauna and the coastal climate in most parts of the northeastern Atlantic: a westerly branch affects the south and west shores of Iceland, warm water is carried into the Greenland Sea, and another branch follows along the Norwegian coast toward the northeast to enter the Barents Sea.

The Baltic Sea lies in an area of low evaporation and is fed by many large rivers. Consequently, there is a large surface outflow of low salinity water through the Kattegat and into the North Sea. The main gate to the Baltic is the Strait of Darss, and because of its shallowness (16 to 18 meters) the heavier, highly saline water in the depths of the Belt Sea is prevented from entering the Baltic. Some intrusion of bottom water does take place, but its salinity does not exceed 15 to 20 °/oo (Segerstråle 1957:754).

Over 100 years ago, Forbes (1859) observed that the naturalists of Scandinavia were indefatigable in the exploration of their neighboring seas and that mention of those who have successfully worked on the marine natural history of the Boreal Province would fill pages with arrays of eminent names. This tradition of Scandinavian excellence in marine biology, which began with Linnaeus, has continued to the present day.

Early Work and Concepts

Immediately to the south, the work of such notable investigators as Alphonse Milne Edwards and Victor Andouin in France; Johannes Müller in Germany; and in England, Edward Forbes, Samuel P. Woodward, and Philip Henry Gosse provided a good start for the study of marine zoology. The outstanding work of the early period was undoubtedly Forbes' (1859) *Natural History of European Seas.* It is not only the first book on marine ecology (Hedgpeth 1957:2) but is a pioneering and still classic publication on marine zoogeography.

The first permanent marine station was founded at Concarneau, France in 1859. The Marine Biological Association of the United Kingdom was organized in 1884, and its station at Plymouth was completed in 1888. Now there are 30 to 40 similar stations that are contributing to the knowledge of the marine life of the Eastern Atlantic Boreal Region. In 1953, Ekman remarked that probably no other oceanic region has been investigated more intensively.

In view of all the work that has been accomplished to date, one might assume that the question of relationship to adjacent areas and the

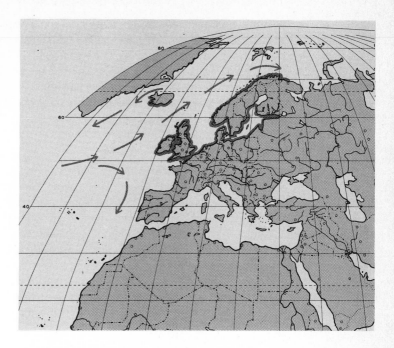

Figure 9-4 Eastern Atlantic Boreal Region.

determination of zoogeographic boundaries had been settled long ago. However, this is not the case. The first distributional maps were published by Forbes and by Woodward both in the same year (1856). Forbes' version indicated a "Boreal Province" extending from northern Norway to the Skagerrak and including the northern parts of the Baltic Sea (the Gulf of Finland and the Gulf of Bothnia). A "Celtic Province" was shown extending from the latter points south to the Lusitanian boundary. Woodward's scheme, based on mollusk distribution, differed mainly in that the entire Baltic was included in the Celtic Province. Later, Forbes (1859) changed his arrangement to agree with that of Woodward.

Modern Work   The concept of two provinces with a dividing line about at the Skagerrak has persisted into the modern literature (Coomans 1962, Mars 1963). Hall (1964:231), who also worked on molluscan distribution, considered the arctic fauna to penetrate into part of the area and divided the rest into Norwegian and Celtic Provinces. However, most modern authors have not recognized such subdivisions, and it simply is not possible to split the region into separate parts because the pattern of endemism will not permit such an action. For example, almost all the boreal fishes range throughout the area (Blegvad 1929–1938, Anderson 1942, Poll 1947, Saemundsson 1949, Bruun and Pfaff 1950, and LeGall and Cantacuzene 1956) and so do the echinoderms (Hyman 1955:673, Ursin 1960), the bryozoans (Ryland 1963), and the gray seal (King 1964). Ekman (1953:101) recognized only a single boreal fauna and observed that the region possessed many endemic elements and that most of them had their center of distribution in the North Sea.

The initial separation of the region into Boreal and Celtic Provinces was almost certainly due to the fact that many eurythermic arctic (arctic-boreal) species manage to range southward along the Norwegian coast as far as the Skagerrak. As they extend southward, many such species undergo isothermic submergence, and the tongue of deep water that projects southward into the Skagerrak apparently forms a convenient pathway for them. The latter is analogous to Cape Cod on the American coast in that it marks the southern boundary of many wide-ranging, arctic-boreal species but does not serve as a major zoogeographic boundary.

It has already been established (p. 203) that the Boreal-Lusitania **Boundaries** boundary lies at about the western entrance to the English Channel. Opinions about the location of the northern limit of the region vary widely. In regard to the fauna in general, Ekman (1953:101) pointed out that, as early as 1851, Michael Sars recognized a faunal change at North Cape, Norway. Forbes (1856, 1859) delineated the boundary at the southern edge of the Lofoten Island chain, Madsen (1936:67) indicated that the arctic (subarctic) fauna extended westward in the Barents Sea as far as North Cape or (in the case of one indicator species) somewhat farther, Geptner (1936:fig. 140) preferred a location about halfway along the northern coast of the Murmansk Peninsula, Schilder (1956:75) drew a line at about North Cape, and Hedgpeth (1957:map) considered the boreal fauna to extend throughout the White Sea and all the way to the eastern end of the Barents Sea.

For mollusks, Woodward (1856) placed the boundary at North Cape, and Coomans (1962:97) agreed, but Hall (1964:231) placed his line considerably south of the Lofoten Islands. For echinoderms, Hyman (1955:673) mentioned the northern limit of Norway (which is North Cape); for the boreal hemichordate species *Rhabdopleura normani*, Burdon-Jones (1954:12) gave the northern temination as North Cape; and for some Norwegian bryozoan species, Ryland (1963:48) recognized the North Cape boundary. King (1964) showed that two species of arctic seals extended through, but not beyond, the Barents Sea, however.

An extensive chapter on the Barents Sea by Zenkevitch (1963:72–178) provided a detailed summary of Russian biological investigations in that area. He emphasized the intensive work that began with the establishment of the Murmansk Biological Station by the St. Petersburg Natural History Society in 1899 and stated that the Barents Sea can now be considered one of the best surveyed areas in the world. In reading the account by Zenkevitch, one receives the impression that the most thorough studies on distribution were carried out by Schorygin (1928) on echinoderms (he recognized an arctic-boreal boundary at about the northern base of the Murmansk Peninsula), by Filatova (1938), and by Brotsky and Zenkevitch (1939). The latter two studies involved comprehensive analyses of the bottom fauna and confirmed Schorygin's findings. However, more recently, Zenkevitch (1963:176) expressed the opinion that the strictly littoral and sublittoral faunas penetrated farther eastward, the first throughout the western part of the White Sea and the second almost to the head of the Murmansk Peninsula.

In his work on the anomuran crustaceans of the U.S.S.R., Makarov (1938) noted the penetration of the Barents Sea by several temperate species. For example, the hermit crab, *Pagurus bernhardus*, an amphi-Atlantic boreal species, was found to extend eastward on the Murmansk coast as far as the Kola Fjord. From the book by Andriashev (1954), it is apparent that about 40 species of common, temperate, shore fishes reach their limits in the western part of the Barents Sea; several others extend further eastward to enter the White Sea.

Despite a considerable amount of conflicting opinion, it may be possible to clarify the problem to some extent: (1) in general, the Russian work demonstrates that an impressive part of the boreal fauna enters the eastern Barents Sea instead of stopping at North Cape; (2) several boreal species extend around the end of the Murmansk Peninsula and into the White Sea, but this pattern seems to be the exception rather than the rule; and (3) the extensive bottom fauna studies and the ranges of the individual fish and invertebrate species lend support to the idea that the greatest faunal change takes place at about the base of the Murmansk Peninsula in the vicinity of the Kola Fjord. Accordingly, it is suggested that the Kola Fjord, the narrow arm of the Barents Sea that leads to Murmansk, be regarded as the most probable location of the arctic-boreal boundary.

Iceland   The marine zoology of Iceland has been under competent investigation by Icelandic and Danish naturalists for about 150 years. Their interest has been devoted not only to the fauna itself but to the questions of its origin and general distribution. A fitting result of their efforts is the fine series entitled *The Zoology of Iceland* that began publication in 1937. To date, more than 80 parts of this series have appeared and many of the marine shore groups have been treated. Consequently, it is now possible to present a reasonably accurate assessment of Icelandic marine zoogeography.

The prosobranch gastropod shore species total 135 (Thorson 1941). Of these, 15 (11 percent) were purely boreal, 20 (15 percent) arctic, 49 (36 percent) arctic-boreal, and 41 (30 percent) eurythermic temperate species that ranged as far south as the Mediterranean and/or northwestern Africa. Einarsson's (1948) report provided information about 74 echinoderm shore species. Of these, only 13 (18 percent) were purely boreal, 15 (20 percent) were arctic, 32 (43 percent) arctic-boreal, and 14 (19 percent) eurythermic temperate. The distribution of the bivalve mollusks (Madsen 1949) was found to be essentially similar to that of the gastropods. The fishes were worked on by Saemundsson (1949), who found a total of 70 shore species. Of these, 41 (59 percent) were purely boreal, 7 (10 percent) arctic, 13 (19 percent) arctic-boreal, and 8 (11 percent) eurythermic temperate.

From the foregoing, it can be seen that Iceland possesses a most interesting faunal mixture. There is a pure boreal component, pure arctic, arctic-boreal, and a considerable number of eurythermic temperate forms. In the fishes, the boreal representation is considerably larger than in the other groups. The most probable reason is that the fishes, being generally the most mobile, have been more successful in migrating from the eastern Atlantic mainland. The distance to the Norwegian coast is about 570 miles, and the time available for colonization by boreal species has been comparatively brief (the surface temperature probably became warm enough to support a boreal fauna about 6,000 to 7,000 years ago).

As far as the boreal fauna in general is concerned, it can be said that the Icelandic relationships are almost entirely with the eastern Atlantic; that is, the only boreal species shared with North America are those that have an amphi-Atlantic distribution. Iceland may well have served as a way station in the westward migration of these amphi-Atlantic forms. The absence of any special American relationship and the almost complete absence of endemics (except for an occasional subspecies) indicates that the island, or at least a portion of it, should be included in the Eastern Atlantic Boreal Region.

Ekman (1953:101) and many of the earlier zoogeographers recognized that Iceland belonged partly to the boreal zone, especially its southern and western regions. Examination of data on the local distribution of both fishes and invertebrates shows a definite pattern. Both the purely boreal species and the eurythermic temperate forms tend to be confined to the south and west coasts, the arctic species are mainly restricted to the north and east, and the arctic-boreal animals are generally found all around the island.

Faroe Islands

The Faroes are a group of 21 volcanic islands located between Iceland and the Shetlands at about latitude 62°N and longitude 7°W. The primary source of our knowledge about the shore fauna and its relationships is the series entitled *The Zoology of the Faroes* which was published in Copenhagen (1928–1942). Stephenson (1937) contributed a series of articles on the various groups of Crustacea and, in general, found that the species were either boreal or arctic-boreal. Also, all were either Eastern Atlantic or transatlantic species, there being no endemism and no evidence of any special relationship to the Western Atlantic. The same kind of distributional picture was indicated by Bronsted (1942) for the sponges and Kramp (1942) for the marine Hydrozoa.

The total number of species seems to be rather small. For example, Stephensen (op. cit.) noted that there were but 29 species of decapods and that the Shetlands (about 160 miles away) possessed about twice as many. It seems safe to assume that the Faroe Islands are a sparsely populated but otherwise typical part of the Eastern Atlantic Region.

Baltic Sea

The Baltic Sea is the largest estuarine area in the world. The salinity is relatively stable and decreases gradually toward the inner end of the long, narrow basin. At the east end of the Skagerrak the salinity is usually about 30 %o, in the Kattegat it drops to 20 %o, and in the Baltic proper it ranges from about 15 %o at the lower end down to about 2 %o at the extremities of the Gulfs of Bothnia and Finland. The distribution of animals along this natural salinity gradient has interested many of the zoologists from the countries that border the Baltic. The combined effects of low salinity and the extreme seasonal fluctuation in temperature make the environment a very rigorous one for marine animals.

As Segerstråle (1957) has pointed out, a number of marine groups either do not penetrate the Baltic or, at most, reach only its southern entrance. Examples are the corals, echinoderms, pteropods, scaphopods, cephalopods, and selachians. Those marine species that do live in the Baltic are often represented by individuals that are much reduced in size.

According to the historical account by Zenkevitch (1963), the Baltic Sea had its beginning about 12,000 years ago when a large freshwater lake was formed by the retreating ice sheet. About 3,000 years later, the basin was invaded by salt water from the Atlantic to form the Yoldian Sea. At this time, the climate was still cold and an arctic fauna occupied the area. An uplift of land in southern Sweden then cut off the basin from the ocean to form the Ancyclus Lake Sea which lasted about 2,200 years. Following this, a subsidence of the land barrier together with a eustatic rise in sea level reestablished the marine connection, forming the Littorina Sea. The latter sea had a higher salinity and a slightly higher temperature than is found at present. During the past 4,000 years, the sea floor at the outlet has been rising again, and this has restricted the exchange of water with the Atlantic, causing the salinity to drop.

In regard to the marine fauna, the Baltic is populated primarily by those boreal Atlantic species that are euryhaline. But there is also an interesting group of arctic relics that are considered to be survivors of the Yoldian Sea fauna. Some of these are the mollusk *Astarte borealis*, the priapulid worm *Halicryptus spinulosus*, the crustacean *Mysis aculata*, the fish *Myoxocephalus quadricornis*, and the seal *Phoca hispida*.

Since the history of the Baltic is relatively brief, there has not been enough time for the formation of endemics (at least at the species level). Thus, the recognition by Mars (1963) of a separate Baltic Region has no real zoogeographic significance. The Baltic Sea is regarded as an impoverished but interesting part of the Eastern Atlantic Boreal Region.

Endemism

As Ekman (1953:101) noted, most of the endemic species of the region have their center of distribution in the North Sea. A survey of the North Sea echinoderms by Ursin (1960) showed that 57 species were present, and of these, 13, or about 23 percent, seemed to be endemic. More than half the total, about 29 species, or 51 percent, had general temperate (eurythermic) distributions extending southward as far as the Mediterranean or northwestern African coast; 33 of the total were shared with Iceland, but only 15 (the arctic-boreal component) reached Greenland.

A study of the fishes of the eastern North Sea–Baltic area by Duncker and Ladiges (1960) provided information of similar value. Of a total of

93 shore species, 27 (about 28 percent) were endemic, 39 (42 percent) were eurythermic temperate, and 13 (14 percent) were arctic-boreal.

For the region in its entirety, the rate of endemism for both fishes and invertebrates may be estimated at 20 to 25 percent.

Amphi-Atlantic Relationships

The relationship of the European boreal fauna to the American was apparently first investigated by the Swedish zoologist Loven (1846). He found that, in the mollusks, about 75 percent of the arctic-boreal species were shared by the two sides of the Atlantic, but only about 8 percent of the purely boreal species were shared. Although more recent knowledge would alter these figures to some extent, the same general magnitude of difference continues to be evident.

Since Iceland is much closer to the North American shores than any other part of the Eastern Atlantic Boreal Region, an examination of its fauna is of considerable help in the determination of amphi-Atlantic relationships. In the prosobranch gastropods (Thorson 1941), 42 of 51 arctic-boreal species are amphi-Atlantic, but only 2 of the 15 purely boreal species are. In the echinoderms (Einarsson 1948), 30 of 32 arctic-boreal species are amphi-Atlantic, but none of the 13 boreal forms are. In the fishes (Saemundsson 1949), 11 of 13 arctic-boreal species are amphi-Atlantic, but only 16 of 41 purely boreal species are. Obviously, there are comparatively more amphi-Atlantic boreal fishes than invertebrates.

On the American side (the Gulf of Maine), 7 of the 31 purely boreal fishes extend to the Eastern Atlantic (Bigelow and Schroeder 1953). Of the eastern North Sea–Baltic fishes (Duncker and Ladiges 1960), 10 of the 41 purely boreal species are amphi-Atlantic. In general, it may be concluded that about 24 percent of the purely boreal Atlantic fishes occur on both sides of that ocean. The percentage among the principal groups of marine invertebrates is certainly much lower.

If one compares the general faunal composition of the southern portions of the Boreal Regions of the Atlantic, south of the Skagerrak in the east and south of Cape Cod in the west, there is only a minor degree of similarity. For example, between Cape Cod and Cape Hatteras about 18 percent of the native mollusks also occur on the European side (Coomans 1962). However, Clarke (1963:10), in comparing the mollusks of Nova Scotia and Maine with those of northwestern Europe, emphasized that virtually all the common, intertidal, rocky coast species were amphi-Atlantic. This distinct trend, in which a strong amphi-

Atlantic relationship is apparent to the north but becomes weak to the south, has been nicely demonstrated for the northwest Atlantic opisthobranch mollusks (Franz 1970).

The general impression of great similarity that can be seen in the northern portions of the two boreal regions is due almost entirely to the presence of large numbers of arctic-boreal species. As we have seen, the great majority of this group have transatlantic distributions. Despite their classification, which implies occupation of both boreal and arctic waters, they tend to inhabit only the northern portions of the boreal regions. It was doubtless this northern faunal similarity, brought about by the presence of so many arctic-boreal species, that caused the earlier zoogeographers to recognize a transatlantic boreal region.

The richest boreal fauna seems to occur on the eastern side of the North Atlantic. In the fishes for example, the cod family (Gadidae) is represented in Danish waters by 13 genera and 18 species (Bruun and Pfaff 1950), but there are only 9 genera and 12 species in the Gulf of Maine (Bigelow and Schroeder 1953). This information, when considered together with the strong European relationship of Iceland, suggests that the principal evolutionary center for the recent Atlantic boreal fauna lies along the eastern side of that ocean. It also indicates that successful migration has been taking place from east to west against the prevailing surface currents.

The Western Pacific Boreal Region

Although detailed surface current charts are not available, the predominant flow into the Yellow Sea is apparently from warm currents that flow northward along the Chinese coast from Formosa Strait and through the East China Sea. In the winter, the rate of flow must be quite slow since the surface temperature of the Yellow Sea drops to 0°C in the north and to about 8°C at the mouth.

The southern Sea of Japan receives a steady flow of warm water from a branch of the Kuroshio Current that runs through the Korean Strait. This branch, called the Tsushima Current, tends to remain along the eastern side of the sea flowing as far north as southern Hokkaido. Mixing with colder water takes place in the northern Sea of Japan, and then a southward flow, called the Linan Current, is established along the mainland coast. As the result, the main circulatory pattern takes the form of a counterclockwise gyre (Zenkevitch 1963:755).

In the Okhotsk Sea, another counterclockwise gyre may be seen. The main inflow is contributed by the East Kamchatka Current. The main outflow takes place through the middle of the Kurile Island chain. Since the East Kamchatka Current originates in the western Bering Sea, the temperature regime of the Okhotsk Sea is remarkably cold. In the winter, the entire area remains below 0°C, while the August mean for most of the surface is less than 12.8°C (Fleming 1955).

As the East Kamchatka Current flows to the southeast, paralleling the Kurile chain of islands, it picks up water from both the Okhotsk Sea and the Western Subarctic Gyre. It is then called the Oyashio Current, and its influence extends along the Japanese coast as far south as Cape Inubo. Opposite northern Honshu, the main flow takes an abrupt turn to the east and, after mixing with Kuroshio water, runs across the North Pacific as the West Wind Drift (Dodimead, Favorite, and Hirano 1963).

Geological History

The geological history of the northeastern Asiatic coastline has still to be clearly outlined. However, it does seem that a series of extensive changes from the late Tertiary through the Quaternary have taken place. Studies of both drowned river valleys and elevated terraces seem to indicate that the sea level may have ranged from as much as 150 to 180 meters above to 200 to 300 meters below its present stage. In 1929, Yabe concluded that, at the beginning of the Pleistocene, the Yellow Sea was dry land, the Sea of Japan was an inland body of water, and the Okhotsk Sea was almost cut off from the North Pacific.

G. V. Lindberg has contributed an extensive series of publications that deal with the history of the seas that border the eastern part of the Soviet Union, and his conclusions were summarized in a paper published in 1953. His evidence was derived mainly from studies of the fish fauna of the rivers and the submarine topography of the shallow seas.

Lindberg's hypothesis may be summarized as follows: For a considerable part of the Pliocene, the land level was relatively high and extended far to the east to include the Japanese Islands, the Ryukyu chain, and Taiwan. The present eastern Asian seas did not exist and the Bering Basin was dry land but supported an ancient Paleoyukon River system. This was followed by a "transgression stage" in which the sea level rose to some 150 to 180 meters above its present state, allowing the ocean waters to penetrate inland for considerable distances. Presumably, the latter event took place in the early Pleistocene.

A "penultimate regression stage" was again characterized by a low sea level so that the Pliocene configuration was essentially restored. But a depression persisted in the southern part of the Bering Sea, and the Sea of Japan remained as an isolated body of water. Also the Okhotsk Sea may have been formed at this time.

The penultimate regression was followed by a "penultimate transgression" in which marine waters rose to penetrate inland. This time, the flooding was not so extensive, since the sea level was only about 80 meters above the present level. The "last regression stage" was characterized by a new low sea level (200 to 300 meters below the present level), and the Japan, Okhotsk, and southern Bering Seas were partially or completely isolated. Finally, the "last transgression" or contemporary stage took place to achieve the present shoreline configuration.

Certainly the major causes of the foregoing transgressions and regressions were the worldwide, eustatic changes in sea level that took place during the glacial and interglacial stages of the Pleistocene. However, the magnitude of the changes postulated by Lindberg are such that considerable tectonic movement must also have been involved.

The Provinces

The general distributional patterns, including the location of endemic areas, indicate that the region is divisible into three provinces: a southern Oriental Province, a northern Kurile Province, and a northwestern Okhotsk Province.

Oriental

It seems to be clear that a faunal break exists at about the location of the Tsugaru Strait between the islands of Honshu and Hokkaido. This is discussed from the standpoint of the fauna of the Sea of Japan (p. 235), and the boundary seems to function equally well for the fauna of the outer coast. The area between Tsugaru Strait and the northern limit of the warm-temperate fauna at Cape Inubo was called the "Sanriku-Zyôban" province by Nomura and Hatai (1936). Hall (1964), also on the basis of the mollusk distribution, split it into two subprovinces, the "Sendaian" and the "Choshian," but this subdivision seems unwarranted.

With fishes, the Tsugaru Strait boundary is best defined by the impressive number of northern species that are usually unable to penetrate any farther south (Ueno 1971). This includes four Pacific salmon, *Oncorhynchus gorbuscha*, *O. tschawytscha*, *O. nerka*, and *O. kisutch* (Hikita 1962); a large number of genera and species belonging to the family Cottidae (Watanabe 1960); and five species of smelt belonging to the

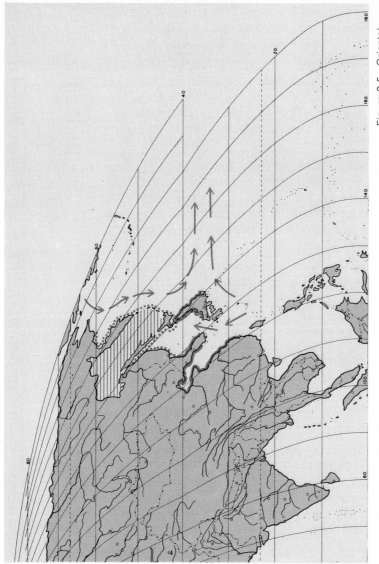

Figure 9-5   Oriental, Okhotsk, and Kurile Provinces.

family Osmeridae (McAllister 1963). Local faunal studies such as that of Sato and Kobayashi (1956) on the bottom fishes of Volcano Bay, Hokkaido, indicate the presence of a species complex that contrasts markedly with that shown by similar studies made at localities on Honshu Island.

The ascidian fauna seems to be very rich, for Tokioka (1963:43) recognized a group of 73 "north cold water species" limited to waters north of Hokkaido and Korea. In regard to marine mammals, it may be noted that the bearded seal ranges from the Arctic to the north coast of Hokkaido and that Stellar's sea lion also extends to Hokkaido (King 1964).

The recognition of boundaries at Tsugaru Strait and at about Chongjin on the Korean coast (p. 271) means that the cold-temperate fauna of the western North Pacific is divided into north-south components that are separated at about latitude 41 to 42°N. The southern portion, which includes the Yellow Sea, the central Sea of Japan, and the Pacific coast of northern Honshu, is here called the Oriental Province.

The close relationship between the cold-temperate faunas of the Yellow Sea and the central Sea of Japan is quite evident. However, little is known about the shore fauna of the eastern side of the Korean Peninsula compared to that of the west coast of Honshu Island. There are also some differences between the faunas of the east and west coasts of Honshu itself, at least as far as the mollusks are concerned (Kuroda and Habe 1952). Further studies may indicate a need for the subdivision of the Oriental Province. Since the faunas of the above areas have not been critically compared, it is difficult to make any estimate of the extent of endemism in the province. In fishes, the level of endemism is about 15 to 20 percent.

There is, unfortunately, little modern information about the distribution **Yellow Sea** of the marine fauna along the northern Chinese coast. As was indicated earlier (p. 236), it is possible that the warm-temperate versus cold-temperate boundary is located in the vicinity of Wenchou, a place considerably south of the mouth of the Yellow Sea. To the east, on the other hand, we know that the tip of Korean Peninsula supports a warm-temperate fauna (p. 235).

Rass (1965) published a list of the fishes of the Yellow Sea and, at the same time, grouped them into four categories: tropical, warm-temperate, cold-temperate, and brackish water. The tropical category was by far the largest, being represented by 174 species. However, it must be kept in mind that almost all of these forms are probably

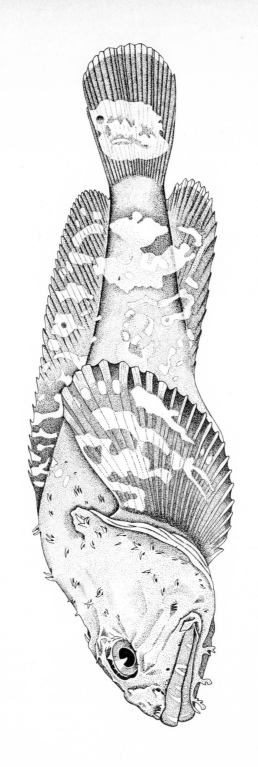

Figure 9-6 A marine cottid (*Eurymen bassargini*). A boreal shelf species apparently confined to the Oriental Province. After Watanabe (1960).

seasonal invaders that penetrate the area during the summer when the surface temperatures reach as high as 30°C. The warm-temperate component is represented by 40 to 44 species, most apparently occurring on the eastern side of the sea near the end of the Korean Peninsula (Mori 1952). But some, such as the croaker, *Pseudosciaena polyactis*, undoubtedly migrate into the sea during the warmer months from the mainland shores to the south.

The resident marine fish fauna of the Yellow Sea apparently consists mainly of cold-temperate species that are also found in the Sea of Japan. Rass (1965) listed 42 such species belonging to the typical cold-water families of the North Pacific such as the Osmeridae, Gadidae, Stichaeidae, Pholidae, Zoarchidae, Hexagrammidae, Cottidae, Liparidae, and Pleuronectidae. The presence of this cold-temperate group was also indicated in the older works of Fang and Wang (1932), Wu (1932), and Koo (1933) as well as the more recent contributions of Mori (1952), Chang (1954), and Lindberg and Legeza (1959–).

In his work on the Polychaeta, Ushakov (1955:59) noted that an appreciable number of the species characteristic of the Far Eastern seas of the U.S.S.R. were also found in the northern part of the Yellow Sea. He concluded that, although the latter sea was separated from the rest of the area by the south coast of Korea, he regarded it as a part of his Far Eastern (cold-temperate) subregion.

The extent of endemism seems to be very low, and this is to be expected, since the Yellow Sea basin is quite shallow and probably was completely dry during the glacial stages of the Pleistocene. The Yellow Sea is accordingly considered to be a somewhat depauperate but integral part of the Oriental Province.

Sea of Japan

The southernmost part of the Sea of Japan supports a warm-temperate fauna with many typical species being found as far north as Shimane Prefecture on Honshu Island and around the tip of the Korean peninsula. To the north, the fauna is cold-temperate, but there are indications that two rather distinct assemblages exist.

In regard to the Mollusca, Nomura and Hatai (1936) recognized a "Noto-San-in" province extending north along the west coast of Japan to the Noto Peninsula; a "Uetu" province from the latter point north to the Tsugaru Strait; and, apparently, a "Hokkaido-Tisima" province north of the Strait. The checklist of Kuroda and Habe (1952) gave the known range of each species, and from this evidence, it would appear that one cold-temperate fauna extends north to about latitude 41 to 42°N where it is replaced by another that continues northward to the Bering Sea. Yoshida, Amio, and Nojima (1957) divided the fauna into three parts at 35°N, 41°N, and 46°N.

The contrasting character of the biota of the northernmost part of the Sea of Japan has been noted by several investigators. Shchapova (1957) emphasized the dominance of the brown algae species compared to the red, Mokievsky (1956, 1959) found a great similarity between the littoral fauna of the northern part of the Tartary Strait and the Okhotsk Sea, and Zenkevitch (1963:778) noted that in Peter the Great Bay the dominant fish species were those belonging to the cold-loving families. The work of Mori (1952) showed that large numbers of north boreal fish species extend about as far south as Chongjin on the northeast coast of Korea. Also, Andriashev (1954) indicated that most of the cold-temperate species found in the western Bering Sea ranged southward to the northern Sea of Japan.

Considering its past history of isolation (p. 265), one might expect that the Sea of Japan would contain a large percentage of endemics. Makarov (1938) did find that 4 out of 25 species of lithodid crabs were endemic, Kuroda and Habe (1952) listed a number of mollusks that seem to occur only on the west coast of Japan, and Ushakov (1955) identified Peter the Great Bay as a center of polychaete endemism. But, it is likely that many species now considered to be endemics will eventually be found outside the Sea of Japan.

The Sea of Japan has a very rich marine fauna, since its waters contain species that belong to four distinct zoogeographic groups. A warm-temperate fauna is found at its southern end, which is also invaded by large numbers of eurythermic tropical species. The greater part of the area is occupied by the cold-temperate (south boreal) fauna of the Oriental Province, and from about 41 to 42°N is found the north boreal fauna of the Kurile Province.

Okhotsk Although the faunas of the northern part of the Sea of Japan and the Okhotsk Sea are very similar, the Okhotsk Sea also possesses considerable endemism. Tokioka's (1963) work on the North Pacific ascidians listed a total of 33 species for the Okhotsk Sea; of these, more than half (18) are apparently endemic. About 40 percent of the pycnogonids are possibly endemic, and for this reason, Hedgpeth (1963:1321) recognized a separate Okhotsk Province. The total number of fish species is about 300 (Shmidt 1950, Zenkevitch 1963:816), with 140 being shared with the Sea of Japan and 112 with the Bering Sea, but around 28 to 30 percent are endemic. Even the polychaetes, a rather slowly evolving group, demonstrate about a 9 percent endemism (Ushakov 1955).

The Pacific salmon genus *Oncorhynchus* may have undergone its initial evolutionary radiation along the Asiatic coast of the North Pacific (Neave 1958). Two of the species, *O. rhodurus* and *O. kawamurae*, are found only in northern Japan; *O. masu* occurs from the Okhotsk Sea to Taiwan, and the remaining five species have amphi-Pacific distributions (Hikita 1962). Although Neave suggested that the Sea of Japan was the most probable place of origin for the genus, the Okhotsk Sea has certainly served as an important evolutionary center and may have played the more significant role.

Various opinions have been expressed about the faunal relationships of the Okhotsk Sea. Vinogradov (1948) recognized three zoogeographic zones within the sea: the "Glacial Okhotsk," the "Western Kamchatka," and the "Southeastern Sakhalin." Pasternak (1957) stated that the boreal forms were absent from the northern part of the sea, but Zenkevitch (1963:794) observed that the littoral fauna was typically boreal.

Since the endemic rate in fishes, ascidians, and pycnogonids is quite high, the Okhotsk Sea is recognized as a distinct province of the Western Pacific Boreal Region. Apparently, one or more periods of isolation or near isolation during the Pleistocene permitted the evolution of many endemics, some of which are still confined to the Okhotsk basin.

Many years ago, Forbes (1856:map) recognized a "Mantchourian" province for Hokkaido, most of Sakhalin Island, and the northern end of the Sea of Japan. Also, he delineated an "Ochhotzian" province from north of Hokkaido to the tip of the Kamchatka Peninsula and including most of the Okhotsk Sea. In the same year, Woodward (1856) recognized an amphi-Pacific "Aleutian" province that extended, on the western side, from Hokkaido to the Bering Strait. In 1936, Nomura and Hatai, in their description of Japanese zoological provinces (based chiefly on molluscan distribution) named a Hokkaido-Tisima province extending from the Tsugaru Strait to the Bering Sea.

More recently, Ekman (1953:178) considered that the northwestern Pacific arctic seas extended from the western Bering Sea to certain layers of the Sea of Japan and included Kamchatka, the northern Kurile Islands, and most of the Okhotsk Sea. The area was called the "Pacific arctic subregion." Schilder (1956:75) followed Ekman, but Hedgpeth

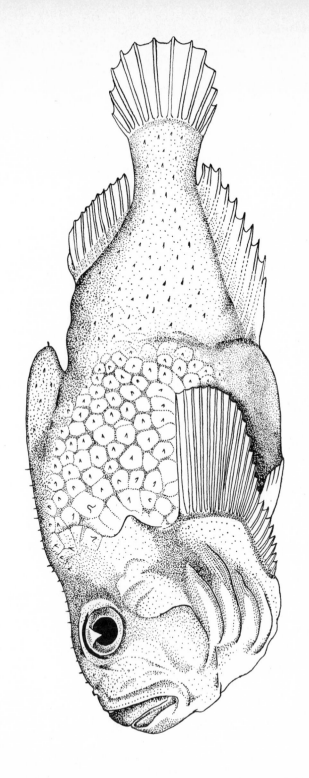

Figure 9-7 A smooth lump-sucker (*Cyclopsis tentacu-laris*). A boreal shelf species confined to the Okhotsk Province. After Lindberg and Legeza (1955).

(1957:map) presented a somewhat different scheme. He depicted an arctic fauna from about Cape Navarin to Cape Olyutorsky and from the tip of the Kamchatka Peninsula through the Okhotsk Sea and south to Vladivostok. The remainder of the area from the Bering Straits to the warm-temperate boundary was considered to be boreal.

There is still very little agreement as to the location of the temperate-arctic boundary in the northwestern Pacific. Shchapova (1948), utilizing the distribution of marine algae, placed the northern part of the Bering Sea (and the Okhotsk Sea) in her "North-Boreal" subdivision. In discussing the invertebrate bottom fauna of the southwestern part of the Chukchi Sea, Ushakov (1952) called attention to the invasion of Pacific boreal forms which comprised about 50 percent of the species. Later, the same author (1953) drew the northern boundary of the boreal region through the Bering Strait.

There are two very good reasons why the boreal-arctic boundary needs to be located somewhere to the south of the Bering Strait. First, the northern Bering Sea is invaded by many arctic species—forms that apparently evolved in the Arctic Basin. Examples among the fishes are *Boreogadus saida*, *Lycodes turneri*, and *Artediellus scaber.* Second, the boreal faunas of the two sides of the North Pacific are quite different. Depending on the group, the fauna may be richer on one side than on the other (see discussion of amphi-Pacific relationship on p. 283). Also, with many of the amphi-Pacific species, there seems to be a northern discontinuity in distribution. The principal cause of these phenomena must be a temperature barrier in the northern Bering Sea.

Where on the Asiatic coast does the boreal-arctic faunal boundary occur? Zenkevitch (1963:745) pointed out that several of the Russian investigators agreed that the area north of St. Lawrence Island, including the greater part of Anadyr Bay, should belong to the Arctic Region (usually called the Low Arctic Subregion). In his ichthyological works, Andriashev (1937, 1954) called attention to the barrier effects of the cold-water area between St. Lawrence Island and the Anadyr Bay. However, he also listed many species that apparently do not extend north of Cape Olyutorsky. In addition, it is interesting to note that in January and February the mean limit of pack ice extends south to about the latter point. Although Cape Olyutorsky now seems to be the most likely boundary location, further investigation may show that it lies to the north, perhaps as far as Cape Navarin.

In summary, the Kurile Province occupies an area extending from the northern Sea of Japan and Hokkaido Island, along the Kurile chain, and ending at about Cape Olyutorsky in the western Bering Sea. A great many of the species occur also in the Okhotsk Sea. Endemism is probably about 30 to 40 percent.

The Eastern Pacific Boreal Region

Until quite recently, comparatively little was known about the oceanography of the North Pacific. Beginning in 1955, data was collected intensively by various agencies from the United States, Canada, and Japan. Reports based on this information are particularly useful for achieving an understanding of the relationship between certain oceanographic features and the distribution of marine animals.

According to the interpretation of Dodimead, Favorite, and Hirano (1963), a West Wind Drift, formed by the mixing of Oyashio and Kuroshio waters, flows eastward across the North Pacific at about 40°N. About 300 miles from the Oregon-Washington coasts, this current divides with one part turning southward to form the California Current and the other flowing northward into the Gulf of Alaska. A minor part of the West Wind Drift also intrudes into the shore waters of British Columbia.

Although the cool, southbound California Current exerts a very important general influence, it has little effect on the shore fauna of the coastal stretch from northern California to British Columbia. Along this coast, the continental shelf is very narrow and the California Current water seems to extend inshore for only a brief period during September and October. At other times, the shore fauna is subjected to upwelling (February to September) or the Davidson Current (November to February) that flows from south to north (Bolin and Abbott 1963).

Dodimead et al. (1963) divided the North Pacific waters into a series of "domains" which were defined mainly by temperature, salinity, and flow characteristics. The West Wind Drift and California Current water was considered to belong to a Transitional Domain. Although it overlaps the Coastal Domain close to shore, the transitional water appears to extend consistently as far north as Dixon Entrance (Dodimead et al. 1963:figs. 221–224). This may be an important reason for the existence of a zoogeographic boundary at this point.

The Gulf of Alaska is under the influence of the counterclockwise Alaskan Gyre. The Subarctic Current brings water into the gyre from the west, and an outward flow takes place to the north through the Aleutian chain. Dodimead et al. (1963) considered the offshore waters of the Alaskan Gyre to belong to a Central Subarctic Domain while the coastal areas of the Gulf of Alaska to the eastern Bering Sea were included in an

Alaskan Stream or a Coastal Domain. The latter two domains seem to have little or no zoogeographic significance.

The distribution of the coastal biota indicates a region divisible into two provinces: a southern Oregon Province and a northern Aleutian Province.

The Provinces

A very rich cold-temperate fauna is found north of Point Conception, California. Forbes (1856:map) distinguished an "Oregonian" province extending north to Puget Sound and a "Stichian" province from there to about Juneau, Alaska. Woodward (1856), on the other hand, recognized a "Californian" province north to Puget Sound and an "Aleutian" province from there all the way to Bering Strait.

Oregon

Modern workers, who have been concerned with the fauna in general, have contributed the following: Ekman (1953) adopted the divisions of Schenk and Keen (1936) who had suggested a radical, new plan based on mollusk distribution. This recognized Californian and Aleutian faunas which were separated by extensive "overlap areas." Ekman's decision was unfortunate, for it had been shown by Newell (1948) that a new analysis was necessary since the actual distribution of the mollusks was closer to the divisions of some of the earlier workers. Hedgpeth (1957) called attention to Newell's work and supported his viewpoint. More recently, Stephenson and Stephenson (1961:239) stated that there appeared to be an Oregonian province extending approximately from Point Conception to the Alaska Peninsula.

In addition to the works of Schenk and Keen (op. cit.) and Newell (op. cit.), the distribution of mollusks in the northeastern Pacific has been examined by several other investigators. Cox (1962) showed that the Abalone genus *Haliotis* (the species *H. kamptschatkana*) extended northward as far as Sitka, Alaska. Hall (1964) recognized Oregonian and Aleutian faunas but considered the barrier between the two to lie at Cape Flattery, Washington. The most recent contribution is that of Valentine (1966) who presented a detailed, numerical analysis of ranges based on the latest data. He also reviewed the schemes of most of the earlier malacologists, and his paper is a good compendium on the subject. Valentine demonstrated that the Oregonian fauna extended to Dixon Entrance with the Aleutian assemblage continuing northward from that area. It was interesting to see that this new, computer-controlled pattern turned out to be virtually identical to that based on a study of the pyramidellid mollusks over 60 years ago by Bartsch (1912).

Invertebrates

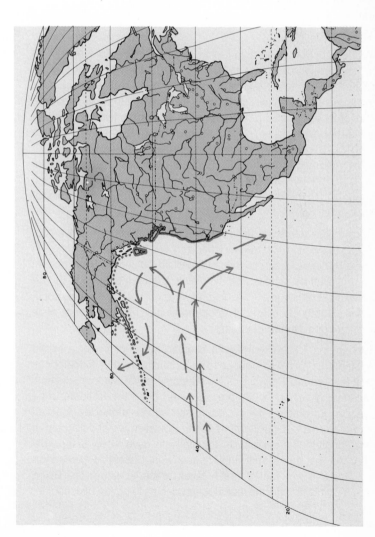

Figure 9-8　Oregon and
Aleutian Provinces.

The distribution of other shore groups has been discussed to some extent. Fisher (1911) described a rich starfish (Asteroidea) fauna extending north to about Sitka. For nemerteans, Coe (1940) listed a "Central" zone extending from California to Puget Sound and a "North Central" zone reaching from there to the Aleutians. Van Name (1945) mentioned that the northern California ascidian fauna probably extended to Cape Flattery, Fraser (1946) referred to the hydroid assemblage ranging from Cape Flattery to the Aleutians as a "single unit," and Osburn (1950–1953) recognized for bryozoans a "cool water area" between Point Conception and the Aleutians. Scagel (1963) spoke of the algal flora as being very uniform from the mouth of the Columbia River as far north as Sitka.

The wonderfully rich, shore fish fauna found north of Point Conception extends into British Columbia waters but rapidly changes character as it reaches southeastern Alaska. In northern California, the rockfishes belonging to the genus *Sebastodes* (family Scorpaenidae) form a large complex of more than 50 related species. More than two-thirds of the species do not extend north of British Columbia or southeastern Alaska (Chen 1971:79). Of the *Sebastodes* species that occur in British Columbia, the majority do not extend as far north as the Gulf of Alaska (Clemens and Wilby 1961). In fact, it can be said that about 50 percent of the entire shore fish fauna of western Canada does not extend north of the Alaskan Panhandle.

Fishes

In regard to the demersal fishes of economic importance, Alverson, Pruter, and Ronholt (1964:168) emphasized that the greatest change in species complexes occurred within the British Columbia, southeastern Alaska, and Gulf of Alaska area. It is also apparent that many Bering Sea species extend into the Gulf of Alaska but no further south (Wilimovsky 1958). These are all good indications of the presence of an important zoogeographic boundary.

Considering the foregoing data, it seems clear that there is in the northeastern Pacific a well-defined lower boreal fauna that is analogous to that of the Oriental Province of the Western Pacific. When the fishes, echinoderms, or marine algae are examined there would seem to be good reason for placing the boundary somewhere along the Alaskan Panhandle, perhaps as far north as Sitka. However, we should bear in mind that, of all the littoral groups, the mollusks have received the most attention and that the current authoritative work (Valentine 1966) indicates that the major faunal break takes place at about the Dixon Entrance. For this reason, the latter locality seems to be the best choice.

Figure 9-9 The smoothhead sculpin (*Artedius lateralis*). A eurythermic temperate shelf species of the north-eastern Pacific. After Bolin (1944).

The name Oregonian Province was first applied by Forbes in 1856 and it has remained in fairly constant use since that time. It is changed to Oregon Province here in order to achieve a standard terminology. The extent of endemism appears to be rather high; for mollusks, it is 29 percent (Valentine 1967:158), and for fishes, it is approximately 50 percent.

Between the Oregon Province and the cold waters of the Arctic Region, lies an upper boreal fauna. Although an "Aleutian" province was first named by Woodward (1856), he thought that it extended from Puget Sound to Bering Strait. In 1939, Andriashev published a zoogeographic map on which he delineated an Aleutian Province that extended from Kodiak Island to about Cape Romanzof. His map has been recently republished (Zenkevitch 1963:747). Ekman (1953:151), following Schenk and Keen (1936), described an Aleutian Province that included the area between Puget Sound and a point north of the Pribilof Islands (58°N). Hedgpeth (1957:map) indicated that the eastern Pacific boreal fauna ranged all the way to Kotzebue Sound, north of Bering Strait. Stephenson and Stephenson (1961:239) referred to the existence of a "Bering" province north of the Alaska Peninsula.

Aleutian

For the Mollusca, Hall (1964) considered the cold or arctic fauna to occupy the entire area from about 56°N to the Arctic Ocean. Valentine (1966) presented a good summary of previous work by malacologists and, using a numerical range analysis, considered the northern boundary for the Aleutian Province to lie at Nunivak Island (about 60°N). Makarov (1938:146) noted, in the southeastern part of the Bering Sea, the presence of a complex of species belonging to the hermit crab genus *Pagurus*—these being species which ranged southward but reached their northern limits at Nunivak Island.

Coe (1940) described a "Northern" zone for the nemertean fauna of the eastern Bering Sea, Mackay (1943) noted that the crab genus *Cancer* (the species *C. oregonensis*) extended as far north as the Pribilof Islands, and Osburn (1950–1953) recognized an "Arctic" area for bryozoans extending from the Aleutians to Point Barrow. Tokioka (1963:149), working on ascidians, referred to a northern, cold-water species group around Alaska and the Aleutian chain in contrast to a more southern, California Current group.

Very little is known about the details of the distribution of fishes along the eastern coast of the Bering Sea. The distributional tables of Andriashev (1954) indicate that many arctic species enter the northern

Bering Sea but are not found in its southern reaches. Conversely, a large number of north boreal Pacific species range northward into the southern Bering Sea but are rare or absent in the northern end. Most of the bottom fishes reported from Bristol Bay by Kobayashi and Ueno (1956) appear to be species that do not occur in the northern Bering Sea. McAllister (1963) has shown that the Eulachon (*Thaleichthys pacificus*) extends northward as far as Bristol Bay and the Pribilof Islands.

Here again, the distribution of the mollusks has been studied more closely than that of the other shore animals so that considerable weight can probably be given to Valentine's (1966) boundary at Nunivak Island. It may also be seen that, according to the U. S. Hydrographic Office map (pub. 225, 1944), the mean limit of the pack ice in January and February lies at about this point. It was noted previously (p. 274) that the Kurile Province–Arctic boundary in the western Bering Sea probably occurs at this same latitude.

Since there is very little information available about distributions within the Bering Sea, it is most difficult to hazard a guess as to the general degree of endemism that exists in the Aleutian Province. Certainly, the fauna is less rich than that of the Oregon Province immediately to the south, and since the geographic area is also considerably less, the endemism is probably smaller. For mollusks, it is about 24 percent (Valentine 1967:158).

Amphi-Pacific Distribution

The Bering Sea is essentially a broad, shallow basin almost completely enclosed to the north and with its southern end bordered by the Alaska Peninsula and the Aleutian chain of islands. In considering the general topography, the absence of obvious barriers might lead one to expect a homogeneous marine fauna. Indeed, the first workers on marine zoogeography (Forbes 1856, Woodward 1856) placed the Bering Sea, along with adjacent areas, in a single province. However, more recent investigators have shown that the eastern and western parts of the southern Bering Sea (below the Arctic zone) have quite different faunas.

Andriashev (1939) recognized an Aleutian Province for the southeastern Bering Sea and an Eastern Kamchatka Province for the southwestern part. Makarov (1938) reported 13 species of anomuran crabs from the Bering Sea but noted that only 5 were common to both sides while 8 were confined to the eastern part. Ushakov (1955) estimated that less than 50 percent of the eastern Bering and Gulf of Alaska polychaete species were shared with the western side, and later (1971) he published a list of 25 amphi-Pacific species. Zenkevitch (1963:841) observed that 52 percent of the decapod crustaceans and 46 percent of the fishes were common to both sides. Of the northern cold-water (upper boreal) ascidians, Tokioka (1963) found that only 29 of 121 species occurred on both sides.

Even though the arctic waters apparently come far enough south to form a temperature barrier at about Nunivak Island on the east and Cape Olyutorsky on the west, it is still surprising to find such a high degree of difference between the two sides of the southern Bering Sea. The gap between the above points is about 800 miles but about half the distance is covered by the shallow waters of the continental shelf. It may be surmised that, for most of the shore species, the deep waters of the northern Aleutian Basin off the Siberian coast form a significant barrier to an east-west dispersal.

As one might suspect, the amphi-Pacific relationship farther south is much less marked. At one time, it was thought that rather strong faunal ties existed between Japan and the American west coast due to the existence of an active migration route along the islands of the Aleutian chain. This was emphasized by some of the early malacologists, but Keen (1940), after an examination of the species involved, concluded that only about 0.3 percent of the mollusks had utilized this route. Makarov (1938) decided that, in the anomuran and pagurid crabs, a limited faunal exchange had taken place via the Aleutians. Gislén (1944) referred to several amphi-Pacific echinoderms that may have utilized this route. But it must be emphasized that these represent only a very small proportion of the fauna involved.

There is a group of eurythermic, arctic-boreal species that have continuous or almost continuous ranges through the northern Bering Sea and extend far to the south on each side of the North Pacific. The presence of such species accounts for most of the similarity that can be found between the Oregon and Kurile faunas. Examples among the fishes are the starry flounder (*Platichthys stellatus*) (Orcutt 1950) and four Pacific salmon species belonging to the genus *Oncorhynchus* (Hikita 1962). In echinoderms, there is *Strongylocentrotus franciscanus*, *Amphiura arcystata*, and *Amphipholis pugetana* (Gislén 1944). Some of the shore birds such as the tufted puffin (*Lunda cirrhata*) exhibit very similar distributional patterns (Udvardy 1963). Among the marine mammals, Stellar's sea lion (*Eumetopias jubatus*) is a good example (King 1964:10).

In most of the shore groups, there is a very high degree of relationship at the generic level, and many geminate or twin species may be identified. There seem to be two important evolutionary centers for the North Pacific fauna, one to the west and the other to the east. The genus *Mya* and other molluscan groups apparently underwent considerable evolution in the Kurile Province and Okhotsk Sea areas (MacNeil 1965). This is probably also true for most of the ascidian genera (Tokioka 1963) and certain fish families such as the Cottidae, Liparidae, and Zoarcidae. In these groups, the majority of the species and most of the endemics are concentrated on the Asiatic side. In contrast, the modern center of evolutionary radiation for the decapod crustaceans (Makarov 1938, Zenkevitch 1963), the asteroid starfishes (Fisher 1911), and other fish families (Osmeridae, McAllister 1963; Scorpaenidae, Phillips 1957) is apparently located on the American side.

In summary, it can be said that, considering the topography of the Bering Basin, the upper boreal faunas on each side are surprisingly distinct. On the average, only about 40 to 50 percent of the fauna appears to be common to both sides. To the south, the Oregon and Oriental faunas are very different, and aside from the wide ranging arctic-boreal species, probably less than 1 percent of the species are shared. At the generic level, the two sides of the North Pacific are very closely related. There seem to be two important evolutionary centers, one to the west and one to the east.

The Arctic Region

Since the Arctic Region extends south into the northwestern Atlantic and also well into the Bering Sea, the general circulation scheme is quite complicated. In the North Atlantic, it is apparent that the cold East Greenland Current, moving out of the Arctic Basin and through the Denmark Strait, is responsible for the consistently cold temperature of eastern Greenland and northeastern Iceland. When this current reaches the southern tip of Greenland, it turns westward into the Labrador Sea. There, it is supplemented by additional cold water from Baffin Bay (via the Davis Strait) and forms the Labrador Current which flows to the southeast.

The relatively warm Norwegian Current is, of course, responsible for the northward penetration of the boreal fauna into the southwestern part of the Barents Sea (see p. 259). Water from this current also has a moderating effect on sea surface temperatures at Novaya Zemlya, Franz Josef Land, and Spitzbergen. The main inflow of Atlantic water into the Arctic Ocean takes place just to the west of Spitzbergen. From there, the main direction of flow is toward the Laptev Sea. Then, the Atlantic water turns to the west to begin a broad, counterclockwise circulation that includes most of the Arctic Basin (Coachman and Barnes 1963).

In the Bering Sea, a northward current flows from about the middle of the Aleutian chain to the Bering Strait staying to the east of the main gyre. The flow of this North Pacific water into the Arctic Ocean is apparently subject to seasonal change. In the summer, the pack ice is pushed back far enough to clear a considerable part of the Chukchi Sea (Zenkevitch 1963:262). In the winter, the pack ice extends down to about 60°N latitude in the Bering Sea, and cold currents have been noted to enter the Bering Sea from the north and to flow southward along the Asiatic side (Zenkevitch, op. cit., 821).

In his earliest work, Forbes (1856:map) depicted a very widespread Arctic Province extending below the Alaskan Peninsula in the Pacific and as far south as Nova Scotia in the Atlantic. Later (1859:map), he presented a more restricted interpretation that showed the arctic fauna extending only a short distance south of the Bering Strait and to the northeastern coast of Newfoundland. Woodward (1856), on the basis of mollusk distribution, showed agreement with Forbes' later view for the Atlantic but considered the Arctic to extend to the Aleutian Islands in the Pacific. Apparently, von Hofsten (1915:203) was the first to subdivide the arctic fauna when he recognized, from echinoderm distribution, High-arctic (Hocharktische) and Low-arctic (Niederarktische) Zones. The two areas were separated on the basis of temperature, the High-arctic being below the 0°C isotherm.

Earlier Work

Gurjanova, Sachs, and Ushakov (1925) advocated a further subdivision into High-arctic, Arctic, and Subarctic Provinces. But, aside from the notation that littoral forms were not found in the High-arctic, the areas were not well defined (according to Zenkevitch 1963). Madsen (1936) maintained that only two regions should be recognized, Arctic and Subarctic. He placed the boundary between the two at the place where the littoral mollusks and the barnacle *Balanus balanoides* disappeared. Gorbunova (1940) divided the High-arctic into two parts, a "brackish water" province as defined by the distribution of the mussel *Portlandia arctica* and a "sea" province characterized by another bivalve *Propeamussium groenlandicum*.

The situation was further complicated when Ekman (1953:176) recognized not only High-arctic and Low-arctic subregions but also stated (p. 178) that the main division of the Arctic shelf was between the

Figure 9-10   Arctic Region.

"Polar-arctic" and the "North-west Pacific Arctic" with the Bering Strait forming a very sharp dividing line. He also (p. 180) identified an "Atlantic-arctic" group of species and a "North Siberian" group. Finally, Zenkevitch (1963:65), following the concepts of Gurjanova (1939), produced a zoogeographic map of the Arctic Region. This showed the Low-arctic occupying only the Barents and White Sea areas with the rest of the shelf being High-arctic. The High-arctic was divided into a North American–Greenland province and a Siberian province with each subdivided into a brackish water and a sea province. According to these most recent views, then, the Arctic Region appears as a complicated jigsaw puzzle.

The boundaries of the Arctic Region were determined during the process of locating the northern boundaries of the four boreal regions. In the eastern Atlantic, the arctic fauna extends south to about Kola Fjord at the base of the Murmansk Peninsula and to the north and east coasts of Iceland. In the western Atlantic, all of Greenland is included in the Arctic Region, and on the mainland coast, it extends south to the Strait of Belle Isle. In the eastern Bering Sea, the arctic fauna extends south to about Nunivak Island and, on the western side, to about Cape Olyutorsky.

Regional Boundaries

As was noted in the discussions on the boreal faunas of the North Atlantic, a significant proportion of the species may be identified as arctic-boreal. While many of these forms have a circumpolar distribution, there is a large group that typically occur from the eastern Canadian Arctic to Novaya Zemlya (sometimes further east) and then southward into boreal waters on each side of the Atlantic. Examples among echinoderms are *Lophaster furcifer*, *Stichaster albulus*, and *Asterias linckii* (von Hofsten 1915). Among fishes, there are *Somniosus microcephalus*, *Raja radiata*, *Lumpenus lumpretaeformis*, and *Anarhichas lupas* (Backus 1957). The presence of such forms led Ekman (1953:180) to recognize the Atlantic-arctic group within the Arctic Region.

Arctic-Boreal Species

The same phenomenon is apparent in the Pacific where many arctic-boreal species are found from about Wrangel Island in the Chukchi Sea eastward to about Point Barrow on the Arctic coast of Alaska (Zenkevitch 1963:268). These species range southward through the Bering Straits and well into the boreal regions. It has been estimated that, in the southwestern part of the Chukchi Sea, about 50 percent of the fauna consists of arctic-boreal species (Zenkevitch, op. cit.). Examples may be found among the fishes (Walters 1955), the polychaetes (Ushakov 1955:28), and pycnogonids (Hedgpeth 1963). Although such species have a considerable influence on the general faunal makeup of the above two portions of the Arctic Region, their presence can be recognized without the necessity of setting up special zoogeographic zones.

Arctic
Subdivisions

There is no doubt that a rather distinct brackish-water fauna exists along considerable portions of the Arctic shelf. According to Thorson (1957), the widespread *Macoma calcarea* community (including *Mya truncata*, *Cardium ciliatum*, and other species) is bound mainly to estuarine conditions. This seems to be also true for the *Portlandia arctica* community (including *Myriotrochus rinki* and *Ophiocten sericeum*). Fishes belonging to this fauna are the arctic flounder, *Liopsetta glacialis*, and the four-horned sculpin, *Myoxocephalus quadricornis* (Walters 1955). The question is, should zoogeographic recognition be given to this brackish water assemblage? Although it seems to occupy a huge geographic area (Zenkevitch 1963:65), there are doubtless many variations that reflect local or seasonal changes in salinity. The question is, therefore, more properly ecological than zoogeographic. For this reason, the brackish-water and sea water provinces of Gorbunova (1940) and Zenkevitch (1963) are not utilized here.

The concepts of High-arctic versus Low-arctic or Arctic versus Subarctic are usually correlated with faunal changes that occur with changes in latitude, although Walters (1955:354) pointed out that the less well known a locality, the more likely it is to be considered High-arctic. Since the major portion of the polar basin is still poorly explored, many localities probably have a much richer fauna than the literature indicates. It has been suggested that a faunal boundary occurs at the position of the 0°C isotherm. However, it seems that, in the summer, the temperature of the shallow waters of areas far to the north rises to 5°C or more. It has also been noted that a sharp contrast exists between the communities found above and below the 50-meter mark, the temperature always being below 0°C in the deeper area (Ellis 1959). At the present state of our knowledge, about the only generalization that can be safely made is that the shore fauna of the polar basin does seem to become poorer in species at the higher latitudes. But since there does not seem to be a significant endemic element at such latitudes, a zoogeographic separation cannot be made.

The recognition of North American–Greenland and Siberian Provinces

(Gurjanova 1939, Zenkevitch 1963:65) does not seem advisable. Both Walters (1955) and McAllister (1962) have shown clearly that the fish fauna of the western Canadian and Alaskan Arctic is virtually identical to that of the Siberian coast. These observations are reinforced by work that has demonstrated the circumpolar distribution of common arctic communities such as the *Macoma calcarea*, *Portlandia arctica*, *Venus fluctuosa*, and *Arca-Astarte crenata* (Thorson 1957, Ellis 1959). Osburn (1955:30) noted that the preponderance of bryozoan species of the Arctic Ocean had circumpolar distributions. Apparently, the great majority (about 85 percent) of New World arctic polychaetes have similarly broad distributions (Pettibone 1954:205, Berkeley and Berkeley 1956). Twelve of the thirteen northeastern Canadian echinoderms that are arctic endemics are now known to be circumpolar (Grainger 1955).

As far as the fish fauna is concerned, there do seem to be two barriers to the circumpolar distribution of species. One is a land barrier formed by the Boothia Peninsula and Somerset Island in the Canadian Arctic, and the other is the open water gap between Spitzbergen and Greenland. McAllister (1962) noted that less than half the shore fishes of the western Canadian Arctic were found in the eastern part and suggested that the fauna from the former area west to the Barents–White Sea be considered a separate "Innuit Fauna." From work done on the fishes of northeastern Canada (Backus 1957) and Greenland (Jensen 1942, 1944, 1948), it is apparent that there is a reduced number of arctic endemics compared to other areas. However, almost all such species that are present apparently have broad arctic distributions. Therefore, the eastern Canadian-Greenland area may be considered somewhat depauperate but not zoogeographically distinct.

Despite the fact that a number of different schemes for zoogeographic subdivisions of the Arctic have been presented, the generally broad distribution of arctic endemics and the corresponding lack of sufficient endemism in the proposed subdivisions shows the presence of an essentially uniform fauna. It needs to be borne in mind, however, that this basic homogeneous pattern may, in certain areas, be highly modified by the presence of euryhaline species or species with restricted arctic-boreal distributions.

Endemism

In his study of the pelecypods of Franz Josef Land, Soot-Ryen (1939:19) noted that about 15 percent of the species were restricted to the Arctic Region. However, in her study of all the Mollusca of Point Barrow,

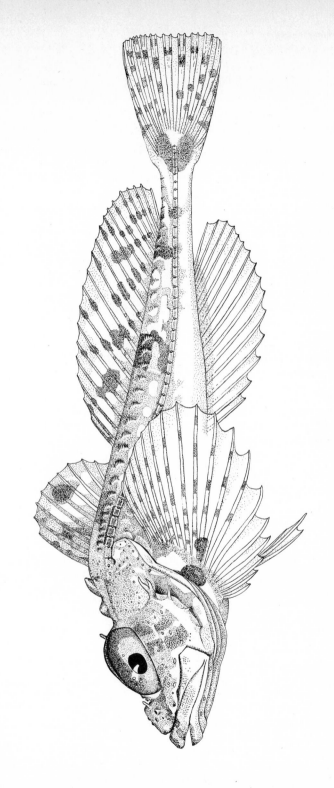

Figure 9-11 The spatulate sculpin (*Icelus spatula*). A cold-water shelf species confined to the Arctic Region. After Gilbert and Burke (1910).

Alaska, MacGinitie (1959) found that only 4 out of 107 species were exclusively arctic. The polychaete worms from Point Barrow demonstrated about a 4.5 percent endemism (Pettibone 1954:205). Of the invertebrate fauna in general from Point Barrow, only the sponges (30 percent) and the amphipods (25 percent) showed appreciable numbers of arctic endemics (MacGinitie 1955). Djakonov (1945) referred to 14 species of echinoderms as being limited to the Arctic shelf. This probably represents about 15 percent of the total shore species.

Along the Arctic coast of Eurasia, about 10 percent of the fishes are confined to the Arctic Region (Andriashev, 1954), some 33 percent of the Spitzbergen fish fauna are arctic endemics (von Hofsten 1919, Hognestad 1961), and about 30 percent of the Canadian arctic fishes are apparently in the same category (McAllister 1960). On the coast of Labrador, about 19 percent of the fishes are true arctic species (Backus 1957:331). Of the seals, only the walrus (*Odobenus rosmarus*) and the harp seal (*Pagophilus groenlandius*) seem to be restricted to Arctic waters (King 1964). In the whales, there are three monotypic, arctic genera, the Greenland whale (*Balaena*), the narwhale (*Monodon*), and the white whale (*Delphinapterus*) (Davies 1963:109).

Although the arctic fauna is comparatively young, it is fairly distinct for two reasons: In some groups, such as the sponges and amphipods and probably the bryozoans (Osburn, 1955), there are large numbers of species that seem to have evolved in the area. Arctic waters also serve as a refuge for geographic and phylogenetic relicts that apparently survive here by avoiding competition with the more advanced forms of temperate waters. The three monotypic whale genera are examples of the latter type.

It is most difficult to arrive at a meaningful figure for Arctic Region endemism. In areas where there are large numbers of arctic-boreal species, such as Labrador or Point Barrow, the percentage of endemics is quite low, but in other localities it can be comparatively high. For fishes, the average would perhaps be 20 to 25 percent. For invertebrates, it would be somewhat less. Valentine (1967:158) gave an overall figure of 14 percent for the mollusks.

Literature Cited

Alverson, D. L., A. T. Pruter, and L. L. Ronholt. 1964. A study of demersal fishes and fisheries of the northeastern Pacific Ocean. *Inst. Fish., Univ. Brit. Col.*, pp. 1–190, 72 figs.

Andersson, K. A. 1942. Fiskar och fiske i Norden Bokförlaget Natur och Kultur, 2 vols., xxiv + 1016 pp., 426 figs., 127 plates.

Andriashev, A. P. 1937. A contribution to the knowledge of the fishes from the Bering and Chukchi Seas. Explorat. des mers de l'URSS., fasc. 25, *Inst. Hydro. Leningrad*, pp. 292–355, figs. 1–27 (Russian with English summary). English translation by Lanz and Wilimovsky, published by U.S. Fisheries and Wildlife Service as a Special Science Report, Fisheries No. 145, in May 1965.

———. 1939. *The fishes of the Bering Sea and its neighboring waters, origin and zoogeography.* Leningrad Univ. Press (not seen; ref. from Zenkevitch 1963).

———. 1954. Fishes of the northern seas of the U.S.S.R. Keys to the fauna of the U.S.S.R. *Zool. Inst. Acad. Sci. U.S.S.R.*, (53):1–617 (English translation Jerusalem, 1964).

Backus, R. H. 1957. The fishes of Labrador. *Bull. Amer. Museum Nat. Hist.*, 113(4):273–338, 2 figs., 2 plates.

Bartsch, P. 1912. A zoogeographic study based on the pyramidellid mollusks of the west coast of America. *Proc. U. S. Nat. Museum*, 42(1906):297–349, plate 40 (map).

Berkeley, E., and C. Berkeley. 1956. On a collection of polychaetous annelids from northern Banks Island, from the south Beaufort Sea, and from northwest Alaska; together with some new records from the east coast of Canada. *J. Fisheries Res. Board Can.*, 13(2):223–246.

Bigelow, H. B., and W. C. Schroeder. 1953. Fishes of the Gulf of Maine. *U.S. Fish Wildlife Serv., Fishery Bull. No. 74*, viii + 577 pp., 288 figs.

Blegvad, H. (editor). 1929–1938. La faune ichthyologique de l'Atlantique Nord. Conseil Permanent International pour l'Exploration de la Mer, A. F. Host, Copenhagen, 418 plates.

Bolin, R. L., and D. P. Abbott. 1963. Studies on the marine climate and phytoplankton of the central coastal area of California, 1954–1960. *Rep. Calif. Coop. Oceanic Fish. Invest.*, 9:23–45, 8 figs.

Bousfield, E. L. 1960. *Canadian Atlantic sea shells.* Ottawa (not seen).

Brondsted, H. V. 1942. Marine Spongia. *Zool. Faroes*, 1(part 1)(3):1–33.

Brotsky, V., and L. Zenkevitch. 1939. Quantitative evaluation of the bottom fauna of the Barents Sea. Tr. V.N.I.R.O., 7 (not seen; ref. from Zenkevitch 1963).

Bruun, A. F., and J. R. Pfaff. 1950. *List of Danish vertebrates. Fishes.* Danish Science Press, Copenhagen, pp. 20–66.

Burdon-Jones, C. 1954. The habitat and distribution of *Rhabdopleura normani* Allman. Universitet I. Bergen, Natur. Rekke, (11):1–17, 3 figs.

Chamberlain, J. L., and F. Stearns. 1963. A geographic study of the clam *Spisula polynyma* (Stimpson). *Ser. Atlas Marine Environ., Amer. Geograph. Soc.*, Folio 3:1–12, 6 plates.

Chang, C., and others. 1954. Atlas of fish commonly seen in the Yellow Sea and the Gulf of Chihli. *Acad. Sinica*, pp. 1–149, 73 figs.

Chen, L. 1971. Systematics, variation, distribution, and biology of rockfishes of the subgenus *Sebastomus* (Pisces, Scorpaenidae, *Sebastes*). *Bull. Scripps Inst. Oceanog.*, 18:1–107, 6 plates, 16 figs.

Clarke, A. H., Jr. 1963. Supplementary notes on pre-Columbian *Littorina littorea* in Nova Scotia. *Nautilus*, 77(1):8–11.

Clemens, W. A., and G.·V. Wilby. 1961. Fishes of the Pacific coast of Canada. *Bull. Fisheries Res. Board Can.*, (68, 2d edition):1–443, 281 figs.

Coachman, L. K., and C. A. Barnes. 1963. The movement of Atlantic water in the Arctic Ocean. *Arctic*, 16(1):9–16, 3 figs.

Coe, W. R. 1940. Revision of the nemertean fauna of the Pacific Coasts of North, Central and northern South America. *Allan Hancock Pacific Expedition*, 2(13):247–323, 8 plates.

Coomans, H. E. 1962. The marine mollusk fauna of the Virginian area as a basis for defining zoogeographical provinces. *Beaufortia*, 9(98):83–104.

Cox, K. W. 1962. California abalones, family Haliotidae. *Calif. Dept. Fish Game, Fish. Bull.*, (118):1–133, 61 figs.

Davies, J. L. 1963. The antitropical factor in cetacean speciation. *Evolution*, 17(1):107–116, 2 figs.

Djakonov, A. M. 1945. On the relationship between the Arctic and the North Pacific marine faunas based on the zoogeographical analysis of the Echinodermata. *J. Gen. Biol.*, 6:125–155 (in Russian with English summary).

Dodimead, A. J., F. Favorite, and T. Hirano. 1963. Review of oceanography of the Subarctic Pacific region. *Bull. Int. Nor. Pacific Fish. Comm.*, (13):1–196, 265 figs.

Duncker, G., and W. Ladiges. 1960. *Die fische der Nordmark.* De Gruyter, Hamburg, 1–432 pp., 145 figs.

Edwards, R. L., R. Livingstone, Jr., and P. E. Harmer. 1962. Winter water temperatures and an annotated list of fishes—Nantucket Shoals to Hatteras. *U.S. Fish Wildlife Serv., Spec. Sci. Rep. Fisheries*, (397):iii + 1–31, 27 figs.

Einarsson, H. 1948. Echinoderma. *Zool. Iceland*, 4(70):1–67, 7 figs.

Ekman, S. 1953. *Zoogeography of the sea.* Sidgwick & Jackson, London, xiv + 417 pp., 121 figs.

Ellis, D. V. 1959. The benthos of soft sea-bottom in Arctic North America. *Nature*, 184(4688):79–80.

Fang, P. W., and K. F. Wang. 1932. The elasmobranchiate fishes of Shangtung coast. *Contrib. Biol. Lab. Sci. Soc. China, Zool. Ser.*, 8(8):213–283, 29 figs.

Filatova, Z. 1938. The quantitative evaluation of the bottom fauna of the south-western part of the Barents Sea. Tr. P.I.N.R.O., 2 (not seen; ref. from Zenkevitch 1963).

Fisher, W. K. 1911. Asteroidea of the North Pacific and adjacent waters. *Bull. U. S. Nat. Museum*, (76):vi + 1–419, 122 plates.

Fleming, R. H. 1955. Review of the oceanography of the northern Pacific. *Bull. Int. Nor. Pacific Fish. Comm.*, (2):1–43, 20 figs.

Forbes, E. 1856. Map of the distribution of marine life. *in* Alexander K. Johnston, *The physical atlas of natural phenomena* (new edition). W. and A. K. Johnston Edinburgh and London, plate no. 31.

———. 1859. The natural history of European seas (edited and continued by Robert Godwin-Austen). John Van Voorst, London, viii + 306 pp., 1 map.

Franz, D. R. 1970. Zoogeography of northwest Atlantic opisthobranch molluscs. *Marine Biol.*, 7(2):171–180, 5 figs.

Fraser, C. M. 1946. *Distribution and relationship in American hydroids.* University of Toronto Press, Toronto, pp. 1–464.

Geptner, V. G. 1936. *General zoogeography.* Blomedgis, Moscow, and Leningrad, 1–548 pp., 140 maps (in Russian).

Gislén, T. 1944. Regional conditions of the Pacific coast of America and their significance for the development of marine life. *Lunds Univ. Arrskrift*, 40(8):1–91, 13 figs., 1 plate.

Grainger, E. H. 1955. Echinoderms of Ungava Bay, Hudson Strait, Frobisher Bay and Cumberland Sound. *J. Fisheries Res. Board Can.*, 12(6):899–916, 8 figs.

———. 1964. Asteroidea of the Blue Dolphin expeditions to Labrador. *Proc. U.S. Nat. Museum*, 115(3478):31–46, 4 figs.

Gorbunova, G. 1940. The bivalve mollusc *Portlandia arctica* as an indicator of the distribution of continental water in the Siberian seas. *Prob. Arctic*, vol. 11 (not seen; ref. from Zenkevitch 1963).

Gurjanova, E. 1939. Contribution to the origin and history of the fauna of the Polar Basin. *Bull. Acad. Sci. U.S.S.R.*, vol. 5 (not seen; ref. from Zenkevitch 1963).

———, J. Sachs, and P. Uschakov. 1925. Comparative survey of the littoral of the northern seas. *Tr. Stat. Biol. Murman Soc. Natur. Leningrad*, 1 (not seen; ref. from Zenkevitch 1963).

Hall, C. A. 1964. Shallow-water marine climates and molluscan provinces. *Ecology*, 45(2):226–234, 6 figs.

Hedgpeth, J. W. (editor). 1957a. Introduction. *in Treatise on marine ecology and paleoecology*. vol. 1. *Geol. Soc. Amer., Mem.*, (67):1–16, 2 figs.

———. 1957b. Marine biogeography. *in Treatise on marine ecology and paleoecology*. vol. 1. *Geol. Soc. Amer., Mem.*, (67):359–382, 16 figs., 1 plate.

———. 1963. Pycnogonida of the North American Arctic. *J. Fisheries Res. Board Can.*, 20(5):1315–1348, 11 figs.

Hikita, T. 1962. Ecological and morphological studies of the genus *Oncorhynchus* (Salmonidae) with particular consideration on phylogeny. *Sci. Rep. Hokkaido Salmon Hatchery*, (17):1–97, 69 figs., 30 plates.

von Hofsten, N. 1915. Die Echinodermen des Eisfjords. *K. Svenska Vetenskapsakad. Handl.*, 54(2):1–282.

———. 1919. Die fische des Eisfjords. *Zool. Res. Swedish Expedition to Spitsbergen 1908*, 2(10):1–129, 1 plate, 20 figs.

Hognestad, P. T. 1961. Contributions to the fish fauna of Spitsbergen. I. The fish fauna of Isfjorden. *Acta Borealia, A. Scientia*, (18):1–36, 5 figs.

Hyman, L. H. 1955. *The invertebrates*, vol. 4: *Echinodermata*. McGraw-Hill, New York, 763 pp., 280 figs.

Jensen, A. S. 1942. Contributions to the ichthyofauna of Greenland 1-3. Spolia Zoologica Musei Hauniensis II, Skrifter Universitetets Zoologiske Museum, Copenhagen II, pp. 1–44, 9 figs., 8 plates.

———. 1944. Contributions to the ichthyofauna of Greenland 4-7. Spolia Zoologica Musei Hauniensis IV, Skrifter Universitetets Zoologiske Museum, Copenhagen IV, pp. 1–60, 15 figs., 8 plates.

———. 1948. Contributions to the ichthyofauna of Greenland 8-24. Spolia Zoologica Musei Hauniensis IX, Skrifter Universitetets Zoologiske Museum, Copenhagen IX, pp. 1–182, 26 figs., 4 plates.

Johnson, C. W. 1934. List of the marine mollusca of the Atlantic coast from Labrador to Texas. *Proc. Boston Soc. Nat. Hist.*, 40(1):1–204.

Keen, A. M. 1940. Molluscan species common to western North America and Japan. *Proc. 6th Pacific Sci. Congr.*, 3:479–483.

King, J. E. 1964. *Seals of the world.* British Museum of Natural History, London, 1–154 pp., 17 figs., 12 plates.

Kobayashi, K., and T. Ueno. 1956. Fishes from the North Pacific and from Bristol Bay. *Bull. Fac. Fisheries Hokkaido Univ.,* 6(4):239–265.

Koo, K. C. 1933. The fishes of Chefoo. *Contr. Inst. Zool. Peiping,* 1(3):1–235, 35 plates.

Kramp, P. L. 1942. Marine Hydrozoa. *Zool. Faroes,* 1(part 1)(5):1–59, 7 figs.

Kuroda, T., and T. Habe. 1952. *Checklist of bibliography of the recent marine Mollusca of Japan.* Hosokawa Printing Co., Tokyo, pp. 1–210, 2 figs.

LeGall, J., and A. Cantacuzene. 1956. Inventaire de la faune marine de Roscoff. Poissons. *Trav. Sta. Biol. Roscoff, suppl.,* 8:1–67.

Lindberg, G. V. 1953. Principles of the distribution of fishes and the geological history of the far-eastern seas. *Akademia Nauk SSSR,* Ikhthiolog. Komm., Moscow-Leningrad. pp. 47–51. (Prelim. transl. by W. E. Ricker.)

——, and M. I. Legeza. 1959–. Fishes of the Sea of Japan and adjacent parts of the Okhotsk and Yellow Seas. *Akademia Nauk SSSR.* (1):1–207, 1959, 108 figs.; (2)1–391, 1965, 324 figs. Series to be continued.

Lovén, S. 1846. Malacologiska notiser. Nagra anmärkningar öfver de Skandinaviska Hafs-Molluskernas geografiska utbredning. Övers K. Svenska. Vet.-Acad. Förhandl., Stockholm. pp. 252–274 (not seen).

MacGinitie, G. E. 1955. Distribution and ecology of the marine invertebrates of Point Barrow, Alaska. *Smithson. Misc. Coll.,* 128(9):iv + 201, 2 figs., 8 plates.

——. 1959. Marine Mollusca of Point Barrow, Alaska. *Proc. U. S. Nat. Museum,* 109(3412):59–208, 27 plates.

Mackay, D. C. G. 1943. Temperature and the world distribution of crabs of the genus *Cancer. Ecology,* 24(1):113–115.

MacNeil, F. S. 1965. Evolution and distribution of the genus *Mya* and Tertiary migrations of Mollusca. *U. S. Geol. Surv. Prof. Paper 483-G,* pp. 1–51, 11 plates.

Madsen, F. J. 1949. Marine Bivalvia. *Zool. Iceland,* 4(63):1–116, 12 figs.

Madsen, H. 1936. Investigations of the shore fauna of east Greenland with a survey of the shores of other arctic regions. *Medd. om Grönland,* 100(8):1–79, 17 figs.

Makarov, V. V. 1938. Crustacea. Anomura. Fauna of U.S.S.R., 10(3), *Zool. Inst. Akad. Nauk SSSR, New Ser.* (16):1–283, 113 figs., 5 plates (English translation, Jerusalem 1962).

Mars, P. 1963. Les faunes et la stratigraphie du Quaternaire Mediterraneen. *Rec. Trav. St. Mar. End., Bull. 28, Fasc.,* 43:61–97, 6 figs.

Marty, Y. Y. 1962. Some similarities and differences under which boreal fish species exist in northeast and northwest Atlantic. *in Soviet fisheries investigations in the northwest Atlantic.* Rybnoe Khozyaistvo, Moscow (English trans. Jerusalem 1963), pp. 55–67, 10 figs.

McAllister, D. E. 1960. List of the marine fishes of Canada. *Bull. Nat. Museum Can., No. 168, Biol. Ser. No. 62,* iv + 1–76 pp.

——. 1962. Fishes of the 1960 "Salvelinus" program from western arctic Canada. *Bull. Nat. Museum Can.,* (185):17–39, 4 figs.

————. 1963. A revision of the smelt family, Osmeridae. *Bull. Nat. Museum Can.*, (191):iv + 53, 14 figs.

Mokievsky, O. 1956. Some characteristics of the littoral fauna of the shore of the Sea of Japan. *Tr. Prob. Thematic Confr., Zool. Inst. Acad. Sci.* U.S.S.R. (Not seen; ref. from Zenkevitch 1963).

————. 1960. The littoral fauna of the northwestern coast of the Japan Sea. *Tr. Prob. Thematic Confr., Zool. Inst. Acad. Sci.* U.S.S.R. (not seen; ref. from Zenkevitch 1963).

Mori, T. 1952. Check list of the fishes of Korea. *Mem. Hyogo Univ. Agri.*, 1(3), Biol. Ser. (1):1–228, 1 map.

Neave, F. 1958. The origin and speciation of *Oncorhynchus*. *Trans. Roy. Soc. Can., Ser. 3, Sect. 5*, 52:25–39.

Newell, I. M. 1948. Marine molluscan provinces of western North America: a critique and a new analysis. *Proc. Amer. Phil. Soc.*, 92(3):155–166, 7 figs.

Nomura, S., and K. Hatai. 1936. A note on the zoological provinces in the Japanese seas. *Bull. Biogeogr. Soc. Japan*, 6(21):207–214, 1 plate.

Orcutt, H. G. 1950. The life history of the starry flounder *Platichthys stellatus* (Pallas). *Fish. Bull., Calif. Div. Fish Game*, (78):1–64, 52 figs.

Osburn, R. C. 1950–1953. Bryozoa of the Pacific Coast of America. *Allan Hancock Pacific Expeditions*, 14(1–3):1–841, 82 plates.

————. 1955. The circumpolar distribution of Arctic-Alaskan Bryozoa. *in Essays in the natural sciences in honor of Captain Allan Hancock.* University of Southern California Press, Los Angeles, pp. 29–38.

Parr, A. E. 1933. A geographic-ecological analysis of the seasonal changes in temperature conditions in shallow water along the Atlantic coast of the United States. *Bull. Bingham Oceanog. Lab.*, 4(3):1–90, 28 figs.

Pasternak, F. 1957. Quantitative distribution and faunistic composition of benthos in the Sakhalin Gulf and adjacent parts of the Sea of Okhotsk. Tr. I. O. A. N., 23 (not seen; ref. from Zenkevitch 1963).

Pettibone, M. H. 1954. Marine Polychaete worms from Point Barrow, Alaska, with additional records from the North Atlantic and North Pacific. *Proc. U. S. Nat. Museum*, 103(3324):203–356, 39 figs.

————. 1956. Marine Polychaete worms from Labrador. *Proc. U. S. Nat. Museum*, 105(3361):531–584.

Phillips, J. B. 1957. A review of the rockfishes of California (family Scorpaenidae). *Calif. Fish Game, Fish Bull.*, 104:1–158, 56 figs.

Pickard, G. L. 1964. *Descriptive physical oceanography.* Pergamon Press, Oxford, viii + 199 pp., 31 figs.

Poll, M. 1947. *Faune de Belgique: Poissons Marins.* Musée Royal d'Histoire Naturelle de Belgique, Bruxelles, 452 pp., 267 figs.

Pulley, T. E. 1952. A zoogeographic study based on the bivalves of the Gulf of Mexico. Ph.D. thesis, Harvard University. 1–215 pp., 8 figs., 19 plates.

Rass, T. S. 1965. Fishery resources of the European Seas of the USSR. *Inst. Oceanology Akad. Nauk. SSSR*, pp. 1–105, 18 figs. (in Russian).

Ryland, J. S. 1963. Systematic and biological studies on Polyzoa (Bryozoa) from western Norway. *Sarsia*, (14):1–59, 14 figs.

Saemundsson, B. 1949. *The zoology of Iceland*, vol. 4, Part 72, Marine Pisces. Ejnar Munksgaard, Copenhagen and Reykjavik, 150 pp.

Sato, S., and K. Kobayashi. 1956. The bottom fishes of Volcano Bay, Hokkaido I. A taxonomical study. *Bull. Hokkaido Reg. Fish. Lab.*, (13):1–19, 12 figs.

Scagel, R. F. 1963. Distribution of attached marine algae in relation to oceanographic conditions in the northeast Pacific. *in* Marine distributions. *Roy. Soc. Can., Special Publ.*, (5):37–50, 11 figs.

Schenk, H. G., and A. M. Keen. 1936. Marine molluscan provinces of Western North America. *Proc. Amer. Phil. Soc.*, 76:921–938, 6 figs.

Schilder, F. A. 1956. *Lehrbuch der Allgemeinen Zoogeographie.* Gustav Fischer, Jena, viii + 150 pp., 134 figs.

Schorygin, A. 1928. Die echinodermen des Barents Meeres. Ber. M. N. I., 2, 1 (not seen: ref. from Zenkevitch 1963).

Schwarzbach, M. 1963. *Climates of the past.* Van Nostrand, London, xii + 328 pp., 134 figs.

Segerstråle, S. G. 1957. Baltic Sea. *in* J. W. Hedgpeth (editor), *Treatise on marine ecology and paleoecology.* vol. 1. *Ecology. Geol. Soc. Amer., Mem.* (67):751–800, 22 figs. 4 plates.

Shchapova, T. 1948. Geographical distribution of the order Laminariales in the northern part of the Pacific Ocean. Tr. I.O.A.N., 2 (not seen; ref. from Zenkevitch 1963).

———. 1957. Littoral flora of the continental coast of the Japan Sea. Tr. I.O.A.N., 23 (not seen; ref. from Zenkevitch 1963).

Shmidt, P. Y. 1950. *Fishes of the Okhotsk Sea.* Academy of Science, U.S.S.R., Moscow, pp. 1–370, 51 figs., 20 plates (in Russian).

Soot-Ryen, T. 1939. Some pelecypods from Franz Josef Land, Victoriaoya and Hopen. *Norges Svalbard-og Ishavs-Undersokelser, Medd.*, (43):1–21, 1 plate.

Squires, H. J. 1964. *Pagurus pubescens* and a proposed new name for a closely related species in the northwest Atlantic (Decapoda: Anomura) *J. Fisheries Res, Board Can.*, 21(2):355–365, 6 figs.

———. 1966. Distribution of Decapod Crustacea in the northwest Atlantic. *Ser. Atlas Marine Environ.*, (12):1–4, 4 figs., 4 plates.

Stephensen, K. 1937. Crustacean groups. *Zool. Faroes*, 2(part 1)(22):1–24, (23):1–40, (24):1–23, (26):1–10, (27):1–9, (29):1–8, (30):1–18.

Stephenson, T. A., and A. Stephenson. 1954. Life between tide-marks in North America. III B. Nova Scotia and Prince Edward Island: the geographical features of the region. *J. Ecol.*, 42(1):46–70.

———, and ———. 1961. Life between tide marks in North America. IV B. Vancouver Island, II. *J. Ecol.*, 49(2):227–243, 2 figs., 2 plates.

Taylor, W. R. 1962. *Marine algae of the northeastern coast of North America.* rev. edition. University of Michigan Press, Ann Arbor, viii + 1–509 pp.

Thorson, G. 1941. Marine Gastropoda Prosobranchiata. *Zool. Iceland*, 4(60):1–150, 15 figs.

———. 1957. Bottom communities (sublittoral or shallowshelf). *in* J. W. Hedgpeth (editor), *Treatise on marine ecology and paleoecology.* vol. 1. *Ecology. Geol. Soc. Amer., Mem.* (67):461–534, 20 figs.

Tokioka, T. 1963. Contributions to the Japanese ascidian fauna. XX. The outline of Japanese ascidian fauna as compared with that of the Pacific coasts of North America. *Publ. Seto Marine Biol. Lab.*, 11(1):131–156.

Udvardy, M. D. F. 1963. Zoogeographical study of the Pacific Alcidae. *in* Gressitt (editor), *Pacific basin biogeography*, a symposium. 10th Pacific Science Congress. Bishop Museum Press, Honolulu, pp. 85–111, 20 figs.

Ueno, T. 1971. List of the marine fishes from the waters of Hokkaido and its adjacent regions. *Sci. Dept. Hokkaido Fish. Exper. Sta.*, (13):61–102, 2 figs.

Ursin, E. 1960. A quantitative investigation of the echinoderm fauna of the central North Sea. *Medd. Danmarks Fish. Hav.*, 2(24):1–204, 96 figs.

Ushakov, P. V. 1952. The Chukchi Sea and its bottom fauna. Extreme Northeast of the U.S.S.R., 2 (not seen; ref from Zenkevitch 1963).

———. 1953. The fauna of the Okhotsk Sea and its life conditions. Ed. Z.I.N., Acad. Sci. U.S.S.R. (not seen; ref. from Zenkevitch 1963)

———. 1955. Polychaeta of the far eastern seas of the U.S.S.R. *Zool. Inst. Akadamie Nauk. SSSR* (English translation, Jerusalem 1965) 419 pp., 164 figs.

———. 1971. Amphipacific distribution of polychaetes. *J. Fisheries Res. Board Can.*, 28(10):1403–1406.

Valentine, J. W. 1966. Numerical analysis of marine molluscan ranges on the extratropical northeastern Pacific shelf. *Limnol. Oceanog.*, 11(2):198–211, 7 figs.

———. 1967. The influence of climatic fluctuations on species diversity within the Tethyan provincial system. *Syst. Ass. Publ.*, (7):153–166, 3 figs.

Van Name, W. G. 1945. The North and South American ascidians. *Bull. Amer. Museum Nat. Hist.*, 84:1–476, 31 plates.

Vinogradov, L. 1948. On the zoogeographical zonation of the far eastern seas. *Bull. T. I. N. R. O.*, 28 (not seen; ref. from Zenkevitch 1963).

Walters, V. 1955. Fishes of Western Arctic America and Eastern Arctic Siberia. *Bull. Amer. Museum Nat. Hist.*, art. 5, 106:255–368.

Watanabe, M. 1960. *Fauna Japonica. Cottidae (Pisces)*. Biogeographical Society, Tokyo, Japan, vii + 1–218 pp., 75 figs., 40 plates.

Wilimovsky, N. J. 1958. Provisional keys to the fishes of Alaska. *Fish. Res. Lab., U.S. Fish Wildlife Serv.*, Juneau, pp. 1–113, illus.

Woodward, S. P. 1851–1856. *A manual of the mollusca*. John Weale, London, viii + 1–486 pp., 24 plates + 24 pp., 1 map.

Wu, H. W. 1932. *Contribution à l'Étude morphologique, biologique et systematique des poissons hétérosomes de la Chine*. Jouve et Cie, Paris, pp. 1–179, 25 figs.

Yabe, H. 1929. The latest land connection of the Japanese Islands to the Asiatic continent. *Proc. Imp. Acad. Tokyo*, 5, 4 (not seen; ref. from Zenkevitch 1963).

Yoshida, H., M. Amio, and S. Nojima. 1957. Shell fish fauna in the Japan Sea off Yoshimi in Yamaguchi Prefecture. *J. Shimonoseki Coll. Fisheries*, 7(1):117–120.

Zenkevitch, L. 1963. *Biology of the seas of the U.S.S.R.* George Allen and Unwin, London, pp. 1–955, 427 figs.

Life in the
Open Sea

Chapter
10

The
Pelagic
Realm

*And Nature, the old nurse, took
the child upon her knee
Saying: "here is a story-book
Thy father has written for thee."*

*"Come, wander with me," she said,
"Into regions yet untrod;
And read what is still unread
in the manuscripts of God."*

Henry Wadsworth Longfellow in The Fiftieth Birthday of Agassiz, *1857*

While it has been possible to survey the fauna of the continental shelves and to separate, on the basis of the amount of endemism, the shallow-water world into zoogeographic regions and provinces, the same kind of approach is not generally practical for the open sea and the deep sea. In both the Pelagic and Deep Benthic Realms, one must rely, rather heavily, on systematic works involving relatively few of the animal groups that are present.

For the two upper layers of the open ocean, I have delineated regions and provinces on the basis of distribution patterns shown by various species in some groups that have been recently investigated, i.e., chaetognaths, euphausiids, squids, pelagic polychaetes, copepods, pteropods, and several fish families, but these do not provide enough information to give meaningful estimates of endemism. In regard to the deeper pelagic waters, the sparse data that are available permit only very general observations and conclusions.

Vertical Divisions of the Open Ocean

There is yet to be a general agreement on the natural, vertical divisions of the open ocean. Most recent workers have followed the schemes outlined by Hedgpeth or Bruun. Hedgpeth (1957:18) depicted an epipelagic zone for about the upper 200 meters, followed by a mesopelagic zone extending to about 1,000 meters, a bathypelagic to about 4,000 meters, and an abyssopelagic occupying the deepest areas. Bruun (1957:742) considered the epipelagic zone to extend downward as far as sunlight penetrates (thought at that time to reach only about 100 to 200 meters). Below this, Bruun defined the mesopelagic zone as that occupying a relatively thin layer extending only 100 to 700 meters beneath the surface, depending on oceanic conditions, the lower limit being the 10°C isotherm. The bathypelagic layer was judged to occur between the 10°C and the 4°C isotherms, and the abyssopelagic below 4°C. It seems, however, that these divisions were premature, since sufficient evidence was not available for the setting of boundaries with this degree of precision.

Epipelagic Zone

Most biologists agree that about the upper 200 meters is, on the basis of its characteristic biota, quite separate from the deeper waters. Ekman (1953:353) has related that the first attempt at a vertical division based on the actual distribution of the animals was probably that of Fr. Dahl in 1894 who divided the copepods of the Atlantic into three groups, one in the upper 200 meters, the second between 200 and 1,000 meters, and

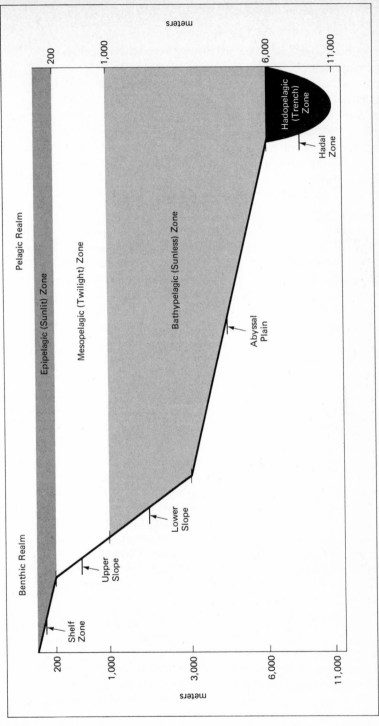

Figure 10-1 Diagrammatic representation of the vertical divisions of the world ocean based on the distribution of the fauna.

the third below 1,000 meters. Glover (1961:207) remarked that the change in the abundance and species composition of the plankton at the 100-fathom (200-meter) contour is so marked that it is frequently possible to detect its position by a visual inspection of the recorder (continuous plankton recorder) silks. In regard to chaetognath distribution, Alvariño (1965:151) observed that, in general, the number of species and the number of individuals per unit volume of water was greatest in the upper 200 meters. Ferguson Wood (1965:111) has shown that, in tropical waters, the phytoplankton maximum usually occurs between 40 and 100 meters.

Not long ago, it was thought that 200 meters was about the limit of light penetration in clear, oceanic waters. But work with the deep-sea photometer (Clarke 1966:80) has detected daylight at depths as great as 800 meters, and the eyes of some pelagic animals are probably even more sensitive. Also, healthy, chlorophyll-bearing plants, once thought to be confined to the epipelagic layer, have been taken as deep as 5,000 meters (Ferguson Wood 1966:177). M. E. Vinogradov (1966:382) concluded that, although surface species of plankton dominated the upper 100 to 200 meters in the tropics, they extended to 200 to 500 meters in the subpolar regions. A further complicating factor is the large number of species that demonstrate diurnal, seasonal, and ontogenetic migrations.

Diurnal (diel) movement has been generally regarded as the normal behavior pattern in pelagic animals. The adaptive value of this habit was investigated by McLaren (1963), who maintained that an animal which feeds in the warm surface waters but rests in the cooler waters below gains an energy bonus. The deep scattering layer is apparently made up of active species that tend to concentrate between 300 and 800 meters in the daytime and come close to the surface at night. Here, myctophid fishes and siphonophores are important, since their gas bladders are of a size to be resonant for the sound impulses generated by the echo sounder (Barham 1966). However Banse (1964:94), in his review of vertical distribution in the zooplankton, concluded that usually less than half the species showed this kind of behavior so that most of the grazers remained in the area of phytoplankton concentration. He also found that diurnal migration is weakly pronounced, if present at all, among the pelagic larvae of bottom invertebrates.

It seems, therefore, that despite a large migratory traffic that moves in and out of the epipelagic layer, the majority of species present are restricted to that environment. It may be concluded that the epipelagic zone is certainly a distinct habitat, although its lower boundary is liable to vary from 100 to 500 meters.

Deep Pelagic
Zones

It is important to note that, in the pelagic habitat, the greatest number of zooplankton species does not always occur in the uppermost zone. In the northwestern Pacific, the 1,000 known species at 2,000 to 3,000 meters is about three times that at the surface (Zenkevitch and Birstein 1956). This phenomenon apparently also occurs in the tropics, at least in some taxa (Banse 1964:91). It is possible that the unusually constant environment of the deeper levels has been conducive to successful evolution (p. 440) among certain zooplankton groups.

It was observed by Banse (1964:90) that the meso- and bathypelagic zones did not seem to be clearly distinguishable. But M. E. Vinogradov (1966, 1968), in discussing vertical changes in the zooplankton, recognized a "mid-depth type" down to 3,000 to 4,000 meters in the subpolar seas and to 1,000 to 2,000 meters in the tropics and an "abyssal type" for the deepest pelagic waters. In this scheme, the mid-depth organisms were considered to be mainly migratory interzonal species and their carnivorous predators. In contrast, the abyssal zone was described as being poor in plankton and not directly affected by the migrations of the interzonal species.

Ekman (1953:356), following the early work of Murray and Hjort (1912), recognized two deep zones based primarily on fish distribution. These were an "upper bathypelagic zone" from 150 to 500 meters and a "lower bathypelagic zone" below 500 meters. However, more recent work has shown that the populations of most deep-pelagic fishes are concentrated between 200 and 1,000 meters. Marshall (1965:300) referred to this area as the twilight zone and noted that it was characterized by many species of lanternfishes (Myctophidae), stomiatoids (eight families), melamphaids (Melamphaidae), alepisaurids (Alepisauridae), etc. Certain other groups of fishes seem to have centers of concentration between about 1,000 and 4,000 meters. Examples are some of the bristlemouths (Gonostomatidae), the gulper eels (Eurypharyngidae), deep-sea eels (three families), and deep-sea anglers (10 families). Furthermore, the species above 1,000 meters commonly have well-developed swimbladders and demonstrate extensive diurnal migrations. Those living below 1,000 meters lack well-developed swimbladders and do not migrate toward the surface at night (Marshall 1960:83, 1965:311).

For the chaetognaths, Alvariño (1964:64) recognized a mesoplanktonic stratum from 200 to 1,000 meters and a bathyplanktonic level below 1,000 meters. Work on the calanoid copepods of the northeast Atlantic has shown sharp changes in species at the 200- and 1,000-meter levels (Grice and Hulsemann 1965). The latter data tend to reinforce the earlier conclusions of Brodsky (1957) who worked on the same group in the North Pacific. Thus, both invertebrate groups show a very good agreement with the fish distribution pattern. It would seem, then, that the concept of a mesopelagic, twilight zone lying beneath the epipelagic layer is probably accurate. Here, there are apparently many herbivorous species and their predators that undergo diurnal migrations. Furthermore, it may be possible to subdivide the mesopelagic fauna into several ecological groups (Ebeling et al., 1970). Beneath the mesopelagic zone, beginning at 1,000 meters (at least for fishes, copepods, and chaetognaths), there is a sparsely populated, bathypelagic, sunless zone where the species undergo little interzonal movement.

Bruun (1957:643) considered that an abyssopelagic zone existed below the bathypelagic and that the boundary between them was the 4°C isotherm (occurring at about 2,000 meters in the Atlantic but shallower in the Pacific and Indian Oceans). Although this zone has been recognized by other workers, the evidence for its existence in terms of a characteristic, endemic fauna is lacking. At present, there seems to be a better case for considering the bathypelagic zone to occupy the entire area from 1,000 to about 6,000 meters (overlying the lower continental slopes and the entire abyssal plain).

It is interesting to see that the three pelagic zones so far discussed fit very well the three major zones of temperature structure. Physical oceanographers refer to an upper mixed zone down to 200 meters with temperatures similar to those at the surface, next a thermocline zone extending to 1,000 meters in which the temperature decreases with depth rapidly and below this a deep zone in which the temperature decreases with depth slowly (Pickard 1964:36). For the huge bathypelagic zone, occupying close to 75 percent of the total volume of ocean water, the major physical characteristics are remarkably constant: the temperature varies from about 1.5 to 5°C and the salinity from about 34 °/oo to 35 °/oo.

Below 6,000 meters, the pelagic fauna of the trenches has been called the "ultra-abyssopelagic" or "hadopelagic." This fauna is still poorly known, and so far, only the Russian vessel *Vitiaz* has operated closing nets below 6,000 meters. One haul in the Kurile-Kamchatka Trench between 6,000 and 8,500 meters yielded 20 species of copepods, 4 of ostracods, and 5 of amphipods (Wolff 1960:104). A total of 21 species of amphipods have been recorded from all the trenches (Dahl 1959, Barnard 1962). Apparently, most of the species taken so far are confined to the trenches and the percentage of endemism seems to be very high, since few are known to be common to more than one trench. All of the animals are reported to be completely unpigmented in contrast to their bathypelagic relatives where a dark red color predominates (Zenkevitch and Birstein 1956). Of the two names so far utilized, the term hadopelagic seems to be the most appropriate.

Summary   It is apparent that the *epipelagic* zone is a distinct habitat. Although there is a good deal of migratory movement in and out of this layer, most of the zooplankton and the great majority of the phytoplankton species seem to be confined to it. Its lower boundary is most likely to occur at about 200 meters, but it can vary from 100 to 500 meters. This tends to be a mixed layer where the temperature is similar to that of the surface.

The *mesopelagic* or twilight zone extends down to about 1,000 meters. Here, there are many zooplankters and a variety of species that prey on them. They live in a thermocline area where the temperature decreases rapidly with depth. The *bathypelagic* zone extends to 6,000 meters and overlies the entire abyssal plain. The fauna is relatively poor, and the decrease of temperature with depth is very slow. On the basis of our present knowledge about the distribution of the deep-pelagic species, an additional abyssopelagic zone cannot be recognized. There seems to be a separate *hadopelagic* or trench fauna existing below 6,000 meters.

Epipelagic and Mesopelagic Zones   It is apparent from the high degree of correlation between the distributional patterns of the epipelagic and mesopelagic animals that, as far as horizontal dispersal is concerned, the two groups can be treated as one. This raises the question of why species that live in the 200- to 1,000-meter layer should show such a close correspondence to surface patterns. As one can see by examination of the temperature charts, the deeper layer is not only considerably cooler, but the average latitudinal temperature changes are much less.

We know that the mesopelagic zone is populated by many species that undergo diurnal migrations. In the lower latitudes, such species are probably quiescent in the cold, mesopelagic waters but become very active upon migrating to the warm, upper layer. Compared to most marine animals, these species are unique in their ability to withstand rapid and extensive temperature changes. For this reason, they are usually considered to be highly eurythermic. However, it seems likely that, in terms of the temperature necessary for them to feed actively and efficiently, their requirements may be just as narrow as those of many forms that are confined to the surface.

It is entirely possible that a mesopelagic predator, such as the lanternfish, *Lepidophanes gaussi,* that is found only in tropical areas, must migrate into an upper layer that is at a temperature of 20°C or more. Otherwise, it may not be able to function well enough to compete successfully. In reference to the lanternfishes and other mesopelagic groups, Bolin (1957:373) concluded that most of them tended to be distributed in relation to the water temperatures of the upper 200 meters. Although the following account pertains to both of the upper pelagic zones, it must be borne in mind that faunas of the two zones are quite distinct.

The Flotsam Environment

Of special interest in the epipelagic zone is the wide variety of creatures that are attracted to floating objects. Flotsam may consist of larger, living organisms or debris such as parts of trees, coconuts, pumice, etc. The latter objects are usually picked up by currents in coastal waters and carried off to sea.

Living Flotsam

Two kinds of living flotsam, jellyfishes and seaweed, are more or less passively carried by currents and provide, in turn, a home for many smaller organisms. Hosts are usually the larger species of scyphomedusae or siphonophores and the floating seaweeds, particularly the various *Sargassum* species. There is now available a large body of literature about the relationships between small fishes and jelly fishes.

The fish-jellyfish symbiosis is usually a temporary association in which the coelenterates are passive hosts and the fishes active opportunists (Mansueti 1963). In most cases, the relationship is apparently commensalistic wherein the fishes receive the benefit of shelter and the jellyfishes are unharmed. However, observations have shown that some fishes, as they grow larger, will begin to feed on their host. This has been documented for the young harvest fish, *Peprilus alepidotus,* and butterfish, *Poronotus triacanthus,* with the scyphomedusae *Chrysaora quinquicirrha* (Mansueti, op. cit.). Also, some symbiotic medusae are known to kill and devour associated fishes.

Almost all the fishes that consort with jellyfishes do so only during their younger stages of development. The only exceptions to this rule seem to be the man-of-war fishes (family Nomeidae) that apparently have developed a lasting, obligatory commensal relationship. Young of the following pelagic fish families have been captured with medusae: Carangidae, Stromateidae, Centrolophidae, Nomeidae, and Tetragonuridae. Many of the species involved have very broad distributions, and it is likely that their dispersal is aided by their habit of residing with jellyfishes.

In contrast to the other kinds of flotsam, the *Sargassum* weed is apt to carry such a rich fauna that it is often considered to be an independent community or ecosystem. Animals commonly associated with this plant in the Western Atlantic include many coelenterates, flatworms, annelids, arthropods (pycnogonids, barnacles, amphipods, decapods, copepods, isopods, and shrimps), mollusks, bryozoans, and fishes (Adams 1960). Another study in the Western Atlantic has shown a *Sargassum* association with 54 species of fishes belonging to 23 families, a number about equal to the associated invertebrate fauna (Dooley 1972). A rich associated fish fauna has also been observed in the Indo-West Pacific by Besednov (1960). While the relationship of most of the fishes is temporary or casual, there are two species that occur only in this environment. These are the sargassum pipefish, *Syngnathus pelagicus,* and the sargassum fish, *Histrio histrio* (Fig. 10-2), which is circumtropical in distribution, indicating that some *Sargassum* weed may occasionally be carried around the Cape of Good Hope.

Debris   Most kinds of nonliving flotsam, such as tree trunks and branches, apparently do not remain in the epipelagic zone long enough to accumulate a wide variety of invertebrates. However, fishes are attracted to such objects very rapidly. Experiments in the tropical Eastern Pacific have shown that a moored, floating object will attract a full complement of fishes after about 5 days (Hunter and Mitchell 1967). This affinity for floating objects is shown by several fish species of commercial importance, and fishermen in some areas have traditionally

Figure 10-2 The sargassum fish (*Histrio histrio*). An obligate member of the sargassum community, this species has a worldwide (circumtropical) distribution.

used moored rafts. The occurrence of juvenile fishes beneath debris is much more frequent than that of adults.

From a zoogeographic standpoint, it may be noted that certain fishes, that do not appear to be well adapted to the pelagic habitat, are apparently able to migrate across the open ocean barriers by accompanying floating debris. Examples are the sergeant major, *Abudefduf saxatilis,* the rough triggerfish, *Canthidermis maculatus,* and the scrawled filefish, *Alutera scripta.* All three have circumtropical distributions.

The Neuston In recent years, it has become apparent that the thin layer of water immediately below the surface film provides a habitat for a special group of small organisms that are called the neuston or the pleuston (David 1965). Sometimes the larger jellyfishes are included (Savilov 1966). Species found in the neuston community represent a wide variety of animal groups (coelenterates, ctenophores, chaetognaths, copepods, amphipods, mysids, stomatopods, decapods, mollusks, and fishes), many of which have been found to produce a blue pigment for protective coloration (Herring 1967).

The North Atlantic ocean
Arctic-Temperate Boundary In the western North Atlantic, the cold East Greenland and Labrador Currents transport water from the Arctic Ocean to maintain arctic conditions far to the south (to the Strait of Belle Isle at about 52°N). In the eastern North Atlantic, the relatively warm water carried by the North Atlantic and Norwegian Currents has a moderating influence that extends all the way to the western Barents Sea (at about 72°N). The boundary area between these warm and cold current systems is an important zoogeographic barrier. This arctic-temperate boundary extends across the North Atlantic in the form of a great sigmoid curve (see inside the covers).

The slope and position of the arctic-temperate boundary is nicely outlined by the distributional patterns of certain pelagic invertebrates. Dunbar (1964) plotted the ranges of two arctic amphipods (*Parathemisto libellula* and *Gammarus wilkitzkii*) that extend southward to but not across the boundary. Alvariño (1965) showed that two widely distributed chaetognaths (*Sagitta hexaptera* and *S. lyra*) extend from the tropics northward to the boundary. This pattern is also shown by the squid *Onychoteuthis banksi* (Clarke 1966:142). Both epipelagic and mesopelagic fishes of general temperate or eurythermic tropical distributions range northward to the boundary. For the mesopelagic fishes, the Greenland-Iceland rise, Iceland-Faroe rise, and the Wyville Thompson ridge form an effective barrier preventing penetration into the Norwegian Sea.

In 1953, Ekman (p. 341) pointed out the remarkable fact that the North Atlantic did not possess an endemic, boreal epipelagic fauna, and Kramp (1959) emphasized that not a single species of the hydromedusae is characteristic of the boreal waters. These observations are upheld by more recent investigations.

The Boreal Waters

The extensive strip of ocean between the arctic and warm-temperate waters is almost entirely occupied by animals with three basic patterns of distribution: arctic-boreal, eurythermic temperate, and broad eurythermic tropical. The arctic-boreal pattern is nicely demonstrated by the chaetognath *Sagitta elegans,* a species that extends from the Arctic Basin south to the western entrance to the English Channel on one side and to Cape Hatteras on the other (Alvariño 1965:166). Other invertebrate examples are the pteropods *Limacina helicina* and *Clione limacina* (Ekman 1953), the euphausiids *Thysanoessa inermis* and *T. raschii* (Dunbar 1964), and several species of pelagic shrimps (Squires 1966). The herring *Clupea harengus* is arctic-boreal, although it is found in neritic as well as pelagic waters (Marty 1962:59). Mesopelagic fishes with this pattern are the paralepid *Notolepis rissoi kroyeri* (Rofen, 1966:287) and the lanternfishes *Hierops arctica* and possibly *Benthosema glaciale* (Bolin 1959).

The eurythermic temperate pattern, involving both cold-temperate and warm-temperate waters, is demonstrated by such forms as the chaetognath *Sagitta tasmanica* (Alvariño 1965:162), the squid *Histioteuthis bonellii* (Clarke 1966:195), and the euphausiid *Nyctiphanes couchii* (Einarsson 1945). Epipelagic fishes are the mackerel shark *Lamna nasus,* the basking shark *Cetorhinus maximus* (Bigelow and Schroeder 1953), and the Atlantic mackerel *Scomber scombrus* (Matsui 1967). Examples among the mesopelagic fishes are the lanternfishes *Hygophum hygomi, H. benoiti, Symbolophorus veranyi, Myctophum punctatum,* and *Lampanyctus intricarius* (Bolin 1959); the stomiatid *Stomias boa ferox* (Morrow 1964:300); and the pomfret *Brama brama* (Mead and Haedrich 1965).

The surface of the boreal North Atlantic is invaded by many broad eurythermic tropical species that range entirely through both equatorial and temperate regions. The chaetognaths *Sagitta lyra* and *S. hexaptera* are good examples (Alvariño 1965). So are two of the large, epipelagic fishes, the bluefin tuna *Thunnus thynnus* and the swordfish *Xiphias gladius*. At the mesopelagic level, there is the chaetognath *Sagitta zetesios* (Alvariño, op. cit.) the melamphaid fish *Scopelogadus beanii* (Ebeling and Weed 1963:39), and the lanternfishes *Diaphus mollis* and *Lobianchia gemellari* (Nafpaktitis 1968). The northward penetration of this group seems to be subject to considerable seasonal variation.

The
Warm-Temperate
Waters

This rich, pelagic area is populated principally by a very large group of eurythermic tropical species, a smaller group of species with general temperate distributions (discussed above), and some forms that are apparently confined to it. In the western Atlantic, warm-temperate waters occupy a rather narrow zone between about 28 and 35°N. But, on the eastern side they extend from about 15 to 50°N. and include the Mediterranean and Black Seas.

A great many of the eurythermic tropical species manage to range well into warm-temperate waters but not beyond. North Atlantic species with such patterns are the chaetognaths *Sagitta bipunctata*, *S. minima*, and *S. enflata* (Alvariño 1965) and the squids *Onychia carribaea*, *Cranchia scabra*, and *Liocranchia reinhardti* (Clarke 1966). Among the many epipelagic fishes are the sharks *Prionace glauca* and *Carcharhinus longimanus*, the albacore *Thunnus alalunga*, the skipjack *Katsuwonus pelamis*, and the dolphin *Coryphaena hippurus*. Mesopelagic fishes are the viperfish *Chauliodus danae* (Haffner 1952:112); the omosudid *Omosudis lowei* (Rofen 1966:479); and the paralepids *Macroparalepis breve*, *Lestrolepis intermedia*, *Paralepis atlantica*, *P. elongata*, *Notolepis rissoi rissoi*, *Lestidiops affinis*, and *L. jayakari* (Rofen op. cit.). It is interesting to see that a large proportion of the species in this category have worldwide distributions.

A number of fishes are apparently restricted to the warm-temperate belt. Epipelagic species are the squaretail *Tetragonurus cuvieri* (Grey 1955:33) and the flying fish *Prognichthys rondeleti* (Bruun 1935:96). Mesopelagic forms are the lanternfishes *Aethopora metopoclampa*, *Lampadena speculigera*, and *Lampanyctus pusillus* (Bolin 1959); the stomiatid *Stomias brevibarbatus* (Morrow 1964:303); the paralepids *Sudis hyalina* and *Macroparalepis affine* (Rofen 1966, Haedrich and

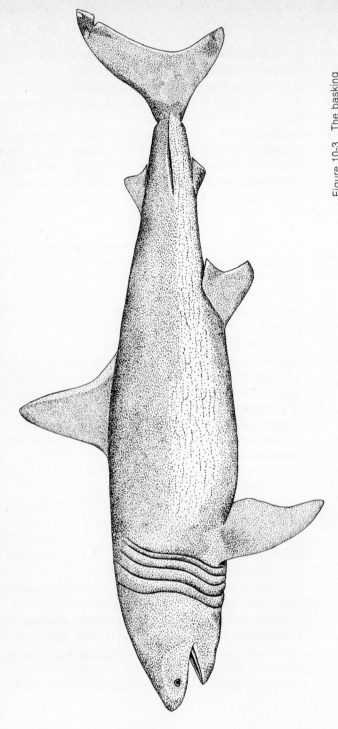

Figure 10-3 The basking shark (*Cetorhinus maximus*). An epipelagic species with a worldwide, antitropical distribution in temperate waters. After Jordan and Evermann (1900).

Nielsen 1966:913); and the evermannellid *Evermannella balbo* (Rofen, op. cit.). Some of these species seem to be endemic, but others have broader, antitropical distributions.

The
Mediterranean

Although somewhat reduced in number of species, the epipelagic fauna of the Mediterranean is very similar to that of the warm-temperate North Atlantic. Representatives of the three basic distributional groups (eurythermic tropical, eurythermic temperate, and warm-temperate) exist in about the same proportion as in the outer ocean. In almost all species, the Mediterranean population appears to be identical to that of the Atlantic. This close relationship is probably maintained by migratory movement through the Straits of Gibraltar. Here, there is a constant, eastward flowing, surface current that is capable of carrying pelagic organisms into the Mediterranean.

Communication at the mesopelagic level is more difficult. The sill in the Straits of Gibraltar has a maximum depth of about 320 meters, but below 200 meters, there is an undercurrent that flows out of the Mediterranean. Also, the mesopelagic layer of the Mediterranean, averaging about 14°C and 39 $^o/oo$, presents a notable contrast to the cooler and less saline water of the eastern Atlantic. As a result, many of the common Atlantic mesopelagic organisms are missing from the Mediterranean. The fishes of this habitat are now quite well known (Tortonese 1963) but only two species, the myctophid *Notoscopelus elongatus* (Bolin 1959) and the paralepid *Paralepis speciosa* (Rofen 1966:220), appear to be endemic to the Mediterranean. The hydromedusa *Solmissus albescens* (Kramp 1959) is also apparently endemic.

Since the Mediterranean contains, primarily, an impoverished continuation of the outer ocean fauna, it belongs in the same pelagic zoogeographic region. Ebeling (1962:137) considered the Mediterranean to comprise a primary zoogeographic region for the deep-pelagic (mesopelagic) fauna, but evidence to support this arrangement is wanting. For the epipelagic fauna, the Black Sea is, to some extent, a depauperate continuation of the Mediterranean. The mesopelagic layer of the Black Sea is uninhabitable for aerobic organisms, since there is no dissolved oxygen below about 180 meters (Pickard 1964:141).

The restricted, nondistinctive character of the Mediterranean epipelagic and mesopelagic faunas (compared to the eastern Atlantic) offer an interesting contrast to the relationship of the shelf fauna. It was found

(p. 203) that the Mediterranean shelf fauna was considerably richer than that of the outer coast and that many endemics were present.

In summary, it can be said that the boreal waters of the North Atlantic are comparatively depauperate in numbers of species and are occupied entirely by organisms that are shared with adjoining zoogeographic regions. The warm-temperate belt is a good deal richer in species due mainly to the presence of large numbers of eurythermic tropical forms. Also, a few of the warm-temperate species are apparently endemic. The distributional patterns of the epi- and mesopelagic organisms are very similar, the only notable difference being in the northeast where there are underwater ridges that prevent the dispersal of most mesopelagic species north of about 65°N.

Summary

In view of the lack of endemism in the boreal portion and the relatively small number of species that appear to be North Atlantic warm-temperate endemics, it seems best to recognize but a single, large temperate region for this part of the upper pelagic world. It extends eastward to include the Mediterranean and the Black Sea.

The Pacific Ocean north of about 42°N has been referred to by oceanographers as the Subarctic Pacific Region. Its most unique physical-chemical feature is undoubtedly the presence of a low salinity layer of less than 34 °/oo at the surface with a permanent halocline existing between 100 and 200 meters. The region is considered to be limited to the south by the Subarctic Boundary which represents the dividing line between the West Wind Drift and the North Pacific Current (Dodimead, Favorite, and Hirano 1963). It is important to note that this southern boundary is subject to seasonal shifting so that it lies south of 40°N in the early spring and at about 42°N in the fall (Bogarov 1958:152). To the north, the region probably extends to about 62°N, the limit of the deep water of the Bering Sea.

The North Pacific Ocean
North Boreal Region

The relatively sharp changes in temperature and salinity that occur at the southern boundary of the region have profound distributional effects. The result is a highly distinct biotic region that has no equivalent in any other part of the world. The pelagic organisms have apparently been influenced by the same environmental conditions that stimulated the evolution of a separate north boreal shore fauna.

Because of the distributional relationship to the shelf fauna of the same area, it seems appropriate to refer to the region as the North Boreal rather than the Subarctic. Many of the Russian workers have called it simply the Boreal Region or Boreal Complex. However, Parin (1968), in discussing the distribution of epipelagic fishes, recognized both North Boreal and South Boreal Zones.

Some of the most distinct and conspicuous elements of the North Boreal pelagic fauna are the fishes. Characteristic and numerous in the surface layer are the Pacific salmon (Larkins 1964). In fact, this may be called the domain of the Pacific salmon with the most abundant species in this group being the chum salmon *Oncorhynchus keta*, the pink salmon *O. gorbuscha,* and the sockeye salmon *O. nerka.* Other common fishes apparently confined to these waters are the Atka mackerel *Pleurogrammus monopterygius* and the walleyed pollock *Theragra chalcogrammus.* Brodsky (1957) depicted a South Bering Sea Province for the oceanic, calanoid copepod fauna of this area, Beklemishev and Semina (1956) described a distinct, boreal phytoplankton assemblage, and Bradshaw (1959) recognized a separate group of planktonic foraminiferans. Fager and McGowan (1963) found that a group of five zooplankton species was consistently present in the area. Parin (1961a) referred to the mesopelagic fish fauna as distinctive with a total of about 40 species including a number of endemic species and genera.

The southern boundary of the North Boreal Region at about 42°N appears to be significant for the mesopelagic fishes. Ebeling (1962:140) presented a latitudinal distribution chart for the eastern Pacific that showed many range terminations at about 42 to 43°N. Pearcy (1964:94) found that, off the Oregon coast between 43°20′N and 46°14′N, the number of species decreased and that the northern limits for some of them occurred within the area. Bussing (1965) referred to the presence of a boundary in the northeastern Pacific at about 42 to 46°N. Also, Becker (1966) found the main latitudinal change in the lanternfishes (family Myctophidae) took place at about 40 to 43°N.

In addition to species that are confined to it, the North Boreal Region contains three other large groups with distributional patterns that seem to be distinct. These are an arctic-boreal group, a general boreal group (inhabiting both North and South Boreal Regions), and a eurythermic temperate group (occupying both the boreal regions and the warm-temperate waters). An arctic-boreal distribution is demonstrated by the pteropod *Limacina helicina* (McGowan 1963), the chaetognath *Sagitta elegans* (Alvariño 1965:166), and the euphausiids *Thysanoessa longipes, T. raschii,* and *T. inermis* (Brinton 1962). Fishes showing this pattern are the herring *Clupea harengus* and the dolly varden *Salvelinus malma.*

The general boreal group is very diverse and contains at least one squid *Moroteuthis robusta* (Clarke 1966:146), a polychaete *Tomopteris pacifica* (Tebble 1962:436), a euphausiid *Tessarabrachion oculatus* (Brinton 1962:150), and five epipelagic fishes: the Chinook salmon *Oncorhynchus tshawytscha,* the coho salmon *O. kisutch,* the skilfish *Erilepis zonifer,* the sablefish *Anoplopoma fimbria,* and the steelhead trout *Salmo gairdneri* (Larkins 1964). Parin (1961a) has called attention to three mesopelagic lanternfishes with this pattern *(Lampanyctus leucopsaurus, L. nannochir,* and *L. regalis).* Ebeling (1962:140) charted the latitudinal range of several other mesopelagic fishes that appear to have general boreal distributions *(Gonostoma gracile, Chauliodus macouni, Diaphus rafinesquei, Lampanyctus jordani).*

The more broadly distributed, eurythermic temperate group includes the euphausiid *Euphausia pacifica* (Brinton 1962:109), the salmon shark *Lamna ditropis* (Strasburg 1958), the basking shark *Cetorhinus maximus,* and the jack mackerel *Trachurus symmetricus.* Mesopelagic fishes with this distribution are apparently the lanternfishes *Diaphus theta, Electrona arctica,* and *Tarletonbeania crenularis;* also the big-scale *Melamphaes lugubris* (Ebeling 1962:140, Paxton 1967:429).

Among the epipelagic fishes, it is apparent that most of the species are members of nonpelagic families (such as the Salmonidae, Hexagrammidae, and Anoplopomatidae), but they have succeeded in dominating the pelagic environment in the North Boreal Region. This seems to indicate that the southern regional barrier is very effective against the primary pelagic groups and that the high productivity of the area has attracted species from families that are usually confined to the shelf. The low salinity of the surface layer may be an important factor in the success of the salmonid species. In addition to the three major distributional groups just outlined, there are a few species that are highly eurythermic, being found all the way from the Bering Sea to the tropics.

It is unfortunate that the South Boreal Region has been generally referred to as a zone of mixing or a transition region. Such names imply that it is simply an area of change without a separate fauna of its own. In fact, Bogorov (1958:150) stated that no distinct plankton fauna occurs between the boreal and tropical communities. However, both the chaetognath *Sagitta scrippsae* (Alvariño 1965:170) and the euphausiid

South Boreal Region

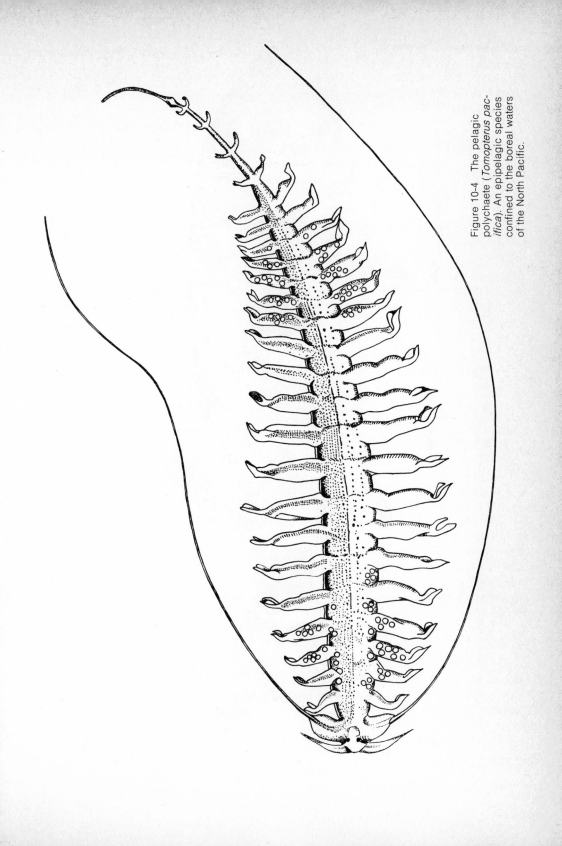

Figure 10-4 The pelagic polychaete (*Tomopterus pacifica*). An epipelagic species confined to the boreal waters of the North Pacific.

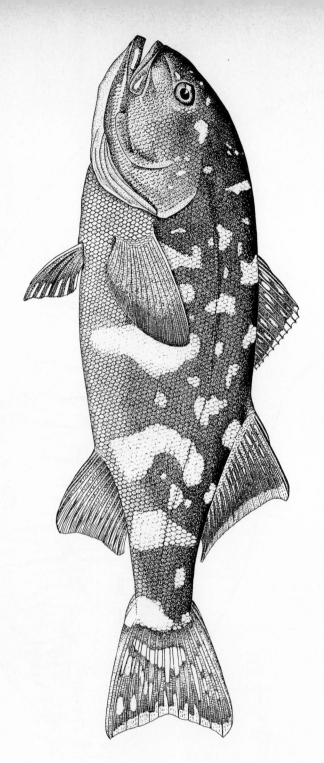

Figure 10-5 The skilfish (*Erilepis zonifer*). An epipelagic species with a general boreal distribution in the North Pacific.

Nematoscelis difficilis (Brinton 1962:153) are good examples of south boreal species. Each extends across the ocean between about 35 and 43°N in waters that are colder than the warm-temperate range. Both are found farther south in the cool California Current, but this should be expected. Almost identical patterns are shown by the mesopelagic lanternfishes *Myctophum californiense* and *M. asperum* (Parin 1961a).

Besides the general boreal and eurythermic temperate groups already discussed (p. 316), the South Boreal Region is populated by an assemblage that is shared only with the adjoining warm-temperate belt. Examples are the euphausiids *Thysanoessa gregaria, Euphausia recurva,* and *E. hemigibba* (Brinton 1962) and the squid *Todarodes pacificus* (Clarke 1966:128). Among the fishes are epipelagic species such as the squaretail *Tetragonurus cuvieri* (Grey 1955), the saury *Cololabis saira* (Parin 1960a), and the pomfret *Brama japonica* (Mead and Haedrich 1965). There is also the mesopelagic bigscale *Melamphaes parvus* (Ebeling 1962:122).

Finally, the South Boreal Region is occupied by a large number of species that have broad, eurythermic tropical distributions (extending all the way from the tropics to the northern boundary of the region at about 42°N). Included are the chaetognaths *Krohnitta subtilis, Sagitta bipunctata,* and *S. minima* (Alvariño 1965); the euphausiids *Euphausia gibboides, E. mutica,* and *Nematoscelis microps* (Brinton 1962); the copepod *Clausocalanus farrani* (Frost and Fleminger 1968); and the pelagic polychaetes *Naiades cantrainaii, Krohnia lepidota,* and *Lopadorhynchus uncinatus* (Tebble 1962). Conspicuous epipelagic fishes are the great blue shark *Prionace glauca* (Strasburg 1958), the albacore *Thunnus alalunga* (Otsu and Uchida 1963), and the bluefin tuna *T. thynnus* (Gibbs and Collette 1967). All three of these fish species are noted for their marked seasonal migrations. This also seems to be a common pattern for mesopelagic fishes; the distributional chart by Ebeling (1962:140) indicates that, in the eastern Pacific, about five species have this distribution.

Warm-Temperate Region

The North Pacific Warm-Temperate Region lies between the latitudes of about 23 to 24°N and 34 to 35°N. The southern boundary is identical to the 20°C isotherm for the coldest month (see inside covers). Distribution within these limits is demonstrated by certain epipelagic fishes. The flying fish *Prognichthys rondeleti* has a worldwide, antitropical distribution in warm-temperate waters. In the North Pacific, it is found off the coasts of both California and Japan (Parin 1960b:62). Two other flying

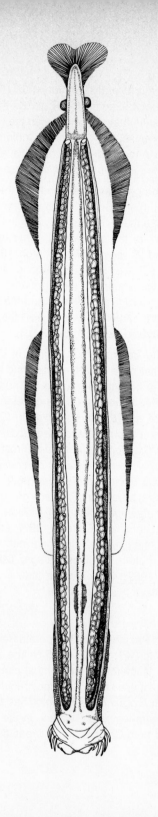

Figure 10-6 A chaetognath (*Sagitta scrippsae*). An epipelagic species apparently confined to the South Boreal Region of the North Pacific.

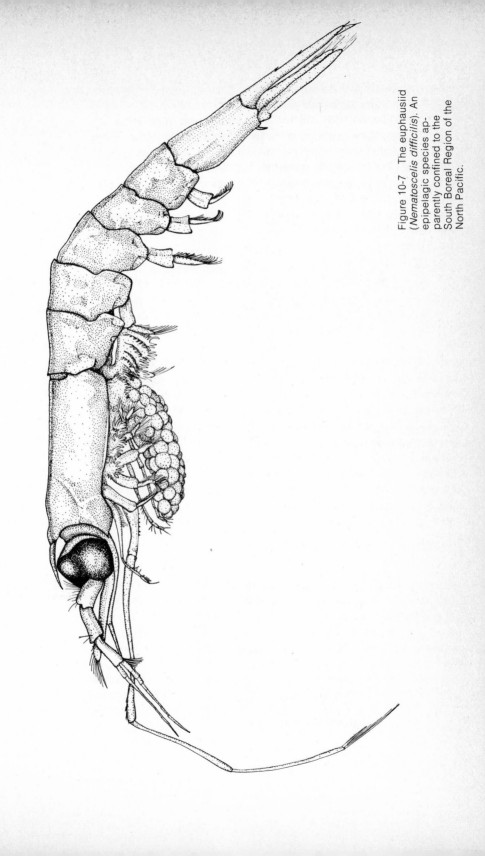

Figure 10-7 The euphausiid (*Nematoscelis difficilis*). An epipelagic species apparently confined to the South Boreal Region of the North Pacific.

fishes, *Cypselurus pinnatibarbatus* and *C. hetururus*, are also broadly distributed in warm-temperate waters and occur on both sides of the North Pacific (Parin 1959, 1961b). In addition, three other flying fish species appear to be endemic to the western side of the ocean (Parin 1961a). Several mesopelagic fishes demonstrate a warm-temperate distribution. These are the lanternfishes *Myctophum affine* (Parin 1961a:262), *Centrobranchus brevirostris* (Becker 1964:58), and *Diogenichthys atlanticus* (Bussing 1965:225); the bathylagid *Bathylagus wesethi;* and the scaly dragonfish *Stomias atriventer* (Ebeling 1962:140).

Besides the warm-temperate endemics, the region is inhabited by the eurythermic temperate group and a group that is shared only with the South Boreal Region (both discussed above). However, the greater number of species probably belongs to the eurythermic tropical category (represented by forms occupying both tropical and warm-temperate waters). Examples are the chaetognaths *Sagitta ferox* and *Pterosagitta draco* (Alvariño 1965); the euphausiids *Euphausia eximia, E. brevis, Thysanopoda monacantha, T. pectinata, T. obtusifrons,* and *T. aequalis* (Brinton 1962); the squids *Liocranchia reinhardti, Cranchia scabra, Octopoteuthis sicula,* and *Onychia carribaea* (Clarke 1966); and several planktonic foraminferans (Bradshaw 1959). Among the many pelagic fishes are the striped marlin *Makaira audax,* the skipjack tuna *Katsuwonus pelamis,* the yellowfin tuna *Thunnus albacares,* and the black skipjack *T. lineatus.* Apparently, about a dozen of the common mesopelagic fishes of the eastern Pacific can be placed in this category (Ebeling 1962:140); also, it is now evident (Rofen, 1966) that at least three paralepid species (*Lestidiops pacificum, Omosudis lowei,* and *Paralepis atlantica)* should be added.

The Tropical Seas

When one examines the longitudinal distribution of the species confined to the tropics, it becomes evident that the major breaks are caused by the New World and Old World Land Barriers. While the former is a complete block to the migration of such species, it is apparent that the Cape of Good Hope region of South Africa is occasionally bypassed. Certainly, many of the circumtropical species manage to maintain their integrity by this means. However, the worldwide patterns of many groups indicate that often the Old World Barrier is very effective. For example, there are only six tropical oceanic chaetognaths. Of these, four *(Sagitta robusta, S. neglecta, S. regularis,*

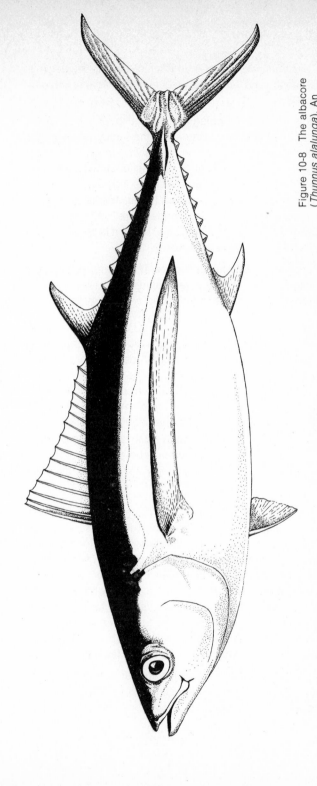

Figure 10-8 The albacore (*Thunnus alalunga*). An epipelagic species with a worldwide (broad eurythermic tropical) distribution extending through tropical, warm-temperate, and cold-temperate waters. After Walford (1937).

and *S. pulchra*) are confined to the Indo-Pacific, one *(S. hispida)* is confined to the Atlantic, and only one *(Krohnitta pacifica)* is circum-tropical (Alvariño 1965).

Worldwide patterns for the epipelagic fishes also emphasize the importance of the Old World Land Barrier. In the family Carangidae, there are approximately 50 species in the entire tropical Atlantic (including those with eurythermic tropical distributions). Of these, only about 10, or some 20 percent, are shared with the Indo-Pacific. About half of the Atlantic flying fishes (Exocoetidae) (Parin, 1960b) and about one-third of the mackerels (family Scombridae) are confined to that ocean. For the mesopelagic fishes, on the other hand, it is probable that the East Pacific Barrier is the more effective (p. 326).

Considering the tropical pelagic fauna in general, it seems appropriate to establish two major regions that are separated by the New and Old World Land Barriers. These are the Indo-Pacific and the Atlantic.

Indo-Pacific Region

The tropical marine region of the Indo-Pacific includes an area that is bounded to the north and south by the Tropical Convergence (marked by the 20°C isotherm for the coldest month). To the east, the region is constricted due to the cool California and Peru Currents that flow toward the equator. In the western Indian Ocean, tropical waters span the distance from the northern Persian Gulf at about 30°N to the Kei River mouth in South Africa at 32°S. Longitudinally, this vast region extends all the way from Colombia to eastern Africa, about two-thirds of the distance around the world! The region may be divided into two provinces.

Eastern Pacific Province

For some of the major groups of pelagic invertebrates such as the chaetognaths, euphausiids, and squids, the eastern Pacific tropics do not seem to harbor a distinct fauna. Also, the larger kinds of epipelagic fishes tend to range entirely across the Pacific. Examples of the latter are the blue marlin *Makaira nigricans* and black marlin *M. indica* (Howard and Ueyanagi 1965), the sailfish *Istiophorus platypterus* (Morrow and Harbo 1969), the whitetip shark *Pterolamiops longimanus* and the silky shark *Carcharhinus falciformis* (Strasburg 1958), and several species in the Carangidae and Scombridae families.

In contrast, many of the smaller fish species appear to be confined to the eastern tropical Pacific. Of the six species of strictly tropical flying fishes, three *(Cypselurus calliopterus, C. xenopterus,* and *Prognichthys tringa)* are apparently endemic (Parin 1960b). About 15 species of carangid fishes belonging to common genera *(Caranx, Seriola, Trachinotus, Oligoplites,* etc.) are usually considered to be endemic.

For the mesopelagic fishes, the faunal independence is very marked since the eastern Pacific group has a very high incidence of endemism. In fact, Ebeling (1962:137) indicated that there was no species overlap between the eastern and western Pacific. Becker (1966), in discussing the distribution of lanternfishes in the Pacific, observed that there is, in the tropical waters of the eastern part, a very discrete fauna of about 10 endemic species. However, there are indications from captures made elsewhere that a few of the strictly tropical, eastern Pacific species have broader ranges. For example, the bigscale *Melamphaes janae* has been taken in the Indian Ocean (Ebeling 1962:123), and the luminescent shark *Isistius brasiliensis* is now known to be worldwide (Parin 1964:171).

In the Eastern Pacific Province, more than in any other part of the world, there seems to be a notable discrepancy between the latitudinal distribution of the epipelagic and mesopelagic animals. Here, relatively cool (warm-temperate) water is carried by the Peru Current as far north as the Gulf of Guayaquil. This has the effect of restricting the range of the purely tropical, epipelagic species from about 3°S to 24°N. However, work on the mesopelagic fishes has shown that many of the tropical species extend southward for a considerable distance. Bussing (1965:224) noted that the Peru-Chile species apparently restricted to waters north of 20°S are tropical forms.

It may be concluded that the tropical waters of the eastern Pacific harbor a distinct, pelagic fauna (at least as far as the fishes are concerned). This Eastern Pacific Province is separated from the Indo-West Pacific by the East Pacific Barrier. At one time, it was thought that this open ocean barrier was operative only for the littoral fauna. But, as the result of a study of the general distribution patterns of the wide-ranging pelagic species, it was found that the barrier was highly effective for many surface and deep-pelagic fishes (Briggs 1960:172).

It seems incongruous that a stretch of open ocean, even a very broad one, should comprise a barrier for active, pelagic animals. Yet, this is clearly the case. We can only surmise that a great many pelagic species are evidently dependent on a certain degree of proximity to land masses or to relatively shallow-water areas. The attraction of areas of high organic productivity that are often associated with coastal upwelling undoubtedly has important distributional consequences. In this province, the degree of endemism for zooplankton species seems to be low, for the smaller epipelagic fishes it appears to be rather high (perhaps 50 percent or more), and for the mesopelagic fishes it is apparently very high (possibly 90 percent).

This broad and bountiful province reaches all the way from the vicinity of Easter and Sala y Gómez Islands at about 105°W to eastern Africa at about 35°E. The question is, does an essentially homogeneous fauna cover the entire area or should it be segmented into a series of smaller provinces?

Indo-West
Pacific
Province

In considering the question of the homogeneity of the Indo-West Pacific Province, it seems appropriate to investigate the distribution of those fish groups that were found to be affected by the East Pacific Barrier. As for many families, the center of abundance for the jacks (family Carangidae) seems to be in the Philippine–East Indies area. A total of 52 species has been recorded for the Philippine Islands (Herre 1953). The work of Williams (1958) on the Carangidae of eastern Africa demonstrated the presence of 30 species. Of the 30, at least 28 range all the way to the western Pacific and 2 are possibly endemic; 25 species have been reported from the vicinity of the Hawaiian Islands (Gosline and Brock 1960), 11 from the Marshalls and Marianas (Woods 1953), and 8 from the Tuamotu Islands (Harry 1953 estimated that this number represented about half the species occurring there). The great majority of species occurring about these Polynesian island groups are also very widespread, most of them extending all the way to the eastern coast of Africa.

Although their systematics and distribution are not as well known, the flying fishes (Exocoetidae) seem to have the same general pattern (Parin 1960b). That is, the greatest number of purely tropical species are found in the Philippine–East Indies area, but those occurring in the more distant localities are usually also found in the central area. The tropical squaretail *Tetragonurus pacificus* is confined to the Indo-West Pacific but ranges from Polynesia to East Africa (Grey 1955:32). For the mesopelagic fishes, it may be noted that the bigscales (Melamphaidae), lanternfishes (Myctophidae), and hatchetfishes (Sternoptychidae) are represented by more species in the tropical Indo-West Pacific than in any other area (Ebeling 1962, Bolin personal communication, Baird 1971). Furthermore, most of the species present seem to be relatively widespread within the area, the only obvious barriers being those of the province itself.

The above data, particularly that on the family Carangidae, indicate that (1) the richest pelagic fauna seems to occur in the Philippine–East Indies area, (2) the species found in the more distant localities are almost all wide-ranging, and (3) in such distant places there is little or no endemism. It seems clear, therefore, that we are dealing with a very extensive but single zoogeographic province.

The Red Sea The Red Sea is separated from the Gulf of Aden by a shallow sill with a maximum depth of about 100 meters. The movement of shallow water between these two basins does not appear to be very restricted, but the deep waters below the sill are effectively separated. The main body of deep water in the Red Sea is known to be very uniform at a temperature of 21.7°C and a salinity of 40.6 ‰ (Pickard 1964:177). This means that a unique mesopelagic environment is present that is much warmer and more saline than other waters.

Considering the contrast in habitat, it is perhaps not surprising that the mesopelagic fauna of the Red Sea should demonstrate some interesting peculiarities. Since the mid-waters of this area have yet to be well explored, these differences are difficult to evaluate. However, Marshall (1963:187) has written that three out of four of the commonest mesopelagic fishes show some morphological differentiation. If further investigations bear out these preliminary findings, it is likely that the Red Sea will need to be considered as a separate province—at least in regard to its mesopelagic fauna.

Atlantic Region As in the Pacific, the marine tropics of the Atlantic are bounded to the north and south by the Tropical Convergences. In the western North Atlantic, the Gulf Stream has a northward displacement effect. The boundary extends from about 28°N off the Florida coast, then swings northward to about 35°N above Bermuda. The presence of a thermal front with profound effects on the latitudinal distribution of mesopelagic fishes off the Florida coast at about 27 or 28°N has been noted by Backus et al. (1969). The line then runs to the southeast to intersect the western coast of Africa at about 15°N. The southward displacement on the eastern side is due to the influence of the relatively cool Canary Current as it flows toward the equator. In the latter area, the boundary for mesopelagic fishes seems to be also located at 15°N (Backus et al., 1965).

In general, the Atlantic Ocean does not possess a rich, strictly tropical, pelagic fauna. This is apparent in both invertebrates and fishes. For example, there are only two such chaetognaths (*Krohnitta pacifica* and *Sagitta hispida*)—one of them an endemic and the other circumtropical (Alvariño 1965). Only two euphausiids (*Thysanopoda tricuspidata* and *Euphausia americana*) are confined to the tropics in the Atlantic—one is an endemic and the other occurs also in the indo-West Pacific (Mauchline and Fisher 1969). Although many species of epipelagic fishes may be found, about four-fifths of them have broad eurythermic distributions. A greater proportion of the mesopelagic fishes seems to be tropical. About half the species of lanternfishes (Myctophidae) found so far in the tropical Atlantic appear to be confined to the tropics (Bolin 1959, Nafpaktitis 1968).

Of the epipelagic fish families in the tropical Atlantic, the jacks (Carangidae) show the greatest degree of difference between the two sides of the ocean. Approximately 30 percent of the species appear to be endemic to each area (it is likely that this apparent difference will be reduced when more systematic work is done). In contrast, there is very little local endemism among the 16 species of tropical Atlantic flying fishes; on the western side there are probably 3 endemics (1 species and 2 subspecies) and on the eastern side only 1 (a subspecies) (Bruun 1935, Parin 1960b, Staiger 1965). Of the 16 species of mackerels (Scombridae) in the western Atlantic, apparently all but 2 or 3 have transatlantic distributions; 5 of the 6 western Atlantic species of needlefishes (Belonidae) extend to the eastern side.

A number of mesopelagic fishes belonging to several different families (Myctophidae, Paralepididae, Melamphaidae, Evermannellidae, Sternoptychidae) are apparently strictly tropical. Some of them appear to be Atlantic endemics: the lanternfishes *Lepidophanes gaussi*, *Lampanyctus cuprarius*, and *Lampanyctus photonotus* (Bolin 1959) and *Diaphus subtilis*, *D. vanhoeffeni*, *D. lucidus*, and *D. splendidus* (Nafpaktitis 1968); the paralepid *Stemonosudis intermedia* (Rofen 1966:430); the hatchetfishes *Polyipnus polli* and *P. laternatus* (Baird, 1971); the bigscales *Melamphaes leprus* and *M. eulepis* (Ebeling 1962); the stomiatoid *Stomias lampropeltis* (Gibbs 1969:15); and the viperfish *Chauliodus schmidti* (Morrow 1964:279). The paralepid and one hatchetfish have so far been taken only on the western side of the Atlantic while the other hatchetfish, one bigscale, the stomiatoid, and the viperfish are presently known only from the eastern side. Five species of lanternfishes belonging to the genus *Diaphus* are apparently confined to the western side (Nafpaktitis 1968).

Backus et al. (1970) studied the relative abundance of the common mesopelagic fishes that are found in the western and equatorial North Atlantic. As a result they were able to designate 10 different geographic regions. Work of this kind, although very important, is primarily of ecological rather than zoogeographic significance.

With the single exception of the fish family Carangidae, the epipelagic animal groups so far investigated appear to be broadly distributed, with the great majority of the species ranging to both sides of the Atlantic. For this reason, the region is not divided into eastern and western provinces. When the patterns of the mesopelagic fishes become better known, it may be necessary to reconsider the question.

The Southern Ocean
Warm-Temperate Region

South of the tropics, the pelagic fauna of the upper layers comes under the influence of the circumpolar West Wind Drift and becomes very widely dispersed. This is true for the warm-temperate belt despite the interruptions caused by the southern tips of South America and Australia. The northern boundary of the warm-temperate (often called subtropical) waters is marked by the position of the 20°C surface isotherm for the coldest month of the year. This line is apparently identical to the Tropical Convergence, or Frontal Zone.

In the South Atlantic, the Tropical Convergence is found at about 20°S in midocean, but toward shore it is affected by the prevailing currents of the South Atlantic Gyre. On the eastern side, it is carried north by the Benguela Current to about 15°S. To the west, it extends to about 23°S as a result of the Brazil Current (see inside covers). In the Indian Ocean, the Tropical Convergence is located somewhat farther to the south. Off southeastern Africa, the Agulhas Current causes a displacement to about 33°S. From there, the line extends in a northward arc that intersects the western coast of Australia at about 26°S (see inside covers). In the South Pacific, the boundary runs very close to 30°S until it is affected by the colder waters of the Peru Current. The latter current causes a major northward shift to about 3°S.

Faunistically, the above boundary is quite well defined, as many tropical organisms appear unable to cross it. Almost all the examples of tropical species that were given for the three oceans above are limited by it. Also, there are two categories of organisms that are generally unable to cross the boundary in a northerly direction: the southern eurythermic temperate group and the warm-temperate group.

The southern boundary of the warm-temperate belt is formed by the

Temperate (Subtropical) Convergence,[1] a line that holds a relatively constant latitudinal position at about 40°S except possibly in the eastern South Pacific (Pickard 1964:114), where it has not been accurately determined. Since there are large numbers of eurythermic tropical organisms (having tropical plus warm-temperate ranges), one might expect to find this boundary well marked. This is indeed the case. For example, Tebble (1960) found that, in the South Atlantic alone, there are about 16 species of pelagic polychaetes that do not occur south of the Temperate Convergence. David (1962) emphasized the importance of this boundary for chaetognaths. Gibbs (1968) listed 17 species of mesopelagic, stomiatoid fishes that extend no farther south.

Of the species that are confined to the warm-temperate belt, either as endemics or with antitropical distributions, the majority seem to have broad, circumglobal ranges. The euphausiids are *Euphausia hemigibba*, *E. recurva*, *E. spinifera*, and *E. lucens* (Baker 1965). Epipelagic fishes are the snipefish *Notopogon fernandezianus* (Mohr 1937:60), the squaretail *Tetragonurus cuvieri* (Grey 1955:33), the stromateid *Palinurichthys antarcticus* (Penrith 1967), and the flying fish *Prognichthys rondeleti* (Parin 1960b). Another flying fish, *Cypselurus pinnatibarbatus*, although somewhat more neritic in its habits, seems to have essentially the same distribution (Parin 1959). The southern bluefin tuna *Thunnus maccoyii* appears to belong to this group (Gibbs and Collette 1967:116). Five mesopelagic lanternfishes may also be strictly warm-temperate; *Hygophum hanseni*, *Loweina interrupta*, *Lampadena chavesi*, *Lampanyctus intricarius*, and *L. australis* (Nafpaktitis and Nafpaktitis 1969). No less than 10 species of mesopelagic, stomiatoid fishes appear to be restricted to these waters (Gibbs 1968).

A conspicuous part of the fauna is the varied group of planktonic and nektonic species that have eurythermic temperate distributions (existing in both warm-temperate and cold-temperate waters) and are thus not restricted by the Temperate Convergence. As was the case with the restricted warm-temperate group, most of the species seem to be very wide-ranging. Examples are the pelagic squids *Desmoteuthis megalops* (Muus 1956) and *Histioteuthis bonellii* (Clarke 1966:195); the chaetognaths *Sagitta tasmanica* and *S. planktonis* (Alvariño 1965); the euphausiids *Euphausia similis* (Baker 1965) and *Thysanoessa gregaria* and *Nematoscelis megalops* (Brinton 1962); and the pteropods *Limacina retroversa* and *Clio antarctica* (Chen 1968). Epipelagic fishes are the trevalley *Caranx georgianus*, the barracouta *Thyrsites atun*, the scaled tuna *Gasterochisma melampus*, the ocean sunfish *Mola ramsayi*, and the basking shark *Cetorhinus maximus*. Mesopelagic fishes

[1] I have taken the liberty of utilizing the term "Temperate Convergence" instead of Subtropical Convergence since it refers to the boundary between the warm-temperate and cold-temperate faunas and therefore seems to be the more descriptive.

with this pattern seem to be *Melamphaes microps* (Ebeling 1962:122); *Protomyctophum tenisoni* (Andriashev 1962); *Sio nordenskjoldii* (Moss 1962); *Idiacanthus niger* (Novikova 1967); and *Protomyctophum subparallelum*, *Lampadena dea*, and *Lampanyctus pusillus* (Nafpaktitis and Nafpaktitis 1969).

Despite the presence of many eurythermic tropical species, eurythermic temperate species, and some species with even broader latitudinal distributions, the warm-temperate zone does have a distinct fauna of its own. This seems to be true for both epi- and mesopelagic layers. It is therefore possible to recognize a circumglobal Southern Warm-Temperate Region that lies between the Tropical and Temperate Convergences.

Cold-Temperate Region

A sharp temperature break may be found where the cold, northward flowing surface water from the Antarctic Continent sinks beneath the surface. This is called the Antarctic Convergence. It is found at about 50°S in the Atlantic and Indian Oceans and about 60°S in the Pacific (Pickard 1964:114). The cold-temperate waters of the Southern Hemisphere are found between the Temperate and Antarctic Convergences. The pelagic fauna of this region consists mainly of eurythermic temperate species (examples of which were given above), an Antarctic–cold-temperate group, and species that are limited to cold-temperate waters. There are also a few species that exhibit very broad, almost cosmopolitan distributions.

It is interesting to note that the research vessel *Anton Bruun*, in conducting a transect of the southern portion of the Peru Current off Chile at 34°S, captured a mixed fauna of mesopelagic fishes. The dominant elements proved to be circumglobal species found associated with, and south of, the Temperate Convergence (Craddock and Mead 1970). Only two species appeared to be endemic to the oceanic waters off Chile. This catch reflects a northward displacement of the cold-temperate fauna which is attributable to the local influence of the Peru Current. The displacement effect was found to be most pronounced to the east of 80°W longitude.

In considering the species that are confined to cold-temperate waters, the distribution of the euphausiid *Thysanopoda acutifrons* is very interesting, since it is one of the few forms that exists also in northern, cold-temperate waters (the North Pacific). There is probably some migration between the southern and northern populations by isothermic submergence, for this species has been found to occur as deep as 4,000 meters (Brinton 1962:91). Apparently endemic are *Calanus australis* and *C. tonsus* (Brodsky 1962) and *Euphausia longirostris* (Baker 1965). None of the epipelagic fishes appear to be endemic. But some of the mesopelagic species probably are. Examples are the lanternfishes *Electrona subaspera* and *Gymnoscopelus bolini* (Andriashev 1962) and *Protomyctophum parallelum* and *Lampadena notialis* (Nafpaktitis and Nafpaktitis 1969).

The Antarctic–cold-temperate group seems to be fairly numerous. There are the euphausiids *Euphausia vallentini* and *E. triacantha* (Baker 1965), the chaetognath *Sagitta gazellae* (Alvariño 1965:170), the salp *Salpa thompsoni* (Foxton 1961), and the pteropods *Limacina helicina* and *Clio sulcata* (Chen 1968). Among the mesopelagic lanternfishes are *Protomyctophum anderssoni* and *Gymnoscopelus braueri* (Andriashev 1962). Finally, the cold-temperate area is invaded by certain broadly distributed eurythermic species that extend all the way from the tropics to the Antarctic Convergence. Examples are the polychaete *Vanadis longissima* (Tebble 1960), and the squids *Brachioteuthis riisei*, *Onychoteuthis banksi*, and *Phasmatopsis cymoctypus* (Clarke 1966).

Antarctic Region

This distinctive and very cold region lies between the continental shelf and the Antarctic Convergence. Here, the surface temperature is between −1.9 in winter and +4°C in summer. Baker (1954) showed that all of the common species of the zooplankton have circumpolar distributions. This pattern was observed earlier for the phytoplankton by Hart (1942). Most of the species, at least among the zooplankton, are apparently endemic to the region.

Among the conspicuous endemics are the Antarctic krill *Euphausia superba* (Marr 1962) and the related *E. frigida* (Baker 1965); the hydromedusae *Calycopsis borchgrevincki* and *Pantachogon scotti* (Kramp 1959); the polychaetes *Rhychonerella bongraini*, *Vanadis antarctica*, and *Tomopteris carpenteri* (Tebble 1960); the copepods *Calanus propinquus* and *C. acutus* (Brodsky 1962); and the chaetognath *Sagitta marri* (Alvariño 1965). Andriashev (1962) noted that 13 to 14 species of lanternfishes belonging to six genera had been taken south of the Antarctic Convergence and that almost half could be considered Antarctic endemics. He also stated that other mesopelagic species that are permanent residents of the area belong to the families Bathylagidae, Gonostomidae, Paralepidae, Macrouridae, and Scopelarchidae. The most abundant species is apparently the lanternfish

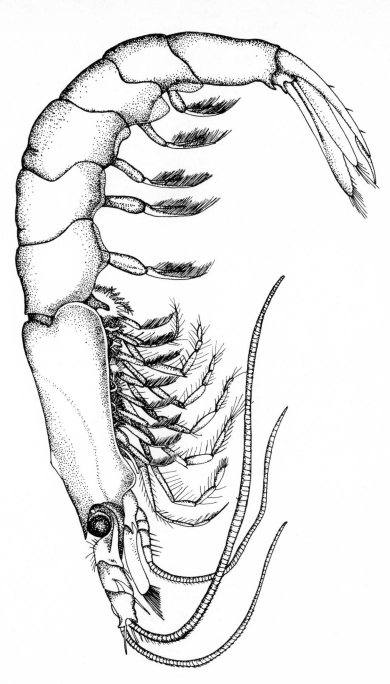

Figure 10-9 The euphausiid (*Thysanopoda acutifrons*). An epipelagic species inhabiting the cold-temperate waters of both hemispheres. A north-south migration probably takes place by means of isothermic submergence. After Marshall (1954).

Electrona antarctica, and Andriashev (op. cit.) considered it to be a valuable indicator species. Marshall (1964a), on the other hand, observed that only three mesopelagic fishes appear to be confined to Antarctic waters: a barracudina *Notolepis coatsi* and two lanternfishes.

In addition to the endemics and the Antarctic–cold-temperate group, there are a few species that have exceedingly broad distributions. The chaetognath *Eukrohnia hamata* has been shown to have a continuous range from the Antarctic to the Arctic (Alvariño 1965:fig. 2). At latitudes greater than 60°, it is found in the epipelagic zone, but at lower latitudes it is found in deeper waters, reaching 1,000 meters or more in equatorial regions. It has been cited as a classic example of bipolar distribution with tropical submergence (David 1958). The mesopelagic euphausiid *Stylochcheiron maximum* has a distribution that is almost as broad, from 63°S in the Pacific to the Gulf of Alaska (Brinton 1962:169). The squid *Galiteuthis armata* has been taken off the Antarctic Continent and northward to about 60°N in the Bering Sea (Clarke 1966:239).

Arctic Region

As is detailed in the final chapter (p. 410), it is possible that by the early Pliocene, sea surface temperatures in the Arctic Basin had dropped to about 4 to 5°C. This probably had the effect of forcing the boreal fauna southward, thus making room for the evolution of a new, colder-water marine fauna. The present pelagic fauna of the Arctic is therefore comparatively young and not very rich in numbers of species.

The greater part of the Arctic Ocean has an almost continuous cover of floating ice occupying in summer about 60 to 80 percent of the surface area. Beneath the ice there is a poor epipelagic biota, but there are local situations where plankton may become very abundant. High concentrations, with phytoplankton species predominating, have been found in the estuaries of tributary rivers and along the fringe of the pack ice (Zenkevitch 1963:50). It has also been noted that, while the greatest volume of zooplankton is concentrated near the surface (the upper 50 meters), there are many more species existing in the deeper waters (Jashnov 1940, quoted by Zenkevitch 1963:49).

An extensive study of the calanoid copepod fauna was completed by Brodsky (1957), who found that 20 of the 40 species inhabiting the epipelagic zone were arctic endemics. Grainger (1963), who studied the distribution of the three *Calanus* species occupying the Canadian Atlantic, concluded that two of them (*C. glacialis* and *C. hyperboreus*) should be considered true arctic species.

All of the nine species of medusae reported from the Chukchi and Beaufort Seas (Hand and Kan 1961) are also known from the North Pacific so they probably should be categorized as arctic-boreal. Only two chaetognaths are found in the Polar Basin, but both are widely distributed outside the Arctic (Alvariño 1965). Dunbar (1964) showed that two amphipods, *Parathemisto libellula* and *Gammarus wilkitzkii*, are confined to Arctic waters and that other common arctic species of amphipods and euphausiids had broader distributions, mainly arctic-boreal.

We noted that the epipelagic environment of the North Boreal Region of the Pacific Ocean was dominated by fish species belonging to families that are usually considered to be nonpelagic. The same situation seems to hold in the Arctic. In the open waters, the herring *Clupea harengus* and the arctic char *Salvelinus alpinus* apparently occur in large numbers. Nielson and Jensen (1967:13) presented evidence to show that the polar cod *Arctogadus glacialis* inhabits pelagic rather than benthic waters. Also, the arctic cod *Boreogadus saida* is said to be usually associated with ice floes (Walters 1955:302). Both cod species have also been found below 500 meters.

Earlier, it had been observed (Marshall 1954:341, 1963:181) that deep-pelagic fishes do not live in Arctic waters. However, we now have some evidence showing that they do. It has been noted (p. 310) that a paralepid and two lanternfishes have well-defined, arctic-boreal distributions. Apparently, one of the lanternfishes, *Benthosema glaciale*, has penetrated well into the Arctic Basin, since it has been reported from Point Barrow, Alaska (Bolin 1959:10). Although the North Atlantic Current flows directly into the Norwegian Sea, relatively shallow underwater ridges seem to be effective in preventing the penetration of most mesopelagic fishes.

It appears that the Arctic pelagic waters, including the Arctic Ocean itself and extending south to encompass northern Iceland, southern Greenland, and southern Labrador, comprise a rather poorly populated but distinct zoogeographic region. Some of the eurybathic species of the Arctic Region also extend over the continental shelf of the Chukchi Sea and southward into the Bering Sea.

Water Masses and Currents
The characteristic occurrence of certain plankton species in particular water masses and the possible use of such species as water mass indicators was first investigated by Bigelow (1926). In modern times,

there has been a formidable increase in the amount of research done on the systematics and distribution of planktonic animals. The underlying theme of much of this work has been the elaboration of Bigelow's concept. That is, most of the attempts have been concerned with the biological definition of the principal water masses.

The introduction of the T-S diagram (a temperature-salinity graph) by Helland-Hansen in 1916 has significantly influenced both descriptive and theoretical oceanographic research for the past 50 years (Neumann and Pierson 1966:478). Many biologists have used T-S diagrams (as water mass indicators) to explain the distribution of plankton. This research has been mainly confined to individual species, but some work has been accomplished on the distribution of zooplankton communities (Fager and McGowan 1963). Bary (1959, 1963a and b, 1964), although convinced that the distribution of zooplankters coincided with that of the water bodies, found that often the distributional patterns of the species were not correlated with temperature and salinity. He therefore proposed that each water mass possessed a unique, unknown property that had an important influence on the distribution of indigenous species.

Johnson and Brinton (1963) published a summary of the subject in their article on biological species, water masses, and currents. In this work they observed that it was to be expected that water masses would produce environments for distinct faunas, and this was indeed the case in many studies which have been made. The ultimate in this trend was achieved by Beklemishev (1966) who made it the basis for a grand, worldwide scheme. This can be briefly summarized: (1) the large oceanic gyres enclose primary water masses with primary pelagic communities, (2) currents sandwiched between the gyres transport mixed water with secondary communities comprised of a mixture of species from the primary communities, and (3) distinct neritic communities are located between the gyres and the coastline.

As the result of all this activity, much of which has been directed toward the goal of discovering indicator species for the various water masses, the literature about pelagic animals is strongly weighted toward the concept that each water mass has its own faunal identity. The question which must be asked is; Can it now be considered true that each major water mass is a distinct environment supporting its own characteristic community of organisms? It is possible to shed some additional light on the matter by examining the distributional patterns of the species in some of the better-known pelagic groups.

The chaetognaths can now be said to be a reasonably well known group, since they have been studied by several investigators in recent years. According to the detailed review by Alvariño (1965), there is a total of about 52 species in the 6 pelagic genera. Of these, about 35 seem to be open ocean rather than neritic forms. However, when the distribution of each species in the group of 35 is examined, we find that *only 3* seem to be confined to a single water mass. The majority are widely distributed in more than one ocean and occur in several different water masses.

Tebble (1962), in his work on the pelagic polychaetes of the North Pacific, discussed a total of 33 species and noted that *all but 3* also occurred in the Atlantic. Of a total of 48 euphausiid species reported from the Pacific, 35 have been recorded from the Atlantic (Ponomareva 1963).

In regard to fishes, the study of Haffner (1952) on the viperfish genus *Chauliodus* indicated that each population was apparently restricted to certain water masses. This report caused considerable speculation about the possible value of fishes as indicator species. However, subsequent investigations have not yielded the same result. In the flying fishes (family Exocoetidae), Parin (1959, 1960b, 1961b) found that the latitudinal distributions were very closely tied to the surface temperature regime while the longitudinal distributions were very broad, often extending entirely across the Atlantic, Indian, and Pacific Oceans. The same kinds of patterns are apparent for other epipelagic groups such as the sharks (Strasburg 1958), the tunas (Blackburn 1965), and the billfishes (Howard and Ueyanagi 1965).

Our knowledge about the distribution of epipelagic fishes in the Pacific has been summarized by Rass (1967) and Parin (1968) who recognized a series of ocean-wide regions that are definable on the basis of temperature. These are (1) Arctic, (2) northern-boreal, (3) southern-boreal, (4) northern-subtropical, (5) tropical, (6) southern-subtropical, (7) notal, and (8) Antarctic. Parin further noted that the above regions can be grouped into three major pelagic provinces and that their boundaries are suitable for planktonic as well as nektonic species.

For the deep-pelagic fishes, the work of Ebeling (1962) is most useful, for in addition to his research on the bigscales (family Melamphaidae) he analyzed the distribution of 13 families, 27 genera, and 135 species. In general, it was found that the distribution of species tended to correlate well with water mass boundaries in a north-south direction, especially where sharp temperature differences were encountered. On the other hand, correlation with the east-west water mass boundaries, where the temperature differences were small, was poor. In fact,

Ebeling recognized a Circumcentral Tropical Zoogeographic Region because so many species proved to have broad tropical distributions. Marshall (1963) observed that certain bathypelagic fishes appeared to be confined within a particular water mass, but others are not so limited.

In answer to the question we have posed, it is still impossible to say just how distinct any given water mass is in terms of its pelagic fauna. This would require a complete assessment of all the species involved together with accurate data on the geographic distribution of each. Judging from the fact that, so far, only a few nektonic or planktonic species are known to be restricted to a single water mass, the case for the distinct fauna concept seems to be rather weak. Even with the deep-pelagic fishes, which appear to show the best correlation, Ebeling (1962) concluded that there was merely a tendency to associate with water masses. Consequently, according to our present knowledge, the worldwide scheme of Beklemishev (1966) does not appear to fit the facts.

North Pacific versus North Atlantic

A problem of interest that has considerable historical significance is the great contrast between the temperate pelagic faunas of the North Pacific and the North Atlantic. The relatively rich fauna of the North Pacific may be divided into North Boreal, South Boreal, and Warm-Temperate Regions. Although overlapping groups of species are found, each region may be recognized by the presence of a significant number of endemic species. In contrast, the comparatively sparse fauna of the North Atlantic fits into one, large temperate region (Briggs 1970:22). There appears to be a total lack of endemism in the North Atlantic waters that have boreal temperature characteristics, and only a few species seem to be confined to the warm-temperate waters.

The relatively rigorous climatic history of the North Atlantic Ocean, compared to that of the North Pacific, has apparently produced a number of faunal peculiarities, among which is the absence of a boreal, pelagic fauna (p. 427).

General Patterns

In examining the general distributional pattern of the upper-layer pelagic invertebrates that have been investigated in modern times, one is struck by the fact that group after group is composed predominantly of eurythermic tropical species with very broad, longitudinal ranges.

Large numbers of such species inhabit a wide, circumglobal belt that extends roughly from about 40°N to about 40°S. This appears to be as true for species belonging to the larger, nektonic groups such as the squids (Clarke 1966) and hydromedusae (Kramp 1959) as it is for the smaller, planktonic chaetognaths (Alvariño 1965), polychaetes (Tebble 1960), copepods (Brodsky 1957), and foraminiferans (Bé 1966).

In addition to the major groups listed above, the worldwide eurythermic pattern seems to dominate in lesser groups, such as the pelagic gastropods (Laursen 1953), octopods (Thore 1949), amphipods (Fage 1960), and heteropods (Tesch 1949). Further, it may be noted that when strictly tropical species do occur, they are usually not circumglobal and are more apt to be confined to the Indo-Pacific rather than the Atlantic. The worldwide, rather homogeneous, warm-water invertebrate fauna presents an interesting contrast to a pelagic fish fauna that tends to have a more restricted longitudinal distribution. It may be that, in general, the fishes, due to their superior mobility, are better able to maintain their positions in certain areas of high productivity.

It is revealing that the pelagic faunas of the two upper layers (epipelagic and mesopelagic) demonstrate latitudinal divisions that are virtually identical to those found for the shore faunas. The only major exception to this general rule is in the North Atlantic where there is a distinct boreal shore fauna but the corresponding pelagic group is missing. This may indicate that, for some reason, the extensive climate changes of the Cenozoic had their most adverse effect on the pelagic species. One may conclude, however, that the marine animals of the upper layers, whether shelf or pelagic, respond to the temperature characteristics of their environment in a very similar way.

The Bathypelagic Zone

We know that the sea occupies about 71 percent of the earth's surface, but it is not generally realized that most of this water layer is relatively deep. In fact, about 88 percent of the world's oceanic area is over 1,000 meters in depth. This means that bathypelagic animals are spread over about two-thirds of the earth's surface. Considering that the average depth of the ocean is about 3,800 meters and that the abyssal plain extends down to about 6,000 meters, it can be seen that the bathypelagic zone occupies an enormous volume of liquid area. If we look at only that portion of our planet into which life has penetrated (the atmospheric, terrestrial, and aquatic habitats) and call it the biosphere, then the bathypelagic zone certainly comprises the majority of it.

How successfully has this major portion of our biosphere been occupied? We need to keep in mind that we are dealing with an area where no food is manufactured by photosynthesis, since it is beyond the reach of sunlight, where the temperature is very cold (1 to 5°C) and where the pressure is high (100 atmospheres or more). Furthermore, the animals must suspend themselves in the water rather than rest on a firm substrate. Also, in general, bathypelagic animals are situated several links down on a food chain that has its beginning in the sunlit, epipelagic zone at least 800 meters above. Considering these factors, the variety of life is impressive.

In contrast to the upper layers that are affected by surface mixing and horizontal, wind-driven currents, the bathypelagic zone is influenced mainly by the relatively slow, thermohaline circulation. Cold, saline water with a high oxygen content originates at the surface primarily in two areas, southeast of Greenland in the North Atlantic and the Weddell Sea in the Antarctic. When such dense water sinks, it becomes distributed to all of the contingent bathypelagic and hadopelagic areas by means of slow, deep currents (Stommel 1958).

The dense water that sinks from the surface serves to replace water that upwells from the depths so that a massive, deep circulation takes place. This ensures a slow but continuous renewal of oxygen, making the greatest depths available to the higher aerobic organisms. A distinction is often made between the deep water and bottom water; also, Antarctic intermediate water and Mediterranean water appear to intrude below 1,000 meters in some places (Pickard 1964:135). So far, however, there is no evidence that these water masses possess endemic faunas.

Marshall (1954:90) has related that when a net is towed horizontally at bathypelagic depths, the crustaceans are usually the most numerous in the catch with the copepods generally far outnumbering the other species. The results from six hauls made off Bermuda between the depths of 730 and 1,650 meters showed that 76 percent of the species were copepods, 15.5 percent other crustaceans, 3.1 percent chaetognaths, 1.9 percent siphonophores, 1.7 percent fishes, 1.2 percent radiolarian protozoans, and the remaining 0.6 percent comprising annelid worms, mollusks, tunicates, echinoderms, jellyfishes, and ctenophores. Such information indicates that we are evaluating an area that could be entitled the "domain of the Crustacea," for this class has been by far the most successful in penetrating this vast but difficult environment. Since about three-fourths of all bathypelagic species are copepods, it is clearly this group that offers the greatest potential for zoogeographic studies.

The work of Grice and Hulsemann (1967:8) in the western Indian Ocean gave some interesting information about the abundance and vertical distribution of the copepods. They recorded 153 bathypelagic species and noted that 122 of them occurred between 1,000 and 2,000 meters, 73 between 2,000 and 3,000 meters, and only 13 between 3,000 and 4,000 meters. Seven species were continually distributed from 1,000 to 4,000 meters. They also found that the mean size of the animals was larger and the total zooplankton volume was greater in the 1,000- to 2,000-meter layer than in those below. The values found were similar to those recorded by the same authors for comparable depth intervals in the northeastern Atlantic.

Adaptations   Species that have succeeded in penetrating the bathypelagic environment have adapted to their surroundings by developing an array of unusual structural modifications. One of the most striking of these is color. The crustaceans (mainly copepods, amphipods, ostracods, and mysids) are apt to be entirely scarlet or else partly red and partly transparent. The deep-sea medusae have beautiful colors; *Atolla* is red, cream, and purple, and *Periphylla* is plum-colored. Even the chaetognaths, nemerteans, and polychaetes display rich blues, yellows, reds, and oranges (Hardy 1957:colored illus.). Most of the fishes are black, but some are dark brown, and a few, such as the whale fish *Barbourisia rufa* (Rofen 1959:plate 2), are red or orange.

Other modifications for life in the bathypelagic zone are also of great interest to the biologist. For example, in the fishes it has been shown that the air bladder is lost or regressed, there is an increase in stored fat, and there is relatively less muscular tissue. In addition, organs such as photophores, eyes, brain, kidneys, and gills are usually reduced (Marshall 1960). The skeleton tends to be poorly developed and weakly ossified. The latter modification was once thought to be the result of a poor environmental supply of vitamin D, but it seems more likely that it is simply one in a syndrome of several adaptive changes. It has been noted that the deep-pelagic species of squid and octopods are apt to be very fragile, with much of their musculature being replaced by gelatinous tissue (Marshall 1954:107, Voss 1967). The bathypelagic nemerteans also possess a large proportion of gelatinous tissue and a relatively feeble musculature (Coe 1946:458).

It is probable that all of the internal modifications listed above can be related to the adoption of a more sedentary existence. Animals that do not have to undergo extensive vertical migrations and compete for food in the well-populated upper layers can perhaps afford to lead a more leisurely existence. Most of the bathypelagic fishes have some means of attracting their prey. The female deep-sea anglerfishes have been described as solitary, floating, baited traps (Mead et al., 1964:588). Most such species employ lighted, moving lures that are marvels of evolutionary design. Some have "fishing poles" (illicia) that can be extended or contracted, and one, *Lasiognathus saccostoma* (Fig. 10-10), is completely outfitted with pole, line, "float," lure, and hooks!

Although the bathypelagic zone does support an interesting variety of animals, food items are scarce and the populations of most species are evidently widely scattered. The problems of mate location and reproduction are, therefore, considerably magnified. In the deep-sea anglers, the predominant fish group, the males are dwarf, actively swimming forms with relatively large eyes and, generally, highly developed olfactory organs. Once an adult female is located, the male can attach himself to her body, either temporarily or as a permanent parasite. This arrangement assures that fertilization will take place at the proper time. Furthermore, the female produces a large number of small eggs which float to the surface and hatch there. The larvae remain in the food-rich epipelagic zone until metamorphosis when they sink into the depths. This reproductive pattern of high fecundity and epipelagic larval development seems to be common to all bathypelagic fishes (Mead, Bertelsen, and Cohen 1964:590).

It was the German Deep-Sea Expedition aboard the *Valdivia* from 1898 to 1899 that brought back the first extensive collections of bathypelagic animals. The captures were made by employing a series of large townets between the surface and the bottom of the ocean. In 1910, the *Michael Sars* expedition utilized a series of townets equipped with closing devices, giving us a much better knowledge of the depth distribution of many different species. The results were reported by Murray and Hjort (1912) in their classic work *The Depths of the Ocean*. These pioneering investigations were followed by the Danish *Dana* expeditions between 1908 and 1930, the *Discovery* expedition of 1925 to 1927, and many others in recent years.

General Distribution

To this day, we know more about the bathypelagic fauna near Bermuda than that of any other part of the world. During the summers of 1929,

1930, and 1931 the Bermuda Oceanographic Expeditions under the direction of Dr. William Beebe, took a total of 1,042 hauls at depths of 1,000 to 2,000 meters through a circular area 8 miles in diameter located about 9 miles southeast of Nonsuch Island. The results gave us, for the first time, some concept of the exceedingly broad geographic distribution and relative scarcity of most bathypelagic species. For example, more than 10,000 specimens of caridean, decapod crustaceans were taken. But of the 36 species identified, only 2 were common and 7 were represented by just a single specimen; 15 of the 25 previously known species had also been taken in the Indian Ocean, 11 in the equatorial Atlantic, 11 in the eastern North Atlantic, 6 off the Cape of Good Hope, 9 in the western Pacific, 6 in the eastern Pacific, 5 off the coast of Iceland, 5 near the Hawaiian Islands, and 4 in the southern Pacific (Chace 1940).

Data on the copepods are particularly valuable, since this group is represented by so many bathypelagic species. The cruises of the *Anton Bruun* in the western Indian Ocean in 1963 and 1964 resulted in the accumulation of important distributional data. Of the 96 species taken in the Indian Ocean between 0 and 10°S, 95 are known to occur also in the North Atlantic. Of the 98 species taken in the Arabian Sea between 10 and 20°N, 90 had been previously taken in the North Atlantic (Grice and Hulsemann 1967:10). Sewell (1948) and Wimpenny (1966:204) considered that the center of distribution for the deep-water copepods was the North Atlantic, but this conclusion has not been supported by the work of Grice and Hulsemann (*op. cit.*). The Antarctic copepod fauna was investigated by Vervoort (1965) who found that *all* the species occurring in the Antarctic bottom water had been taken in the deep water of all the great oceans. On the other hand, Brodsky (1957:42) reported 30 copepod species for the deep water of the Arctic Basin and noted that the fauna consisted mainly of endemics and Atlantic species with only 11 occurring in the Pacific. Dunbar and Harding (1968:323) took 7 zooplankton species in Arctic deep water (2,000 to 3,000 meters) and concluded that all but 1 were arctic endemics.

In his work on the Pacific euphausiids, Brinton (1962) recognized three bathypelagic species, all with broad distributions. *Bentheuphausia ambylops* and *Thysanopoda cornuta* were considered to be cosmopolitan, while the third species, *T. egregia*, was found to occur around the world between 40°N and 55°S. The more restricted distribution of the latter was explained by the fact that its larvae are apparently confined to comparatively shallow water (280 to 500 meters) with a warmer temperature regime. For the euphausiids, Brinton considered the upper limit of the bathypelagic zone to lie at about 1,500 meters.

Vampyroteuthis infernalis is a black and purplish deep-sea cephalopod with light organs and other interesting modifications. In the Atlantic, it is found between 1,000 and 2,000 meters and in the Pacific from 1,500 to 2,000 meters. Its distribution is worldwide beneath tropical and warm-temperate waters, and there is evidence that it is quite sensitive to water density (Pickford 1946). One species of pycnogonid, *Pallenopsis calcanea*, is bathypelagic and apparently has a worldwide distribution (Hedgpeth 1962).

The bathypelagic chaetognath fauna seems to be limited to four species belonging to the genus *Eukrohnia*. *E. hamata* provides a good example of isothermic submergence, since it occurs below 900 meters in the tropical Pacific but in much shallower water toward the poles. The other three species are apparently confined to depths below 1,000 meters: *E. fowleri* is considered to be cosmopolitan, *E. bathypelagica* may be restricted to the Pacific and Indian Oceans, and *E. bathyantarctica* may be an Antarctic endemic, since it apparently has not been taken north of the Temperate Convergence (Alvariño 1965). However, David (1965:307) stated that the horizontal distribution of the latter species is not yet known.

Fishes

Of about 1,200 known species of pelagic fishes, approximately 130 to 150 appear to be typical of the bathypelagic habitat. The great majority of the latter group, perhaps as many as four-fifths (Marshall 1963:182), are deep-sea anglers belonging to 10 families in the suborder Ceratioidei (Bertelsen 1951). Other bathypelagic species belong to the nemichthyoid eel group (families Nemichthyidae, Serrivomeridae, and Cyemidae), the gulper eels (families Saccopharyngidae, Eurypharyngidae, and Monognathidae), the whale fishes (families Cetomimidae, Barbourisiidae, and Rondeletiidae), and the families Gonostomatidae, Macrouridae, Myctophidae, Searsiidae, Alepocephalidae, Aphyonidae, and Chiasmodontidae.

The monograph on the ceratioid anglerfishes by Bertelsen (1951) gives some interesting distributional information about this fascinating group. Although a total of 80 species are recognized, 31 are known only from single specimens, and only 16 are known from more than 20 specimens. Of the 16 better-known species, 14 have been taken from all three oceans. While very broad distributions may prove to be the rule in this group, the fact that 18 species are still known only from the tropical eastern Pacific (Grey 1956) may indicate the existence of a significant level of endemism in this area. Bertelsen (1951) also observed that, although the adults have been taken from the sub-Arctic to the sub-Antarctic, the larvae are restricted to the warmer surface waters between 40°N and about 35°S. Apparently, adults living at higher latitudes, both north and south, are expatriates that do not belong to the reproductive population.

Although all the adult nemichthyoid eels appear to be primarily bathypelagic, only two species, *Nemichthys scolopaceus* and *Borodinula infans*, are very common. Both are probably worldwide in distribution (Briggs 1960, Castle 1961). Only one species of gulper eel, *Eurypharynx pelecanoides* (Fig. 10-11), is relatively common, and it also has been found all over the world (Grey 1956, Briggs 1960).

Most of the 18 species referred to the family Searsiidae by Parr (1960) as well as some of the related Alepocephalidae (Mead Bertelsen, and Cohen 1964:571) are probably bathypelagic. However, all of the species are so poorly known that meaningful conclusions about geographic distribution cannot be reached. Four species of the genus *Cyclothone* (family Gonostomatidae) appear to be bathypelagic: *C. microdon*, *C. acclinidens*, and *C. obscura* are common species represented by populations centered at levels of 1,000 meters or more (Marshall 1960:101) and *C. pacifica* was recently described (Mukhacheva 1964). One of the four is worldwide in distribution, two are fairly widespread elsewhere but do not reach the eastern Pacific, and one is apparently restricted to the temperate North Pacific. Another bathypelagic gonostomatid fish, *Gonostoma bathyphilum*, seems to occur throughout the North and South Atlantic but not elsewhere (Grey 1964:183).

Marshall (1966) has shown that a few species of rat-tails (family Macrouridae) have become adapted to the bathypelagic environment. Specifically, he noted that *Macrouroides inflaticeps* and two other species were very widespread, having been taken in all three oceans. A fourth species, *Cynomacrurus piriei*, may be confined to Antarctic and sub-Antarctic waters (Marshall 1964b:88). In the chiasmodontid fish genus *Kali*, three of the four species appear to be cosmopolitan below tropical and warm-temperate waters, but the fourth is endemic to the eastern Pacific (Johnson 1969).

A few species of lanternfishes (Myctophidae) may typically inhabit the bathypelagic instead of the mesopelagic zone. An apparent example is

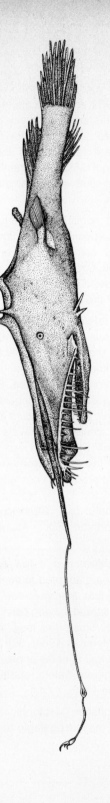

Figure 10-10 The deep-sea angler (*Lasiognathus sac-costoma*). A bathypelagic species known so far only from the Western Atlantic. Fishing equipment consists of extensible rod, filament, "float," and an illuminated lure with hooklike denticles. This wonder hath evolution wrought! After Bertelsen (1951).

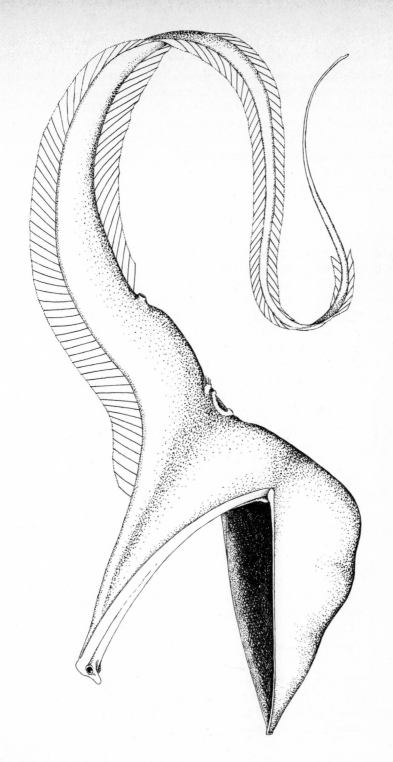

Figure 10-11 The deep-sea gulper (*Eurypharynx peleca-noides*). A bathypelagic species with a worldwide distribution beneath tropical and temperate waters.

Taaningichthys bathyphilus, a circumtropical species that evidently lives below 800 meters (Bolin 1959:26). The whalefishes comprise a small group that is known, so far, from relatively few specimens. There are 13 species recognized in the family Cetomimidae, and 1 each in the Barbourisiidae and the Rondeletiidae. The best-known species in the group is *Rondeletia bicolor*, which is now considered to be circumtropical (Rofen 1959:259). Data on some of the other species suggest that widespread distributions may be the rule in all three families.

Considering the present state of our knowledge about the distribution patterns in the bathypelagic zone, one cannot set up zoogeographic regions and provinces with any degree of confidence. However, it may be useful to state that, so far, it appears likely that one huge Bathypelagic Region may suffice for most of the world. This would include all waters between 1,000 and 6,000 meters except the Arctic Basin. Since Brodsky (1957:42)and Dunbar and Harding (1968:323) reported that the deep-water copepod fauna of the latter area included a large proportion of endemic species, there probably should be a separate Arctic Region. If preliminary data about some of the fishes is indicative, the eastern tropical Pacific may need to be given some kind of distinction.

Conclusion

In summary, it can be said that the bathypelagic zone harbors a fauna that is distinct in the following ways:

General Characteristics of the Fauna

1 There is a high degree of diversification, but the crustaceans are the dominant group, including more than 90 percent of all bathypelagic species.

2 The 1,000- to 2,000-meter layer contains a greater volume of zooplankton, a greater number of species, and the individuals of some groups attain a larger size than in the deeper layers.

3 Morphological adaptations for a bathypelagic existence often include certain bright colors, replacement of muscular tissue by gelatinuous tissue, and a reduction in the relative size of eyes, photophores, and other organs. Most of the modifications seem to be related to the adoption of a more sedentary mode of life.

4 Work on the copepods, decapod crustaceans, euphausiids, cephalopods, chaetognaths, and fishes gives evidence that very broad geographic distributions predominate. The number of species inhabiting bathypelagic depths below the tropical and warm-temperate regions of the surface is much

greater than below the colder surface waters. This may reflect a dependency upon warm waters for a critical part of the life cycle. For example, all bathypelagic fishes apparently have epipelagic larval stages that are restricted to the warmer waters.

5 While many of the species inhabiting the low latitude regions of the world appear to be circumglobal, there is some evidence that the fauna of the eastern tropical Pacific may be distinct enough to be given special recognition. This is particularly true of the fishes where the East Pacific Barrier seems to affect the distribution of such groups as the deep-sea anglers and the bristlemouths (genus *Cyclothone*).

6 With the exception of the Arctic Basin, where there appears to be a significant degree of endemism in the zooplankton, the high latitude regions of the world appear to have a depauperate but otherwise indistinct bathypelagic fauna. Some species that are epipelagic in subpolar regions become bathypelagic in the tropics because of isothermic submergence.

The Hadopelagic Zone

As was indicated in the discussion of the vertical divisions of the open ocean (p. 305), the pelagic fauna of the trenches is very poorly known, since only one research vessel has operated closing nets below 6,000 meters. It appears that the crustaceans dominate this environment as well as the bathypelagic depths for the species so far reported are all copepods, amphipods, or ostracods. If further collecting and systematic work bears out the initial indications of very high rates of endemism, then each trench that extends well below 6,000 meters will probably have to be considered a separate Hadopelagic Region.

Literature Cited

Adams, J. A. 1960. A contribution to the biology and postlarval development of the sargassum fish, *Histrio histrio* (Linnaeus) with a discussion of the *Sargassum* complex. *Bull. Marine Sci.*, 10(1):55–82, 5 figs.

Alvariño, A. 1964. Bathymetric distribution of chaetognaths. *Pacific Sci.*, 18(1):64–82, 12 figs.

———. 1965. Chaetognaths. *Oceanog. Marine Biol. Ann. Rev.*, 3:115–194, 17 figs.

Andriashev, A. P. 1962. Bathypelagic fishes of the Antarctic I. Family Myctophidae. *Akad. Sci. USSR, Zool. Inst., Expl. Fauna Seas*, 1(9):216–294, 36 figs.

Backus, R. H., J. E. Craddock, R. L. Haedrich, and D. L. Shores. 1969. Mesopelagic fishes and thermal fronts in the western Sargasso Sea. *Marine Biol.*, 3(2):87–106.

———, ———, ———, and ———. 1970. The distribution of mesopelagic fishes in the equatorial and western North Atlantic Ocean. *J. Marine Res.*, 28(2):179–201, 17 figs.

———, G. W. Mead, R. L. Haedrich, and A. W. Ebeling. 1965. The mesopelagic fishes collected during cruise 17 of the R/V Chain, with a method for analyzing faunal transects. *Bull. Museum Comp. Zool. Harvard*, 134(5):139–157, 9 figs.

Baird, R. C. 1971. The systematics, distribution and zoogeography of the marine hatchetfishes (family Sternoptychidae). *Bull. Museum Comp. Zool. Harvard*, 142(1):1–128, 80 figs.

Baker, A. de C. 1954. The circumpolar continuity of Antarctic plankton species. *Discovery Rep.*, 27:201–218, 5 figs.

———. 1965. The latitudinal distribution of *Euphausia* species in the surface waters of the Indian Ocean. *Discovery Rep.*, 33:309–334. 7 figs.

Banse, K. 1964. On the vertical distribution of zooplankton in the sea. *Progr. Oceanog.*, 2:53–125, 4 figs.

Barham, E. G. 1966. Deep scattering layer migration and composition: observations from a diving saucer. *Science*, 151:1399–1403, 3 figs.

Barnard, J. L. 1962. South Atlantic abyssal amphipods collected by R. V. *Vema* in J. L. Bernard, R. J. Menzies, and M. L. Băcescu, *Abyssal Crustacea.*, Columbia University Press, New York, pp. 1–78, 79 figs.

Bary, B. M. 1959. Species of zooplankton as a means of identifying different surface waters and demonstrating their movements and mixing. *Pacific Sci.*, 13(1):14–54, 20 figs.

———. 1963a. Temperature, salinity and plankton in the eastern North Atlantic and coastal waters of Britain, 1957. II. The relationships between species and water bodies. *J. Fisheries Res. Board Can.*, 20(4):1031–1065, 3 figs., 17 plates.

———. 1963b. Distributions of Atlantic pelagic organisms in relation to surface water bodies in "marine distributions," *Roy. Soc. Can., Spec. Pub.*, (5):51–67, 12 figs.

———. 1964. Temperature, salinity and plankton in the eastern North Atlantic and coastal waters of Britain, 1957. IV. The species relationship to the water body; its role in distribution and in selecting and using indicator species. *J. Fisheries Res. Board Can.*, 21(1):183–202, 7 figs.

Becker, V. E. 1964. Slendertailed myctophids (genera *Loweina*, *Tarletonbeania*, *Gonichthys* and *Centrobranchus*) of the Pacific and Indian Oceans. Systematics and distribution. *Trudy Inst. Okeanologii*, 73:11–75, 25 figs.

———. 1966. The pattern of distribution of the lantern fishes (Fam. Myctophidae) in the Pacific Ocean. *Abstr., 2nd Int. Oceanog. Congr.*, Moscow, pp. 27–28.

Bé, Allan W. H. 1959. Fluctuations in the faunal boundary between temperate and subtropical planktonic foraminifera in the North Atlantic. *Int. Oceanog. Cong. Preprints, Amer. Ass. Adv. Sci., Washington, D. C.*, pp. 134–137, 2 figs.

Beklemishev, C. W. 1966. Large-scale pattern of distribution of oceanic plankton communities. *Abstr., 2nd Int. Oceanog. Congr.*, Moscow, pp. 28–29.

Beklemishev, K. V., and G. I. Semina. 1956. On the structure of the biogeographical boundary between the boreal and tropical regions in the pelagial of the northwestern Pacific Ocean. *Doklady Akademii Nauk SSSR*, 108(6):1057–1060.

Bertelsen, E. 1951. The ceratoioid fishes. *Dana-Rep.*, (39):1–276, 141 figs., 1 plate.

Besednov, L. N. 1960. Some data on the ichthyofauna of Pacific Ocean flotsam. *Trudy Inst. Okeanol.*, 41:192–197.

Bigelow, H. B. 1926. Plankton of the offshore waters of the Gulf of Maine. *Bull. U. S. Bur. Fish.*, 40(2):1–509.

———. and W. C. Schroeder. 1953. Fishes of the Gulf of Maine. *U. S. Fish Wildlife Serv., Fishery Bull.*, (74):viii + 577 pp., 288 figs.

Blackburn, M. 1965. Oceanography and the ecology of tunas. *Oceanog. Marine Biol. Ann. Rev.*, 3:299–322.

Bogorov, B. G. 1958. Biogeographical regions of the plankton of the Northwestern Pacific Ocean and their influence on the deep sea. *Deep-Sea Res.*, 5(2):149–161.

Bolin, R. L. 1957. Deep-water biological provinces of the Indo-Pacific. *Proc. 8th Pacific Sci. Congr.*, 3:373–376, 1 map.

———. 1959. Iniomi. Myctophidae. *Rep. Sci. Res. "Michael Sars" North Atlantic Deep-Sea Expedition 1910*, 4(part 2)(7):1–45, 7 figs.

Bradshaw, J. S. 1959. Ecology of living planktonic foraminifera in the North and equatorial Pacific Ocean. *Contr. Cushman. Found. Foram. Res.*, 10:25–64.

Briggs, J. C. 1960. Fishes of worldwide (circumtropical) distribution. *Copeia*, (3):171–180.

———. 1970. A faunal history of the North Atlantic Ocean. *Syst. Zool.*, 9(1):19–34, 3 figs.

Brinton, E. 1962. The distribution of Pacific euphausiids. *Bull. Scripps Inst. Oceanog.*, 8(2):51–170, 126 figs.

Brodsky, K. A. 1957. *The copepod fauna (Calanoida) and zoogeographical divisions of the North Pacific and adjoining waters.* Izdat. Akad. Nauk. SSSR, 222 pp. (English translation of selected portions by Donald A. Thomson, 1961).

———. 1962. On biogeographical division of pelagical zones of Southern Hemisphere according to the distribution of *Calanus* species (Copepoda). *Deep-Sea Res.*, 10(4):535 (abstract).

Bruun, A. F. 1935. Flying-fishes (Exocoetidae) of the Atlantic, systematic and biological studies. *Dana-Rep.*, (6):1–106, 30 figs., 7 plates.

———. 1957. Deep sea and abyssal depths. *in* J. W. Hedgpeth (editor), *Treatise on marine ecology and paleoecology*, vol. 1. Mem. Geol. Soc., Am. (67):641–672, 9 figs. 3 plates.

Bussing, W. A. 1965. Studies of the midwater fishes of the Peru-Chile trench. *Biol. Antarctic Seas II, Antarctic Res. Ser.*, (5):185–227, 16 figs.

Castle, P. H. J 1961. Deep-water eels from Cook Strait, New Zealand. *Zool. Pub., Victoria Univ., Wellington*, (27):1–30, 6 figs.

Chace, F. A., Jr. 1940. Plankton of the Bermuda Oceanographic Expeditions. IX. The bathypelagic caridean Crustacea. *Zoologica*, 25(2):117–209, 45 figs.

Chen, C. 1968. Zoogeography of the costomatous pteropods in the west Antarctic Ocean. *Nautilus*, 81(3):94–101.

Clarke, G. L. 1966. Comparative studies of light penetration, transparency, and biolumenescence in oceanic areas. *Abstr., 2nd Int. Oceanog. Congr. Moscow*, p. 80.

Clarke, M. R. 1966. A review of the systematics and ecology of oceanic squids. *Adv. Marine Biol.*, 4:91–300, 59 figs.

Coe, W. R. 1946. The means of dispersal of bathypelagic animals in the North and South Atlantic Oceans. *Amer. Naturalist*, 80(793):453–469.

Craddock, J. E., and G. W. Mead. 1970. Midwater fishes from the eastern South Pacific Ocean. *Anton Bruun Rep.*, (3):1–46, 10 figs.

Dahl, E. 1959. Amphipoda from depths exceeding 6,000 meters. *Galathea Rep.*, 1:211–241, 20 figs.

David, P. M. 1958. The distribution of the Chaetognatha of the Southern Ocean. *Discovery Rep.*, 29:199–228, 13 figs. 1 plate.

———. 1962. The distribution of Antarctic Chaetognaths. *Deep-Sea Res.*, 10(4):536 (abstract).

———. 1965a. The Chaetognatha of the Southern Ocean. *in* J. Van Mieghem and P. Van Oye (editors), *Biogeography and ecology in Antarctica*. W. Junk, The Hague, pp. 296–323, 16 figs.

———. 1965b. The surface fauna of the ocean. *Endeavor*, 24:95–100, 17 figs.

Dodimead, A. J., F. Favorite, and T. Hirano. 1963. Review of oceanography of the Subarctic Pacific region. *Bull. Int. Nor. Pac. Fish. Comm.*, (13):1–196, 265 figs.

Dooley, J. K. 1969. The fishes associated with the pelagic *Sargassum* complex. M. A. thesis, University of South Florida, pp. 1–87, 11 figs.

Dunbar, M. J. 1964. Euphausids and pelagic amphipods. *Ser. Atlas Marine Environ. Amer. Geograph. Soc.*, Folio 6:1–2, 8 plates.

———, and G. Harding. 1968. Arctic Ocean water masses and plankton—a reappraisal. *in* J. E. Sater (editor), *Arctic drifting stations*. Arctic Inst. North America, pp. 315–326, 4 figs.

Ebeling. A. W. 1962. Melamphaidae I. Systematics and zoogeography of the species in the bathypelagic fish genus *Melamphaes* Gunther. *Dana-Rep.*, (58):1–164, 73 figs.

———, and W. H. Weed, III. 1963. Melamphaidae III. Systematics and distribution of the species in the bathypelagic fish genus *Scopelogadus Vaillant*. *Dana-Rep.*, (60):1–58, 23 figs.

———, R. M. Ibara, R. J. Lavenberg, and F. J. Rohlf. 1970. Ecological groups of deep-sea animals off southern California. *Bull. Los Angeles Co. Museum Nat. Hist.*, (6):1–43, 9 figs.

Einarsson, H. 1945. Euphausiacea. I. Northern Atlantic species. *Dana-Rep.*, (27):1–185, 84 figs.

Ekman, S. 1953. *Zoogeography of the sea*. Sidgwick & Jackson, London, xiv + 417 pp., 121 figs.

Fage, L. 1960. Oxycephalidae. Amphipodes pelagiques. *Dana-Rep.*, (52):1–145, 79 figs.

Fager, E. W., and J. A. McGowan. 1963. Zooplankton species groups in the North Pacific. *Science*, 140(3566):453–460, 5 figs.

Ferguson Wood, E. J. 1965. The vertical distribution of phytoplankton in tropical waters. *Ocean Sci. Ocean Engineering*, pp. 111–115, 5 figs.

————. 1966. Plants of the deep oceans. *Z. für Allg. Mikrobiol.*, 6(3):177–179.

Foxton, P. 1961. *Salpa fusiformis* Cuvier and related species. *Discovery Rep.*, 32:1–32, 10 figs., 2 plates.

Frost, B., and B. Fleminger. 1968. A revision of the genus *Clausocalanus* (Copepoda: Calanoida) with remarks on distributional patterns in diagnostic characters. *Bull. Scripps Inst. Oceanog.*, 12:1–235, 11 figs., 67 plates.

Gibbs, R. H., Jr. 1968. *Photonectes munificus*, a new species of melanostomiatid fish from the South Pacific subtropical convergence, with remarks on the convergence fauna. *Contrib. Sci., Los Angeles Co. Museum*, (149):1–6, 1 fig.

————. 1969. Taxonomy, sexual dimorphism, vertical distribution, and evolutionary zoogeography of the bathypelagic fish genus *Stomias* (Stomiatidae). *Smiths. Contrib. Zool.*, (31):1–25, 6 figs.

————, and B. B. Collette. 1967. Comparative anatomy and systematics of the tunas, genus *Thunnus. Fish. Bull. U.S. Fish Wildlife Serv.*, 66(1):65–130, 36 figs.

Glover, R. S. 1961. Biogeographical boundaries: the shapes of distributions. *in* Mary Sears (editor), *Oceanography. Amer. Ass. Adv. Sci., Pub. 67*:201–228, 10 figs.

Gosline, W. A., and V. E. Brock. 1960. *Handbook of Hawaiian fishes.* University of Hawaii Press, Honolulu, ix + 1–372 pp., 277 figs., 2 plates.

Grainger, E. H. 1963. Copepods of the genus *Calanus* as indicators of eastern Canadian waters. Marine distributions. *Roy. Soc. Can., Special Pub.*, (5):68–94, 12 figs.

Grey, M. 1955. The fishes of the genus *Tetragonurus* Risso. *Dana Rep.*, (41):1–75, 16 figs.

————. 1956. The distribution of fishes found below a depth of 2,000 meters. *Fieldiana, Zool.*, 36(2):75–337.

————. 1964. Gonostomatidae. *in* H. B. Bigelow (editor), *Fishes of the Western North Atlantic.* Sears Foundation, New Haven, Conn., Mem. 1, part 4, pp. 78–240.

Grice, G. D., and K. Hulsemann. 1965. Abundance, vertical distribution and taxonomy of calanoid copepods at selected stations in the northeast Atlantic. *J. Zool.*, 146(2):213–262, 26 figs.

————, and ————. 1967. Bathypelagic calanoid copepods of the western Indian Ocean. *Proc. U. S. Nat. Museum*, 122(3583):1–67, 319 figs.

Haedrich, R. L., and J. G. Nielsen. 1966. Fishes eaten by *Alepisaurus* (Pisces, Iniomi) in the southeastern Pacific Ocean. *Deep-Sea Res.*, 13:909–919, 1 fig.

Haffner, R. E. 1952. Zoogeography of the bathypelagic fish, *Chauliodus. Syst. Zool.*, 1(3):113–133, 14 figs.

Hand, C., and L. B. Kan. 1961. The Medusae of the Chukchi and Beaufort Seas of the Arctic Ocean including the description of a new species of *Eucodonium* (Hydrozoa: Anthomedusae). *Arctic Inst. North Amer., Tech. Papers*, (6):1–23, 9 figs.

Hardy, A. 1957. *The open sea.* Houghton Mifflin Co., Boston, xv + 1–335 pp., 300 figs., 48 plates.

Harry, R. R. 1953. Ichthyological field data of Raroia Atoll, Tuamotu Archipelago. *Atoll Res. Bull.* No. 18, Pacific Science Board, Washington, D.C., 190 pp., 7 figs.

Hart, T. 1942. *Rhizosolenia curvata* Zacharias, an indicator species in the Southern Ocean. *Discovery Rep.*, 16:413–446.

Hedgpeth, J. W. 1962. A bathypelagic pycnogonid. *Deep-Sea Res.*, 9:487–491, 2 figs.

———. 1957. Classification of marine environments. *in* J. W. Hedgpeth (editor), *Treatise on marine ecology and paleoecology.* vol. 1. Mem. Geol. Soc. Am. (67):17–27, 5 figs.

Herre, A. W. 1953. Check list of Philippine fishes. *U.S. Fish Wildlife Serv., Res. Rep.*, (20):1–977.

Herring, P. J. 1967. The pigments of plankton at the sea surface. Aspects of marine zoology. Symposium. *Zool. Soc. London*, (19):215–235, 7 figs.

Howard, J. K., and S. Ueyanagi. 1965. Distribution and abundance of billfishes (Istiophoridae) of the Pacific Ocean. *Stud. Trop. Oceanog.*, (2):x + 134 pp., 37 figs., atlas, 38 charts.

Hunter, J. R., and C. T. Mitchell. 1967. Association of fishes with flotsam in the offshore waters of Central America. *Fish. Bull. U.S. Fish Wildlife Serv.*, 66(1):13–29, 7 figs.

Johnson, M. W., and E. Brinton. 1963. Biological species, water masses and currents. *in* M. N. Hill (editor), *The sea.* vol. 2. Interscience, New York, pp. 381–414, 13 figs.

Johnson, R. K. 1969. A review of the fish genus *Kali* (Perciformes: Chiasmodontidae). *Copeia*, (2):386–391, 2 figs.

Kramp, P. L. 1959. The hydromedusae of the Atlantic Ocean and adjacent waters. *Dana-Rep.*, (46):1–283, 335 figs., 2 plates.

Larkins, H. A. 1964. Some epipelagic fishes of the North Pacific Ocean, Bering Sea, and Gulf of Alaska. *Trans. Amer. Fish. Soc.*, 93(3):286–290, 1 fig.

Laursen, D. 1953. The genus *Ianthina.* A monograph. *Dana-Rep.*, (38):1–40, 41 figs., 1 plate.

Mansueti, R. 1963. Symbiotic behavior between small fishes and jelly fishes, with new data on that between the stromateid, *Peprilus alepidotus*, and the scyphomedusa *Chrysaora quinquecirrha. Copeia*, (1):40–80, 5 figs.

Marr, J. W. S. 1962. The natural history and geography of the Antarctic krill (*Euphausia superba* Dana). *Discovery Rep.*, 32:33–464, 157 figs., 1 plate.

Marshall, N. B. 1954. Aspects of deep-sea biology. Hutchinson's Publications, London, 380 pp., 103 figs., 4 plates.

———. 1960. Swimbladder structure of deep-sea fishes in relation to their systematics and biology. *Discovery Rep.*, 31:1–122, 47 figs., 3 plates.

———. 1963. Diversity, distribution and speciation of deep-sea fishes. *Syst. Ass. Publ.*, (5):181–195.

———. 1964a. Fish. *in* R. Priestley, R. J. Adie, and G. de Q. Robin (editors), *Antarctic research.* Butterworths, London, pp. 206–218, 4 figs.

———. 1964b. Bathypelagic macrourid fishes. *Copeia*, (1):86–93, 3 figs.

———. 1965. Systematic and biological studies of the macrourid fishes (Anacanthini-Teleostii). *Deep-Sea Res.*, 12(3):299–322, 9 figs.

———, and Å. V. Taaning. 1966. The bathypelagic macrourid fish *Macrouroides inflaticeps* Smith and Radcliffe. *Dana-Rep.*, (69):1–6, 1 fig.

Marty, Y. Y. 1962. Some similarities and differences under which boreal fish species exist in northeast and northwest Atlantic. *in Soviet fisheries investigations in the northwest Atlantic.* Rybnoe Khozyaistvo, Moscow (English translation, Jerusalem 1963), pp. 55–67, 10 figs.

Matsui, T. 1967. Review of the mackerel genera *Scomber* and *Rastrelliger* with description of a new species of *Rastrelliger. Copeia*, (1):71–83, 6 figs.

Mauchline, J., and L. R. Fisher. 1969. The biology of euphausiids. *Adv. Marine Biol.*, 7:ix + 1–454 pp., 137 figs.

McGowan, J. A. 1963. Geographical variation in *Limacina helicina* in the North Pacific. *Syst. Ass. Publ.*, (5):109–128. 14 figs.

McLaren, I. A. 1963. Effects of temperature on growth of zooplankton and the adaptive value of vertical migration. *J. Fisheries Res. Board Can.*, 20(2):685–727, 19 figs.

Mead, G. W., and R. L. Haedrich. 1965. The distribution of the oceanic fish *Brama brama. Bull. Museum Comp. Zool.*, 134(2):29–67, 8 figs.

———, E. Bertelsen, and D. M. Cohen. 1964. Reproduction among deep-sea fishes. *Deep-Sea Res.*, 11:569–596.

Mohr, E. 1937. Revision der Centriscidae (Acanthopterygii Centrisciformes). *Dana-Rep.*, (13):1–69, 33 figs., 2 plates.

Morrow, J. E. 1964. Chauliodontidae, Stomiatidae, and Malacosteidae. *in* H. B. Bigelow (editor), *Fishes of the Western North Atlantic.* Sears Foundation, New Haven, Conn., Mem. 1, part 4:274–310, 523–549.

———, and S. J. Harbo. 1969. A revision of the sailfish genus *Istiophorus. Copeia*, (1):34–44, 15 figs.

Moss, S. A. 1962. Melamphaidae II. A new melamphaid genus *Sio*, with a redescription of *Sio nordenskjoldii* (Lonnberg). *Dana-Rep.*, (56):1–10, 4 figs.

Mukhacheva, V. A. 1964. On the genus *Cyclothone* (Gonostomidae, Pisces) of the Pacific Ocean. *Trudy Inst. Okeano., Akad. Nauk. S. S. S. R.*, 73:93–138, 17 figs.

Murray, J., and J. Hjort. 1912. *The depths of the ocean.* Macmillan and Co., London, xx + 821 pp., 575 figs., 9 plates.

Muus, B. J. 1956. Development and distribution of a North Atlantic pelagic squid, family *Cranchiidae. Medd. Danmarks. Fisk. Hav.*, 1(15):1–15, 9 figs.

Nafpaktitis, B. G. 1968. Taxonomy and distribution of the lanternfishes, genera *Lobianchia* and *Diaphus*, in the North Atlantic. *Dana-Rep.*, (73):1–131, 69 figs., 2 plates.

———, and M. Nafpaktitis. 1969. Lanternfishes (family Myctophidae) collected during cruises 3 and 6 of the R/V *Anton Bruun* in the Indian Ocean. *Bull. Los Angeles Co. Museum*, (5):1–79, 82 figs.

Neumann, G., and W. J. Pierson, Jr. 1966. *Principles of physical oceanography.* Prentice-Hall, Englewood Cliffs, N. J., xii + 545 pp., illus.

Nielsen, J. G., and J. M. Jensen. 1967. Revision of the arctic cod genus *Arctogadus* (Pisces Gadidae). *Medd. Om Gronland*, 184(2):1–28, 7 figs.

Novikova, N. S. 1967. Idiacanthids of the Indian and Pacific Oceans (Pisces, Idiacanthidae). *Trudy Inst. Okeano., Akad. Nauk. S.S.S.R.*, 84:159–208, 5 figs.

Otsu, T., and R. N. Uchida. 1963. Model of the migration of albacore in the north Pacific Ocean. *Fish Bull., U.S. Fish. Wildlife Serv.*, 63(1):33–44, 10 figs.

Parin, N. V. 1959. On the similarity found in the geographical distribution of the sardine and the subtropical flying fish. *Rep. Acad. Sci. U.S.S.R.*, 124(5):1130–1132.

———. 1960a. The range of the saury (*Cololabis saira* Brev.—Scomberesocidae, Pisces) and effects of oceanographic features on its distribution. *Doklady, Akad. Nauk S.S.S.R.*, 130(3):649–652, 3 figs.

———. 1960b. The flying fishes (Exocoetidae) of the northwest Pacific. *Trudy Inst. Okeano., Akad. Nauk. S.S.S.R.*, 31:205–285, 25 figs.

———. 1961a. The distribution of deep-sea fishes in the upper bathypelagic layer of the subarctic waters of the northern Pacific Ocean. *Trudy Inst. Okeano., Akad. Nauk, S.S.S.R.*, 45:259–278, 7 figs.

———. 1961b. Contribution to the knowledge of the flyingfish fauna (Exocoetidae) of the Pacific and Indian Oceans. *Trudy Inst. Okeano., Akad. Nauk, S.S.S.R.*, 42:41–90, 19 figs.

———. 1964. Data on biology and distribution of pelagic sharks, *Euprotomicrus bispinatus* and *Isistius brasiliensis* (Squalidae, Pisces). *Trudy Inst. Okeano., Akad. Nauk, S.S.S.R.*, 73:163–184, 4 figs.

———. 1968. *Epipelagic ichthyofauna of the oceans.* Nauka, Moscow, pp. 1–185, 56 figs.

Parr, A. E. 1960. The fishes of the family Searsidae. *Dana-Rep.*, (51):1–108, 73 figs.

Paxton, J. R. 1967. A distributional analysis for the lanternfishes (family Myctophidae) of the San Pedro Basin, California. *Copeia*, (2):422–440, 16 figs.

Pearcy, W. G. 1964. Some distributional features of Mesopelagic fishes off Oregon. *J. Marine Res.*, 22(1):83–102, 7 figs.

Penrith, M. J. 1967. The fishes of Tristan da Cunha, Gough Island and the Vema Seamount. *Ann. S. African Museum*, 48(22):523–548, 2 figs., 1 plate.

Pickard, G. L. 1964. *Descriptive physical oceanography.* Pergamon Press, Oxford, viii + 199 pp., 31 figs.

Pickford, G. E. 1946. *Vampyroteuthis infernalis* Chun. An archaic dibranchiate cephalopod. I. Natural history and distribution. *Dana-Rep.*, (29):1–40, 8 figs.

Ponomareva, L. A. 1963. Euphausiids of the North Pacific. Their distribution and ecology. *Inst. Okeano., Akad. Nauk S.S.S.R.* (English translation, Jerusalem 1966), pp. 1–154. 40 figs.

Rass, T. S. 1967. *Biology of the Pacific Ocean.* Book III. *Fishes of the open waters.* Academy of Science, U.S.S.R., 275 pp.

Rofen, R. R. 1959. The whale-fishes: families Cetomimidae, Barbourisiidae and Rondeletiidae (Order Cetunculi). *Galathea Rep.*, 1:255–260, 2 plates.

———. 1966. Families Paralepididae and Omosudidae. *in* G. W. Mead (editor), *Fishes of the western North Atlantic.* Mem. 1, part 5:205–481, 118 figs.

Savilov, A. I. 1966. Pleuston of the Pacific Ocean. *Abstr., 2nd Int. Oceanog. Congr., Moscow*, pp. 316–317.

Sewell, R. B. S. 1948. The free-swimming planktonic Copepoda. Geographical distribution. *Sci. Rep. John Murray Expedition*, 8(3):317–592.

Staiger, J. C. 1965. Atlantic flying fishes of the genus *Cypselurus*, with descriptions of the juveniles. *Bull. Marine Sci.*, 15(3):672–725, 19 figs.

Stommel, H. 1958. The circulation of the abyss. *Sci. Amer.*, 199(1):85–90, 6 figs.

Strasburg, D. W. 1958. Distribution, abundance, and habits of pelagic sharks in the Central Pacific Ocean. *U.S. Fish Wildlife Serv., Fishery Bull.*, 58(138):iv + 335–361, 20 figs.

Squires, H. J. 1966. Distribution of Decapod Crustacea in the North-west Atlantic. *Ser. Atlas Marine Environ.*, (12):1–4, 4 figs., 4 plates.

Tebble, N. 1960. The distribution of pelagic polychaetes in the South Atlantic Ocean. *Discovery Rep.*, 30:161–300, 52 figs.

———. 1962. The distribution of pelagic polychaetes across the North Pacific Ocean. *Bull. Brit. Museum* (Nat. Hist.), 7(9):373–492, 55 figs.

Tesch, J. J. 1949. Heteropoda. *Dana-Rep.*, (34):1–53, 44 figs., 5 plates.

Thore, S. 1949. Investigations on the "Dana" Octopoda. Part I. Bolitaenidae Amphitretidae, Vitreledonellidae, and Alloposidae. *Dana-Rep.*, (33):1–85, 69 figs.

Tortonese, E. 1963. Elenco riveduto dei Leptocardi, Ciclostomi, pesci cartilaginei e ossei del mare Mediterraneo. *Ann. Mus. Civico Stor. Nat., Genova*, 74:156–185.

Vervoort, W. 1965. Notes on the biogeography and ecology of free-living marine Copepoda. *in* J. Van Mieghem and P. Van Oye (editors), *Biogeography and ecology in Antarctica*. W. Junk, The Hague, pp. 381–400, 6 figs.

Vinogradov, M. E. 1966. Characteristic features of the distribution of oceanic meso- and macroplankton. *Abstr., 2nd Int. Oceanog. Congr., Moscow*, pp. 382–383.

———. 1968. *Vertical distribution of the oceanic zooplankton*. Nauka, Moscow, pp. 1–320, 84 figs.

Voss, G. L. 1967. The biology and bathymetric distribution of deep-sea cephalopods. *Studies Trop. Oceanog. Miami*, 5:511–535, 4 figs.

Walters, V. 1955. Fishes of Western Arctic America and Eastern Arctic Siberia. *Bull. Amer. Museum Nat. Hist.*, 106(5):255–368.

Williams, F. 1958. Fishes of the family Carangidae in British East African waters. *Ann. Mag. Nat. Hist.*, 1(6):369–430, illus.

Wimpenny, R. S. 1966. *The plankton of the sea*. Faber and Faber, London, pp. 1–426, 20 plates, 102 figs.

Wolff, T. 1960. The hadal community, an introduction. *Deep-Sea Res.*, 6:95–124.

Woods, L. P. 1953. Family Carangidae. *in* L. P. Schultz and collaborators, Fishes of the Marshall and Marianas Islands. *Bull. U.S. Nat. Museum*, (202), 1:504–520, 2 plates.

Zenkevitch, L. A. 1963. *Biology of the seas of the U.S.S.R.* George Allen and Unwin, London, pp. 1–955, 427 figs.

———, and J. A. Birstein. 1956. Studies of the deep water fauna and related problems. *Deep-Sea Res.*, 4:54–64.

Life in the
Deep Sea

The
Deep
Benthic
Realm

No matter how strange or rigorous the inanimate or noncompetitive environment may be, the evolutionary products of such environments are not as thoroughly refined by competition, and are thus not as well prepared for widespread success, as the products of highly competitive associations.

Hobart M. Smith *in* Evolution of Chordate Structure, *1960*

The Deep Benthic Realm extends from the edge of the shelf, at about 200 meters for most parts of the world, to the bottoms of the deepest trenches at almost 11,000 meters. Here, the greater part of the surface of the globe is covered by a submerged sandy or muddy layer that provides a habitat for a variety of burrowing animals. This group is called the *infauna*, and it is generally much richer in number of species than the *epifauna*, which is found on or just above the substrate.

Some authors recognize a separate "benthopelagic zone" inhabited by various species of animals that swim just above the sea floor. In this work, the term "benthic" is used in the broad sense to include all animals associated with the bottom whether they swim above it, sit upon it, or burrow within it. In general, the deep benthic fauna is not as well known as that of the shelf, and the various zoogeographic areas are suggested on the basis of relatively little data.

Until recent years, our knowledge about the general pattern of vertical distribution in the deep-sea benthic fauna was extremely fragmentary. Ekman (1953:267) recognized only an "archibenthal" fauna extending from the outer edge of the shelf to about 1,000 meters and an "abyssal" fauna for the greater depths. Bruun (1956) termed the fauna from the shelf edge to 2,000 meters "bathyal," that from 2,000 to 6,000 meters abyssal, and coined the term "hadal" for the fauna of the trenches (below 6,000 meters). In his chapter on the classification of marine environments, Hedgpeth (1957:18) depicted a bathyal fauna extending from the shelf to about 4,000 meters, an abyssal assemblage from about 4,000 to 6,000 meters, and a hadal fauna for the trenches.

Vertical Distribution

Vinogradova (1962) wrote a review article which analyzed data on the vertical zonation of the benthic animals that were taken mainly by the Russian vessel *Vitiaz*, the Danish *Galathea*, and the Swedish *Albatross*. The analysis included a total of 1,144 species belonging to most of the major marine invertebrate groups. She found that the changes in the systematic composition of the bottom fauna were most pronounced at a depth of about 3,000 meters. Here, a large number of species, genera, and even families characteristic of the slope disappear and are replaced by groups peculiar to the greater depths. For this reason, she identified this depth as the upper limit of the abyssal zone.

Another, less abrupt, change in the fauna led Vinogradova (op. cit.) to subdivide the abyssal zone into two subzones, an "upper-abyssal" extending from 3,000 meters to about 4,500 meters and a "lower-abyssal" from the latter depth to about 6,000 meters. For the abyssal zone in its entirety, a preliminary estimate of the endemism indicated that 58.5 percent of the species were confined to that area. For depths greater than 6,000 meters, an "ultra-abyssal" (hadal) zone was recognized. Vinogradova stated further that these faunal changes were found to occur at similar depths in the Atlantic, Pacific, and Indian Oceans and that the greatest changes coincided with the sharpest changes in the hypsographic curve (profile) of the earth's crust, i.e., at about 200, 3,000, and 6,000 meters. The foregoing scheme has found general acceptance among the Russian scientists (Zenkevitch 1963:714).

Wolff (1960) published the first detailed summary on the marine life of the trenches (the hadal zone) in which he noted that a total of 127 species had been identified. Of these, about 74, or 58 percent, appeared to be endemic to such depths. A more recent evaluation has been provided by Belyaev (1966a and b), who brought the total of known hadal species to 280 and found the rate of endemism to be about 68 percent. Menzies and George (1967) suggested that the recognition of a distinct hadal fauna may not be justified. Without doubting the high rates of endemism, they observed that the trench species did not seem to possess pecularities that were not also found in abyssal species. They further suggested that comparable rates of endemism may be characteristic of many of the shallower ocean basins.

Invertebrates   The epibenthic sled (Hessler and Sanders 1967) has proved to be a remarkably efficient collecting device for the small animals comprising both the infauna and epifauna of soft bottoms. A series of samples taken at various depths along a line between Massachusetts and Bermuda have yielded some very interesting results (Sanders, Hessler, and Hampson 1965): the outer continental shelf was found to support 6,000 to 13,000 individual animals per sq meter, the upper continental slope 6,000 to 23,000, the lower continental slope 1,500 to 3,000, and the abyssal depths generally only a few hundred. The most abundant group was the Polychaeta which was represented by numerous species at all collecting stations.

The polychaetes of the above transect were described in detail by Hartman (1965), who found no less than 266 species belonging to 50 families. She observed that none of the species was ubiquitous and that most were restricted to one or a few stations, demonstrating a distinct horizontal layering. The following complexes were identified: a New England slope fauna from 200 to 2,000 meters, a New England abyssal fauna from 2,469 to 2,762 meters, and a Bermuda slope fauna from 1,000 to 2,500 meters. The benthic amphipods from the same series of samples were analyzed by Mills (1966). He found a marked restriction to narrow depth ranges and recognized five deep-water zones: continental slope 200 to 300 to 1,000 meters, slope–deep-sea transition 1,000 to 2,000 meters, abyssal 2,000 to 2,800 meters, abyssal–deep-abyssal transition 2,800 to 4,000 meters, and deep-abyssal below 4,000 meters.

Phleger (1960) reviewed the data on the depth distribution of the benthonic species of Foraminfera for various parts of the world. He concluded that rather widespread boundaries occurred at 20, 50, 100, 200 to 300, 400 to 500, 1,000, and 2,000 meters. He further noted that the major faunal boundary at about 100 meters coincided with the approximate depth for the seasonal water layer and presumed that the boundary at approximately 1,000 meters represented the bottom of the permanent thermocline. The latter two boundary estimates were based on observations made in the northwestern Gulf of Mexico. Alton (1966) studied the bathymetric distribution of sea stars (Asteroidea) off the northern Oregon coast. He recognized an outer sublittoral fauna at about 100 to 200 meters, an upper bathyal at 240 to 500 meters, a lower bathyal at 520 to 950 meters, and an abyssal at 1,700 to 2,000 meters.

In his study on the benthic caridean shrimps of the western North Atlantic, Thompson (1963) found that catches from the east coast of the United States, the northeast coast of South America, and the east coast of Central America demonstrated similar depth patterns even though the levels occupied by individual species varied from place to place. The 200-meter isobath was recognized as the boundary separating the shelf species from those of the upper and middle slopes. This depth was also observed to be the upper limit of constant year-round temperature. An increase in species and in trawlable biomass began at about 360 to 400 meters where the temperature average was about 10°C. This assemblage continued downward to about 900 to 1,200 meters, where another change in faunal composition occurred.

Fishes

The research vessel *Oregon*, operated by the U. S. Bureau of Commercial Fisheries, has made two exploratory fishing surveys of the upper slope in the western Caribbean Sea (Bullis and Struhsaker 1970). A characteristic association of fish species was found to occupy the upper slope. The upper limit occurred in the 150- to 250-meter depth range where typical upper-slope families were replaced by shelf families. The lower limit at about 700 meters was marked by a drastic reduction in the total number of families and a replacement by lower-slope families. The upper-slope fauna was found to be distributed approximately between the limits of the permanent thermocline layer (19°C down to 7°C). The greatest number of species and individuals occurred at about 450 meters in the vicinity of the 10°C isotherm.

Summary   Although some of the above data seem to be in conflict, possibly due to different responses of the species in various animal groups to the effects of such environmental factors as temperature and depth, one can discern a general pattern that may serve as the basis for a standard distributional outline. The abrupt faunal change that takes place at the edge of the shelf (at about 200 meters) is now so well documented that it has been almost universally accepted. So far, there appears to be quite a good correlation between the permanent thermocline zone from 200 to about 1,000 meters and the occurrence of the *upper-slope* fauna. It should be noted that in the Western Atlantic the thermocline is sometimes considered to be more restricted so that its lower boundary may lie at about 700 meters. The usual thermocline depth limits of 200 to 1,000 meters coincide nicely with the boundaries of the mesopelagic, or twilight, zone of the Pelagic Realm (p. 305).

At about 3,000 meters, there is a break in the hypsographic curve of the ocean basins, and this appears to be accompanied by a very distinct faunal change. We can therefore refer to a *lower-slope* fauna from 1,000 to about 3,000 meters. The abyssal plain generally extends from 3,000 to about 6,000 meters, but there is some evidence for another faunal break at about the midway point—4,500 meters. However, at present, it would be so difficult to justify the recognition of separate upper-abyssal and lower-abyssal faunas that it seems best to stay with the concept of a single *abyssal zone*. The very high endemism (68 percent) so far demonstrated by the benthic fauna of the trenches indicates that these areas belong to a distinct *hadal zone*.

Upper Slope   It has been observed that most species of deep-sea fishes are bound narrowly to the continental slopes where food is more plentiful than to the abyssal reaches (Marshall 1967:477). Furthermore, it is apparent that there is not only a sharp taxonomic difference between the upper and lower slope environment, but the former is far richer in numbers of species, genera, and families. The number of fish families that are confined or virtually confined to the upper slope is surprisingly large. In some areas, the upper slope apparently supports a more diverse fish fauna than the outer shelf.

These phenomena of surprising richness and great systematic distinctiveness for the upper slope are probably also true for the benthic invertebrates. This has been demonstrated for the colder waters of the western North Atlantic by work with the epibenthic sled (Sanders et al. 1965) and has been indicated for the tropical western Atlantic by work on the caridean shrimps (Thompson 1963). Within the confines of the upper slope, a good many of the fish species show an increase of size with depth. It has been suggested that the young of such species are epipelagic and that they only gradually move downward as they grow older (Bullis and Struhsaker 1970, Marshall 1967:477).

Horizontal Distribution
Western Atlantic Region

Because of its uneven topography and generally narrow width, the slope is a difficult environment to sample adequately with trawls and dredges. However, in recent years a number of trawling surveys have been undertaken with the hope of discovering populations of commercially exploitable fishes. Consequently, there is more information available about the horizontal distribution of the fishes than the invertebrates. In the western Atlantic, it is instructive to compare the results of the trawls made in the western Caribbean by the *Oregon* (Bullis and Struhsaker 1970) with those made between Virginia and Nova Scotia by the *Cap'n Bill II* (Schroeder 1955).

Beneath the tropical waters of the western Caribbean, there is a rich slope fish fauna consisting of about 115 species belonging to 50 families. Along the slope beneath the cold-temperate waters to the north are about 75 species belonging to 35 families. Although the difference in diversity is marked, the great contrast between the two areas lies in the systematic relationship of the faunas. Of the 75 species reported from the Mid-Atlantic Seaboard, only 8 were also found to occur in the western Caribbean. Thus, the slope fish faunas of the two areas are vastly different, probably as much so as the shelf faunas.

The next important question is: Where is the zoogeographic boundary (or boundaries) between the lower and higher latitudes of the Western Atlantic? Certain groups of fishes are relatively easy to capture by deep trawling. Among these are the batoid fishes (skates and rays) belonging to the families Rajidae and Anacanthobatidae. Bigelow and Schroeder (1962, 1965a, 1968) have shown that in the tropics there are nine slope species belonging to the genus *Raja*, and using the same distributional criteria that we applied to the shallow-water species, all could be termed either tropical or eurythermic tropical, i.e., many do not extend north of Florida and none north of Cape Hatteras. The same pattern is found for the five species in the genus *Breviraja* and the four belonging to *Cruriraja*. *Dactylobatis armatus* is so far known only between Cape Kennedy and South Carolina and *Springeria folirostris* is apparently confined to the northern Gulf of Mexico.

To the north off the Canadian Atlantic coast, all of the slope species of *Raja* have boreal or arctic-boreal distributions (Leim and Scott 1966). A report on the batoid fishes of the cold-temperate waters of the western South Atlantic has shown that all five species (in three different genera) are probably confined to such waters (Bigelow and Schroeder 1965b). In the western North Atlantic, the catsharks of the family Scyliorhinidae are a typical upper-slope group, since 95 percent of the specimens have been taken between 183 and 914 meters (Springer 1966). Three of the five species of *Scyliorhinus* are restricted to the tropics, one ranges from the western Caribbean to New England, and one from the Lesser Antilles to Argentina. Another member of the family, *Galeus arae*, extends from northern Florida to Colombia and demonstrates a depth segregation within 300 to 760 meters based on sex and maturity (Bullis 1967).

Many of the squaloid sharks (family Squalidae) are more widely distributed, with five species that are apparently bipolar in temperate waters and one (*Squalus acanthias*) that is virtually cosmopolitan (Garrick 1960). However, in the squaloid genus *Etmopterus* there appear to be two species confined to the northern Gulf of Mexico (Bigelow and Schroeder 1957).

Although many bony fish families are typical of the upper slope, the distribution of the various species is, in general, not as well known as that of the elasmobranch species mentioned above. A survey of the deep sea bottom fishes of the Gulf of Mexico below 350 meters (mainly at slope depths) was conducted by Bright (1968). He concluded that this area contained an extension of the Caribbean fauna; 25 of the 219 species listed (11.4 percent) are apparently endemic to the Gulf of Mexico.

The order Zeiformes, including six families (Parazenidae, Macrurocyttidae, Zeidae, Grammicolepidae, Oreosomatidae, and Caproidae), is almost entirely upper slope. Other slope families of importance in the

Western Atlantic are the Argentinidae, Chlorophthalmidae, Bathypteroidae, Gadidae, Moridae, Macrouridae, Apogonidae, Percophididae, Gemypylidae, Trichiuridae, Scorpaenidae, Peristediidae, Draconettidae, Brotulidae, Bothidae, Triacanthodidae, and Ogcocephalidae. The eelpouts (family Zoarcidae) are almost entirely temperate or polar, the northern species being boreal, arctic-boreal, or arctic (Leim and Scott 1966) and the South American species being apparently confined to the cold-temperate waters of that area (Norman 1937).

In regard to the invertebrates, the tropical caridean shrimps were investigated by Thompson (1963), who showed that the great majority of the species were either confined to the tropics or else extended into warm-temperate latitudes. Very few were found to range into cold-temperate waters.

In summary, it can be said that most of the better-known tropical Western Atlantic slope species apparently conform to one of three basic distributional patterns as defined by the temperature of the surface waters. They may be restricted to the tropics (tropical species), extend also into warm-temperate waters (eurythermic tropical species), and a few may extend even further into cold-temperate waters (broad eurythermic tropical species). On the other hand, the cold-temperate areas such as those of New England or Argentina possess slope faunas that are highly distinctive with very few species that also range to the tropics. So, at this time, we can certainly recognize in the Western Atlantic tropical and temperate assemblages but, with the exception of the northern Gulf of Mexico, evidence for distinct slope faunas in the warm-temperate latitudes is scarce.

It appears that the locations of greatest faunal change for the upper slope of the Western Atlantic may coincide with the latitudes of the boundaries between the warm-temperate and cold-temperate faunas of the shelf; that is, off Cape Hatteras in the north and off the mouth of the Río de la Plata in the south.

Eastern Atlantic Region

Blacker (1962) reported a total of 81 benthic fish species from the slopes to the west of the British Isles. It is interesting to compare his catches with those of Schroeder (1955) taken from the opposite side of the North Atlantic. Both surveys were confined to slopes beneath cold-temperate waters, and each consisted of over 200 hauls. Schroeder took a total of 75 species, but the significant difference is that only 10 species were taken in both areas. This is a good indication that the slope fish faunas of the two sides of the boreal North Atlantic are markedly different.

We can also take the opportunity to compare the slope fish fauna of the Eastern Atlantic tropics with that of the boreal zone on the same side of the ocean. Here again there is a marked contrast similar in degree to that between the tropical and boreal Western Atlantic. In 1948 and 1949, the Belgian Oceanographic Expedition to West Africa worked with otter trawls on both the shelf and slope in the tropical waters between Sénégal and Mossâmedes. Only 12 of the slope species reported by this expedition (Poll 1951, 1953, 1954, 1959) were also taken by Blacker (1962) off the British Isles.

When the slope fishes of the tropical Eastern Atlantic are compared to those of the tropical Western Atlantic (Bullis and Struhsaker 1970), the contrast is also very striking. Of the 70 upper-slope species found in the Eastern Atlantic, only 7 have been taken on the western side. It may be recalled (p. 109) that about 25 percent of the tropical Eastern Atlantic shelf fishes are transatlantic in distribution, so the slope faunas are apparently considerably more distinct than the shelf faunas. Some of the invertebrate groups are not this well separated. Thompson's (1963) study on the caridean shrimps of the tropical Western Atlantic showed that 15 of the total of 33 species had transatlantic distributions.

Between the Eastern Atlantic tropics and the cold-temperate waters of northwestern Europe, there probably exists a distinct warm-temperate slope fauna. Dr. G. E. Maul of Madeira has described, in a series of papers published in the *Boletim do Museu Municipal do Funchal* from 1946 to date, an impressive number of slope fishes that are so far known only from Madeira or from Madeira and the adjacent waters of the Eastern Atlantic. In this area, as in most parts of the world, the ahermatypic corals are more numerous on the upper slope than on the shelf or at greater depths. According to the classic revision by Vaughan and Wells (1943:85), more than half of the Eastern Atlantic warm-temperate coral species are apparently endemic. Also, most of the slope species of the Mediterranean appear to be endemic to that area.

South Africa   In recent years, a good deal of deep-water trawling has been carried on off the South African coast by the *Africana II* of the South African Division of Sea Fisheries and also by the *Vema* from the Lamont Geological Observatory of Columbia University in the United States. So far, large numbers of new species have been described (mainly in the

Annals of the South African Museum) with particular attention paid to the mollusks and polychaetes. This information, although far from conclusive, appears to indicate the presence of a distinctive slope fauna. Earlier, Vaughan and Wells (1943:87) had reported a deep-water, ahermatypic coral fauna of 30 species; of these, 16 were considered to be endemic, 9 shared with the North Atlantic, and 6 shared with the Indo-West Pacific. Mortensen (1933) listed 96 species of upper-slope echinoderms (Asteroidea and Ophiuroidea) taken between Cape Province and Natal; 31 are apparently endemic and restricted to the warm-temperate waters between the Cape of Good Hope and the Kei River mouth.

In regard to the southern part of the western Pacific, more deep-water trawling has been done in New Zealand than in any other area. In his extensive paper on the molluscan fauna of the New Zealand slope, Dell (1956) recorded a total of 595 species but noted that complete collections had been made only as deep as 330 fathoms. About one-fifth of the total were considered to be shelf species that continued over the edge to various depths on the upper slope. The very steep slopes or heads of submarine canyons seemed to have unusually large numbers of shelf species. Dell described 91 new species in this work, and since the majority of those already known appear to be confined to New Zealand, the rate of endemism is probably very high. The author concluded that the slope fauna was in large part derived from the shelf and noted that a number of Tertiary genera, which had previously been considered extinct in the New Zealand area, were now known to be still living on the continental slopes.

Indo-West Pacific

The ahermatypic corals of New Zealand were found to be related to those of the Indo-West Pacific, with 10 of the 18 species being endemic (Vaughan and Wells 1943:88). Fell (1958) reported that the 91 species of slope echinoderms represented a mixture of endemics and more widely distributed forms from the Indo-West Pacific, North Pacific, and Sub-Antarctic. Almost all of the squaloid sharks (family Squalidae) from New Zealand slopes are very broadly distributed (Garrick 1960), and the same appears to be true for the three species of benthic eels (Castle 1961). Also, most of the holothurians apparently range broadly in the Pacific or are cosmopolitan (Pawson 1965).

Vaughan and Wells (1943:88) noted that a very distinct fauna of ahermatypic corals was found in the waters off the Great Australian Bight and through Bass Strait to Brisbane; of 50 species, only 14 were known from other areas. They also found that the deep-water coral fauna of the East Indies is, like that of the reef corals, the richest in the world with a total of 49 genera and 152 species and varieties. About half the East Indian species are apparently endemic. The fish fauna of the upper slope in the Indo-West Pacific is also very rich. For example, Norman (1939) reported on a collection of 276 species belonging to 37 families taken mainly from the slopes of the northern Indian Ocean by the John Murray Expedition. Knudsen (1967), in his article on the bivalves of the same expedition, found 34 slope species and noted that they seemed to be part of a general Indo-West Pacific fauna having the same distribution as the adjacent shallow-water fauna. He noted further that the fauna was distinct from that of neighboring areas such as New Zealand–Tasman Sea, the Great Australian Bight, and South Africa.

To the north in Japanese waters, the slope fish fauna is apparently still very rich. Kamohara (1964) has maintained records for many years on the occurrence of offshore bottom fishes at Kochi Prefecture on the east coast of Kyushu Island. Evidence from this paper and from revisionary works such as those on the Champsodontidae (Matsubara, Ochiai, Amaoka, and Nakamura 1964) and the Hoplichthyidae (Matsubara and Ochiai, 1950) indicate that a separate warm-temperate fish fauna exists in Japanese slope waters.

At the Hawaiian Islands, the slope fauna has yet to be thoroughly investigated. In regard to the fishes, many of the slope species originally described by Gilbert (1905) have never been taken elsewhere. There are 37 Hawaiian species of ahermatypic corals; of these, 26 are endemic, 3 cosmopolitan, and the rest are shared with other parts of the Indo-West Pacific (Vaughan and Wells 1943:89).

Since most of the Okhotsk Sea is less than 2,000 meters in depth and almost all of it less than 3,000 meters, the detailed analysis of the deep-water fish fauna by Shmidt (1950) provides some useful information about slope distribution in the boreal waters of the northwestern Pacific. He found the benthic fish fauna to consist of 48 species belonging to 7 families; 15 of the species are shared with the Sea of Japan immediately to the south and only 13 with the Bering Sea. So far, it appears that there is a large number of indigenous species, with the rest extending to adjacent parts of the North Pacific.

Eastern Pacific Very little information has been published on the slope fauna of the American west coast. Considerable work has been done in southern

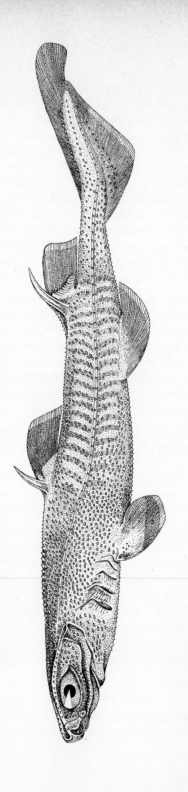

Figure 11-1 The deep-water squaloid shark (*Centroscyllium nigrum*). This species is so far known from the upper slopes of the Hawaiian Islands to the tropical Eastern Pacific. After Garman (1899).

California by the *Velero IV* of the Allan Hancock Foundation, but the material so far published deals mainly with the isolated invertebrate faunas of certain deep basins and submarine canyons. Among the basins, which range in depth from 627 to 2,107 meters, many species of polychaetes were found to have a wide vertical range in the Eastern Pacific and to extend horizontally into far northern waters (Hartman and Barnard 1958:17). Twenty-five isopod species were collected from the slopes of the submarine canyons; of these, 15 were described as new, 6 were previously known only from California, 1 from the North Pacific, and 3 were considered to be cosmopolitan (Schultz 1966:1). The canyon amphipod fauna appeared to have its main relationship toward the North Pacific, but the author (Barnard 1966:28) emphasized that the broader distributions of this group were very poorly known.

A comparative study of the macroinvertebrates of the Gulf of California and of the continental slope off Mexico was undertaken by Parker (1964). He found that the northern Gulf basins and troughs, extending from 230 to 1,500 meters, had a molluscan fauna comprised of California province shelf species. In contrast, the slopes off southern California and Mexico were populated by deep-water species. The California slope seemed to be occupied by northern benthic animals which had migrated south, while the Mexican slope appeared to be occupied by tropical species that had moved northward. Parker suggested that shallow ridges cutting across the slope into Baja California may serve to separate the two faunas. Also, he found the fauna of the lower Mexican slope to be richer and more complex than anywhere else in the deep sea.

At the Galápagos Islands, 10 deep-water, ahermatypic coral species have been taken; 7 are endemic and 3 cosmopolitan (Vaughan and Wells 1943:89).

Arctic   Zenkevitch (1963:61) presented a summary of the earlier work of G. Gorbunov on the benthic fauna of the slopes of the Arctic Basin. It was observed that the slope ("bathyal") fauna was about four times as rich as that of the abyssal plain. Gorbunov identified a total of 429 species from the slopes of the Siberian sector. Of these, 312 were shared with the shelf, 37 with the abyssal zone, and 80 were considered to be typical slope species; 59 of the 80 slope species were listed as being endemic to the Arctic Ocean. In general, this fauna was considered to be related to the abyssal fauna of the Atlantic but quite separate from that of the Pacific. An impressive number of the mollusks of the slope appear to be confined to or to extend not far beyond the Arctic Ocean (Clarke 1963).

Very little is known about the pattern of upper-slope distribution in the Antarctic. For the Atlantic sector, Vinogradova (1964) observed that about 70 percent of the benthic invertebrate species occurring between 200 and 2,000 meters were shared with other parts of the Atlantic. In the Pacific sector, about 60 percent of the species were found in other parts of that ocean. At greater depths, the extent of Antarctic endemism was found to be much higher.

Antarctic

It is clear that the upper slope, extending from about 200 to 1,000 meters, is an environment that supports a rich and distinctive fauna. There are not only many species characteristic of this habitat but also many genera and families. Although the amount of collecting has been limited, one can find in many places highly characteristic faunas, suggesting that most upper-slope species are closely tied to their particular habitat.

Summary

For the few parts of the world where extensive trawling surveys have been undertaken, the available information appears to indicate:

1 The richest slope faunas in terms of numbers of species are located beneath the tropical surface waters. As far as the fishes are concerned, the fauna of each tropical region seems to be highly distinctive, more so than the shelf fauna. The tropical Indo-West Pacific probably possesses the most diverse upper-slope fauna, followed by that of the tropical Western Atlantic. The upper-slope faunas of some tropical, oceanic islands (Hawaii and the Galápagos) appear to be highly endemic.

2 Exploration in the northeastern Atlantic and off South Africa, New Zealand, Southern Australia, and Japan have revealed the presence of distinctive upper-slope faunas beneath the warm-temperate waters of these areas. However, separate warm-temperate slope faunas have yet to be identified in the Western Atlantic and Eastern Pacific. There are indications from the presence of endemic species that a distinct fauna may exist in the northern Gulf of Mexico.

3 The upper-slope fauna beneath the cold-temperate or boreal waters appears to be quite separate for each region, in terms of its relationship to adjoining faunas, and there seems to be very little affinity among regions. For example, we know that the boreal shelf faunas of the eastern and western North Atlantic

share a good many species, yet the slopes of these areas share very few. In the northwestern Pacific, the Okhotsk Sea appears to be an important area of endemism for slope species.

4   The Arctic Basin seems to support a very distinctive fauna, with the majority of the slope species being endemic. The fauna of the Antarctic slopes may be somewhat less distinctive than that of the Arctic, but so little work has been done that an accurate evaluation is not yet possible.

5   In some parts of the world at least, there are indications that the horizontal distribution of the upper-slope faunas closely follows that of the shelf, and we can recognize distinct tropical, warm-temperate, cold-temperate, and cold (Arctic and Antarctic) assemblages. If these initial indications are reinforced by further investigations, we will be able to recognize a relationship very similar to that found for the Pelagic Realm where the horizontal distributional patterns of species in the 200- to 1,000-meter layer almost exactly duplicate those of the surface fauna. For the present, it seems appropriate to use for the fauna of the upper slope the same regional system that has been outlined for the shelf.

Lower Slope   In fishes, the remarkable contrast between the upper-slope and lower-slope faunas is nicely demonstrated by the report of Bullis and Struhsaker (1970) on their trawling survey in the western Caribbean. A diverse fauna including representatives of 50 families was found on the upper slope. This was reduced to only 8 families at 850 to 950 meters. Furthermore, it was found that the great majority of the species on the lower slope (about 68 percent) belonged to a single family, the Macrouridae. The other benthic fish families found on the lower slope were the Alepocephalidae, Congridae, Brotulidae, Rajidae, Halosauridae, Bathypteroidae, and Synaphobranchidae. Off the Oregon coast, where the fauna is considerably less diverse, an upper-slope complex of about 27 species was reduced to only 3 species at 1,350 to 1,829 meters (Day and Pearcy 1968).

On examining the reports of deep-sea expeditions and other publications in which lower-slope fishes have been identified, one can see a basic difference at the family level between the cold-temperate and tropical faunas. For example, in the depths of the Okhotsk Sea, 37 of the 52 reported species belonged to two families, the Liparidae and the Zoarcidae (Shmidt 1950), but in the tropical seas it is the Macrouridae and Brotulidae that are dominant (Grey 1956).

The grenadier fishes of the family Macrouridae comprise a large group of more than 300 species, and at least 90 percent are restricted to the slope environment (Marshall 1965). With the exception of a few bathy-pelagic species (p. 345), almost all have very limited horizontal distributions. Of 30 species known from the western North Atlantic, only 4 also occur along the slopes of the eastern North Atlantic; of the latter 4, 2 are found in the northernmost part of the Atlantic where there are undersea ridges that could facilitate migration between the two sides of the ocean, and 1 has been found to inhabit the abyssal plain as well as the slope.

The Macrouridae is the largest family of deep-sea bottom fishes, and since the great majority of the species appear to be confined to the lower slope, investigations into the evolutionary and geographic radiation of this group should reveal important information about its environment. As things now stand, both the family and its habitat are poorly known. Utilizing the information available at the time, Bolin (1957) presented a map of the deep-water biological provinces of the Indo-Pacific based on the distribution of the Macrouridae. He recognized Antarctic and Antiboreal Provinces to the south and a Boreal Province for the far north. The tropical and warm-temperate areas were divided as follows: Californian, Panamanian, and Peruvian Provinces for the Eastern Pacific; Japanese, Hawaiian, Indo-Australian, and Australo-Pacific Provinces for the Western Pacific; and a single Indian Ocean Province for that area.

As was noted previously, the two dominant, deep-benthic fish families in the cold waters of the Northern Hemisphere are the Zoarcidae (eelpouts) and the Liparidae (sea snails). In the North Pacific, these families are represented by numerous genera and species and comprise a conspicuous part of the fauna in both the shelf and slope habitats. In comparison, there are relatively few species of each group in the North Atlantic so that it seems apparent that the North Pacific has functioned as the primary center of evolution and distribution. Both families were probably exclusively North Pacific in the early stages of their development.

Penetration into the Arctic Basin and thence to the North Atlantic was probably achieved by the zoarcids and liparids when the Bering Land Bridge was inundated in the late Miocene and/or the late Pliocene (p. 417). In addition, accommodation to the depths of the lower-slope environment made it possible for both groups to extend their ranges southward along the American Pacific coast without encountering prohibitively high temperatures. The discovery by Garman in 1899 of four zoarcid and three liparid species on or close to the lower slope

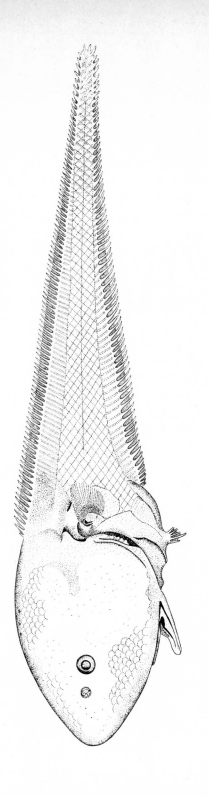

Figure 11-2 A grenadier fish *Squalogadus intermedius.* This peculiar species is known so far only from the lower slope of the Gulf of Mexico. After Grey (1959).

beneath the tropical waters of the Eastern Pacific provided visible evidence of isothermic submergence. Proceeding farther south, and upon reaching the cold waters of the Southern Hemisphere, species of these families were again able to occupy both deep and shallow waters. Now they are numerous about the southern tip of South America (Norman 1937) and have also reached the Antarctic Continent (Rofen and DeWitt 1960) and the Cape of Good Hope region of South Africa (Smith 1949). The species inhabiting each of these widely separated areas are almost all distinct, a possible indication that the original migrations took place long ago.

Despite the foregoing indications, one cannot assume that all of the benthic fishes on the lower slope have highly restricted distributions. The deep-water eels of the family Synaphobranchidae are, with the exception of one abyssal species, almost entirely confined to the lower slope; five of the seven slope species have been taken in more than one ocean (Castel 1960, 1961, 1964, 1968). Two species, *Antimora rostrata* in the family Moridae (Grey 1956:160) and *Acanthonus armatus* in the Brotulidae (Nielsen 1965), have worldwide (circumtropical) distributions. In the family Ipnopidae, two of the three lower-slope species appear to be widely distributed (Nielsen 1966): *Ipnops agassizi* extends from the western Indian Ocean to the Eastern Pacific, and *Bathytyphlops marionae* from the Western Atlantic to the western Indian Ocean.

Information about the distribution of invertebrate groups typical of the lower slope environment is very fragmentary. The primitive, seastar family Porcellanasteridae is found only at depths greater than 1,000 meters (Madsen 1961). Three monotypic genera belonging to this family are confined to the lower slope and seem to be restricted to the Indian Ocean and Malay Archipelago. Almost all the other genera occur at greater depths and are more widespread.

In his work on the taxonomy and zoogeography of the deep-sea holothurians, Hansen (1967) found that the species tended to be widely distributed, but he also observed that, in contrast to the abyssal species, none of the slope species appeared to be cosmopolitan. Thompson (1963), who studied the caridean shrimps of the slope in the tropical Western Atlantic, presented data which indicated that about half the lower-slope species were endemic to that side of the ocean.

In reporting on the slope and abyssal Isopoda Asellota, Wolff (1962) observed that there were about 140 typical slope species (including most of the 15 species which penetrate beyond 2,000 meters) and that almost 60 percent were restricted to the Atlantic Ocean and practically all of them to one section of it. In examining Barnard's (1961) work on the gammaridean amphipods taken from 400 to 6,000 meters, one is struck by the large number of species with apparently limited geographic ranges. The bivalve genus *Xylophaga* is a group essentially confined to the subthermocline regions, and so far, the horizontal distribution of the species appears to be highly restricted (Knudsen 1961:203).

In the deep sea, the primitive gastropod order Archaeogastropoda and the primitive pelecypod order Protobranchiata are well represented, while in shallow waters, they are outnumbered by species belonging to more advanced groups (Clarke 1962a). In the archaeogastropods, it is interesting to see that no less than 102 of the 126 deep-water species occur on the slopes and that 56 species are apparently confined or almost confined to the lower slope (Clarke 1962b). Similarly, in the protobranch pelecypods about 94 of the 136 deep-water species are found on the slopes with 48 virtually restricted to the lower slope. In general, these lower-slope mollusks appear to have remarkably limited horizontal distributions, almost all being recorded from a single area within one ocean.

Finally, Southward and Southward (1967) have determined that the phylum Pogonophora inhabits mainly the continental slope and rise. Their data also seem to suggest that these animals are most numerous on the lower part of the slope. It may be significant that, although 40 species have been described from the northwestern Pacific, only 4 of them have been taken at more than one locality.

In summary, it can be said that the lower-slope fauna, occupying a zone from 1,000 to about 3,000 meters, is much less diverse than that of the upper slope. In regard to its zoogeography, we can only say with certainty that most of the species, with the exception of the holothurians and some of the fishes, appear to have very limited horizontal distributions and that the high-latitude fauna seems to be quite different than that of the lower latitudes.

Why should there be such a multitide of short-ranged species? If one examines a relief map of the floor of the world ocean, it can be seen that the lower-slope habitat is by no means confined to the continental margins. There are shallow rims of deep basins, oceanic rises, mid-ocean ridges, and the bases of islands and sea mounts. It may be that each isolated or semi-isolated elevation that extends above 3,000 meters has developed its own faunal complex with a significant amount

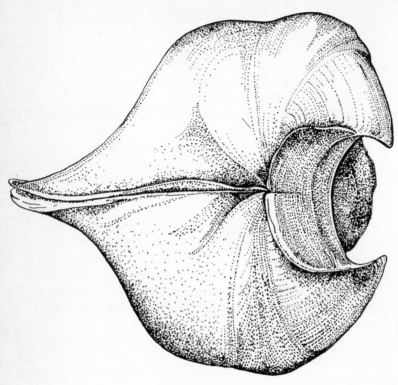

Figure 11-3   A primitive, wood boring bivalve (*Xylophaga concava*). Taken by the *Galathea* Expedition from the lower slope off the Gulf of Panama. After Knudsen (1961).

of endemism. If so, we will eventually need to recognize an enormous number of lower-slope provinces.

Abyssal Zone

Although our knowledge about the abyssal plain has increased considerably during the past 20 years, especially due to the publications resulting from collections taken by the *Galathea* (Danish), *Ob* (U.S.S.R.) *Vitiaz* (U.S.S.R.), *Eltanin* (U.S.A.), and *Vema* (U.S.A.), we still know very little about this enormous and interesting area. As Bruun (1956) has pointed out, the abyssal plain occupies an area of about 273 million km. When compared to the total size of the earth, which is about 510 million km, we see that we are attempting to evaluate an area that occupies more than one-half the earth's surface. Madsen (1961:206) estimated that the total area covered by dredging operations could, at the very most, be about 5 sq km. This means that we have explored only about *one fifty-millionth* of the habitat occupied by the abyssal fauna!

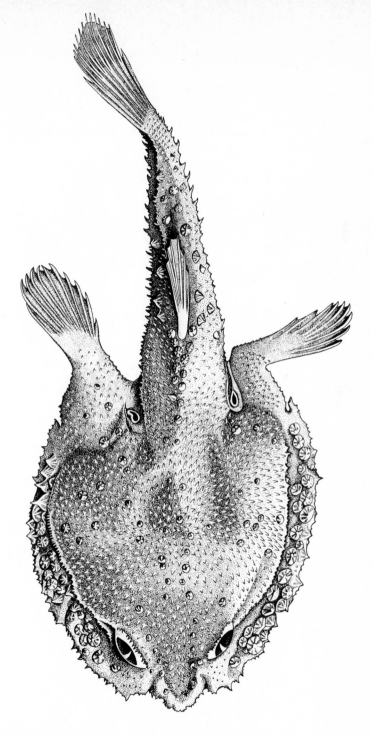

Figure 11-4 An ogcoceph-
alid (*Dibranchus spinosa*). A
lower-slope species confined
to the tropical Eastern Pacific. After Garman (1899).

In 1844, Edward Forbes, having completed work on the vertical distribution of benthic animals in the Aegean Sea, noted a marked decrease in diversity with depth and expressed the opinion that a "zero of animal life" would probably be found somewhere around 300 fathoms. This observation provided an impetus for biologists to test Forbes' theory by sampling greater depths. Subsequently, animal life was recorded from progressively greater depths, but not until the dredging cruises of H.M.S. *Porcupine* and H.M.S. *Lightning* in the summers of 1868, 1869, and 1870 were organisms taken from truly abyssal depths. As the result of this work, Wyville Thomson (1873:4) was able to state with assurance that "the bed of the deep sea, the 140,000,000 square miles of which we have now added to the legitimate field of natural history research, is not a barren waste."

The encouraging results of the work of the *Porcupine* and *Lightning* in the North Atlantic provided the necessary stimulus for the undertaking of a far more ambitious task, the outfitting of a ship to explore the deep seas of the world. A surplus English Navy ship, H.M.S. *Challenger*, a medium-size sailing vessel of 2,306 tons with an auxiliary steam engine, was made available. She left Portsmouth on December 30, 1872 with a scientific staff of six and a navy crew.

The epoch voyage of the *Challenger* is considered to mark the beginning of modern oceanography. She dredged the abyssal plain to more than 5,000 meters and found life everywhere; provided first-hand knowledge about deep-sea sediments; and established the main contours of the ocean basins, the major current systems, and the principal temperature patterns. She returned to Portsmouth on May 24, 1876, having spent more than 3 years at sea—a record that still stands for the longest continuous scientific expedition. Over 4,500 new species were described, and the scientific results of the expedition were eventually published in 50 large, quarto volumes, 40 of them devoted to zoology. This work still provides a foundation for every important library on oceanography.

The success of the *Challenger* inspired many other expeditions, some of the most notable being those of the American vessels the *Blake* and the *Albatross* (1877 to 1910), the work of the *Investigator* in the Indian Ocean (1885 to 1900) sponsored by the Asiatic Society of Bengal, the German *Valdivia* expedition (1898 to 1899), the French expeditions on the *Travailleur* and the *Talisman* (1880 to 1883), the expeditions of the Prince of Monaco on the *Hirondelle* and the *Princesse-Alice* (1892 to 1915), the Dutch *Siboga* expedition (1899 to 1900), a series of Danish expeditions in the 1920s and 1930s on the *Ingolf* and the *Dana*, the

Norwegian on the *Michael Sars* (1910), the English John Murray Expedition (1933 to 1934), and the Swedish Deep-Sea Expedition on the *Albatross* (1947 to 1948).

Despite the enormous geographic area covered by the abyssal plain, the physical characteristics at its surface are remarkably uniform. There is no light except that which is produced by the organisms themselves, the temperature is usually between 1.0 and 2.5°C, except for the Arctic and Antarctic regions where it is a little colder; dissolved oxygen is usually 3.5 to 6.0 milliliters per liter, salinity about 34.9 ‰, and pressure from 300 to 600 atmospheres. The substrate is, apparently almost everywhere, composed of fine particles forming a soft mud; at 3,000 to about 4,500 meters it is mainly a calcareus ooze (from skeletal remains of pteropods, foraminiferans, and coccoliths), and at 4,500 meters to the greatest depths it is mainly a red clay (apparently as the result of the solution of the carbonates in the skeletons of the planktonic animals).

In regard to general characteristics of the animals, Menzies (1965) has pointed out that the average size of all members of an abyssal population appears to be smaller by an order of magnitude compared to an equivalent population from the intertidal zone. He also observed that most were pale in color and that over 80 percent of the isopods were white. It was also noted that abyssal animals seemed to be mainly deposit feeders (over 90 percent of the isopods fall into this category).

Considering the uniformity of the physical surroundings, the presence of a soft mud substrate almost everywhere, and the predominance of a single feeding type among the animals, one might conclude that species—once having become adapted to such an environment—would become exceedingly widespread. The early naturalists did expect to find a monotonous, wide-ranging abyssal fauna, and as recently as 1957, Bruun predicted that the future would show that in general the true abyssal species are cosmopolitan because the abyssal zone formed an ecological unit with no barriers, except for isolated areas like the Sulu Sea.

In 1959, Vinogradova published a general article on the distribution of the abyssal fauna of the world and provided a map that outlined a series of different areas, subareas, and provinces. Madsen (1961:209) felt that Vinogradova's scheme was too complicated and that it was based on data from too many eurybathic species. Madsen proposed a much simpler arrangement in which he recognized only four regions (Arctic, Atlanto-Indian-West Pacific, East Pacific, Antarctic). Vinogradova (1966) reiterated her scheme for the Pacific Ocean and emphasized the relationship between the vertical and horizontal distribution patterns. She noted that, while the whole of the Pacific deep-sea fauna exhibited a 50.4 percent endemism, the animals confined to the upper abyss were 70.3 percent endemic, and those confined to the lower abyss were 92.6 percent endemic. The same effect was demonstrated in a comparison of the abyssal faunas on the eastern and western sides of the Pacific: 30 to 40 percent of the eurybathic species were found on both sides, but only 3 to 12 percent of the upper-abyssal forms were shared, and *none* of the lower-abyssal species were shared.

Although Vinogradova's works do distinguish between upper-abyssal and lower-abyssal faunas, this concept is difficult to apply to the individual groups of abyssal animals. Our information is still so fragmentary that one cannot with assurance identify this pattern within any given group. In fact, the two groups (the asellote isopods and the bivalve mollusks) upon which detailed analyses of vertical distribution have been made (Wolff 1962, Knudsen 1970), do not appear to support the idea of two vertically separated, abyssal faunas.

The following abyssal animal groups have, to some extent, been recently investigated:

Fishes

The valuable and useful compilations of widely scattered records on deep-sea fishes by Grey (1956) and Nybelin (1957) and a systematic work on the family Ipnopidae by Nielsen (1966) have been most helpful in providing a general picture of distribution on the abyssal plain. To date, a total of about 71 species of benthic fishes have been taken from 3,000 to 6,000 meters. Considering the vast extent of the world abyssal plain, the fish fauna so far appears to be rather depauperate.

Of the 71 species taken on the abyssal plain, about 20 appear to be primarily lower-slope species that have extended their vertical ranges only a short distance below 3,000 meters (to less than 3,660 meters). The remaining 51 can be considered as typical inhabitants of the abyssal plain. Of the 51 abyssal species, 14 are apparently quite eurybathic, since they have been taken on the lower slope as well as at considerable depths on the abyssal plain. This leaves 37 species which can, at least temporarily, be placed in a stenobathic category since they have been taken only at abyssal depths.

One would expect the eurybathic group to contain the more widely distributed species, since they should be able to surmount most underwater barriers with little difficulty. There are some indications that this is true. Most of the 14 eurybathic fishes have been taken in more than one deep basin, and one, *Bathysaurus mollis* (Mead 1966:111), has achieved a worldwide (circumtropical) distribution. Other widely distributed members of this group are *Nematonurus armatus* which extends from the western Atlantic to the western Pacific, and *Lionurus filicauda* from the South Atlantic to the southeastern Pacific. Six of the 14 have been taken at lower-abyssal (below 4,500 meters) as well as upper-abyssal depths.

It is very difficult to make any useful observations about the distribution of the stenobathic group because they are so poorly known; 28 of the 37 species are known only from captures of 1 or 2 specimens. It may be significant that 15 of them have been taken, so far, only from the lower part of the plain from 4,267 to 6,000 meters. However, since we know almost nothing about the true distribution of the individual species, the recognition at this time of a separate, lower-abyssal fish fauna would be premature.

Although one would not expect species belonging to the stenobathic group to range very far, some of them do. It is interesting to see that, in the family Ipnopidae (Nielsen 1966), *Ipnops meadi* extends from the western Indian Ocean to the eastern Pacific, *Bathymicrops regis* from the western Atlantic to off southeast Africa, and *B. brevianalis* from the western Indian Ocean to the western Pacific.

In comparison to the lower slope where the fauna is much more diverse and dominated, apparently over most of the world, by the family Macrouridae, the fishes restricted to the abyssal plain reveal so far a quite different complex. Here, the family Brotulidae is dominant with 18 species, next is the Macrouridae with 6, the Ipnopidae with 4, the Zoarcidae with 3, Liparidae with 3, Alepocephalidae with 2, and Bathypteroidae with 1. Considering that 71 species have been taken on the abyssal plain and that 37 may be restricted to it, the overall degree of endemism may be calculated at about 52 percent. But until we learn more about the fishes of this environment, this figure must be regarded as highly speculative and used with caution. Finally, although one can predict that the major ocean basins are likely to have their own characteristic fish faunas, including endemic species, the available data simply do not permit any definite decision on the matter.

Invertebrate
Groups

Clarke (1962a), in his review of the abyssal mollusks of the world, indicated that a substantial degree of endemism exists with respect to individual ocean basins. He listed a total of 1,087 species (not including cephalopods) that are found below 1,800 meters (1,000 fathoms) and noted that, although there are 47 recognized ocean basins, the mean distributional spread was only about 2.0 basins per species. In nearly 90 percent of the cases where more than one basin was occupied by a species, the basins proved to be contiguous. In contrast, a study by Knudsen (1970) on the abyssal Bivalvia (193 species) indicated generally broad distributions for this group with three of the species being considered cosmopolitan.

The great majority of deep-water isopods belong to the Asellota, a group that has been revised by Wolff (1962). Of 202 species occurring deeper than 2,000 meters, none are cosmopolitan and only 3 (1.5 percent) are known from more than one ocean. The percentage of relatively widespread species appears to be much higher among the eurybathic species than among the truly abyssal species. A high rate of endemism is indicated by the large number of new species that are taken by each new expedition. A revealing contrast was made by Menzies (1965) who compared the abyssal isopods from the adjacent Argentine and Cape Basins in the South Atlantic: of a total of 58 species, only 14 percent occurred in both areas and *none* of the species were known from other oceans! Menzies also observed that although many of the abyssal isopod genera were widespread, some were restricted, i.e., *Serolis* in Antarctic waters and *Mesidothia* in the Arctic.

In discussing the distribution of the abyssal, benthic amphipods, Barnard (1962) observed that in the low latitudes of the world, 50 of the 58 species were known from a single record, a single basin, or a single region. Only 8 species were considered to be widely distributed, and 4 of them were known to be eurybathic. Barnard considered that the most striking aspect of the evidence supporting regional endemism was the high rate of recovery of new species and the low rate of recovery of previously known species by each succeeding expedition.

The Porcellanasteridae is a primitive sea star family that is exclusively deep sea (below 1,000 meters) and primarily abyssal. The *Galathea* Expedition took specimens of this family in 27 dredge hauls spread along its whole circumnavigation route, yet not a single species found was new to science (Madsen 1961). Six of the total of 14 species known from more than two localities (43 percent) were considered to be cosmopolitan (occurring in the main deep-sea basin of all three oceans). Madsen also observed that, for echinoderm species in general occurring below 3,000 meters, about 5 percent of the asteroids and about 10 percent of the ophiuroids and holothurians are cosmopolitan.

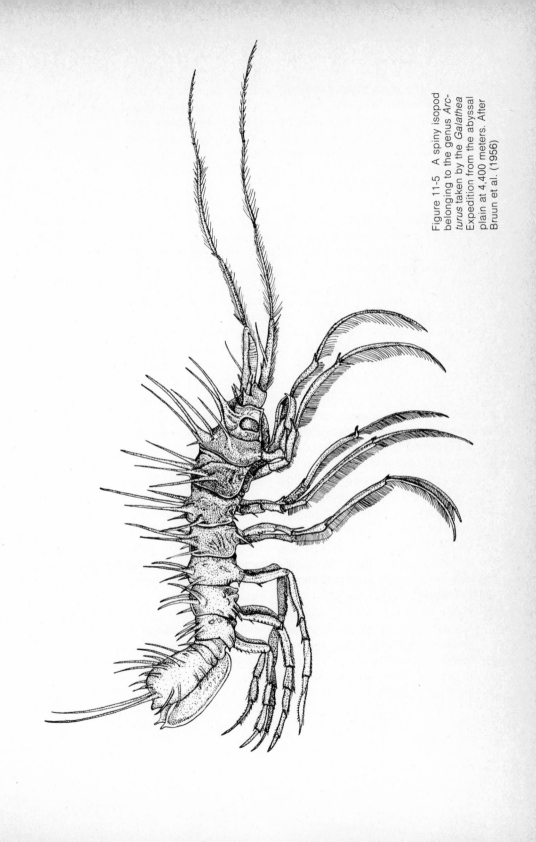

Figure 11-5 A spiny isopod belonging to the genus *Arcturus* taken by the *Galathea* Expedition from the abyssal plain at 4,400 meters. After Bruun et al. (1956)

In his work on the deep-sea holothurians, Hansen (1967) found a good correlation between systematic groups and bathymetric distribution. Examination of the *Galathea* material showed that more species were widely distributed than previously realized, and the author observed that in abyssal species a worldwide distribution may even prove to be the rule. Hansen also noted that a distributional barrier seemed to be present separating the Eastern Pacific fauna from the other parts of the abyss.

For the ascidians, Millar (1970), in discussing the abyssal species collected by the *Vema*, noted that the rates of endemism for the Atlantic (72 percent), Pacific (68 percent) and Indian (43 percent) Oceans gave support to the view that deep-sea faunas show clear zoogeographic divisions. Levi (1964) published a complete list of all sponges taken below 2,000 meters. In the hyalosponge group, there are 95 species occurring at abyssal depths, but only 5 are reported from more than one ocean or ocean basin. In the demosponge group, there are 39 abyssal species, but only 1 appears to occur in more than one ocean. The many new species described in this work, and the large percentage of previously described species that are recorded from only one locality, indicate that the deep-sea sponges are still very poorly known.

In reporting on the hydroid collections taken by the *Galathea*, Vervoort (1966) compiled a list of all species taken below 2,000 meters. Of the 20 species recorded from 3,000 to 6,000 meters, 13 have been taken from more than two oceans or major ocean basins. Almost all of the broadly distributed species are also known from slope depths, and several occur on the shelf. The general impression is one of broad distributions for the species. Zenkevitch (1966) has reported on the abyssal Echiuroidea but considered this group to be too poorly known for systematic or zoogeographic generalizations.

The Swedish Deep-Sea Expedition brought back only three species of sea pens (pennatularians) all belonging to the genus *Umbellula* (Broch 1957). However, it is interesting to see that all three are widely distributed. One is found in the Atlantic, Indian, and Pacific Oceans and the other two in the Indian, Pacific, and Antarctic areas. The reports of the Swedish expedition also include competent summaries on the following groups of abyssal invertebrates: Bryozoa (Silén 1957), Polychaeta (Eliason 1957), Sipunculidae (Wesenberg-Lund 1957) and Cirripedia (Nilsson-Cantell 1957). However, these latter four groups are so poorly known with such a large percentage of the species being recorded from single finds, that one cannot perceive any definite distributional patterns.

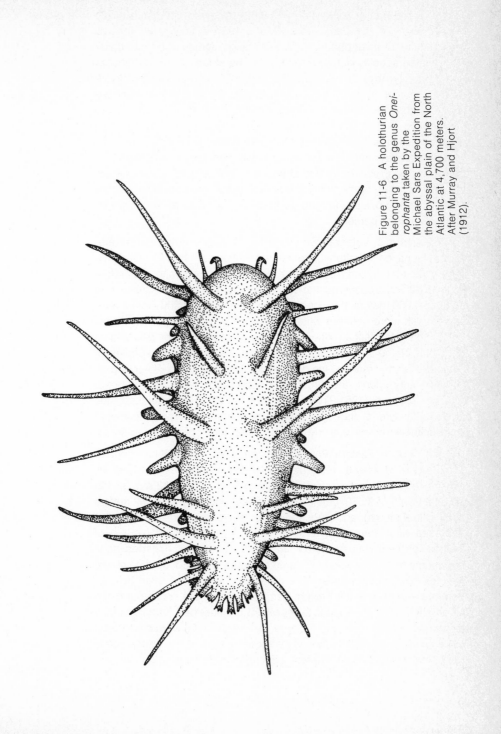

Figure 11-6 A holothurian belonging to the genus *Oneirophanta* taken by the Michael Sars Expedition from the abyssal plain of the North Atlantic at 4,700 meters. After Murray and Hjort (1912).

For the Arctic Basin, Zenkevitch (1963:62) reviewed the results of the Russian work on the *Sadko* from 1935 to 1937. A total of 94 species representing 14 invertebrate phyla were taken. Although a good many of the listed species actually came from the slope, 27 were listed as being "purely abyssal." All but 3 of these 27 were considered to be Arctic Ocean endemics. Filatova (1957) reported 10 species of bivalve mollusks from abyssal depths in the Arctic and considered 5 of them to be endemic. Clarke (1963) recorded a total of 63 Arctic mollusk species living at 1,800 meters or deeper and noted that 81 percent were shared with the Norway and Greenland Seas. He also found that only eury-bathic species (occurring also at less than 558 meters) were found in the Atlantic south of the Wyville Thomson Ridge. In discussing isopod distribution, Menzies and George (1967) referred to an Arctic Basin endemism of 87 percent.

For the Antarctic, Vinogradova (1964), in her work on the deep-water bottom fauna, called attention to the fact that endemism increased with depth. She recognized a separate zoogeographic region for the bottom south of 40°S and observed that only 6 percent of the species taken from 3,000 to 4,000 meters and *none* of those taken below 4,000 meters were shared with more northern waters. Furthermore, she stated that below 3,000 meters the fauna of the Pacific-Indian Ocean sector was distinct from that of the Atlantic sector and that there were no species in common. For this reason, she divided her Antarctic Deep-Water Region into two subregions, the Antarctic-Indian-Pacific and the Antarctic-Atlantic. Madsen (1961:210) noted that the three Antarctic species of porcellanasterid starfishes (known only from the type specimens) were unknown elsewhere and may be especially cold-adapted forms.

For the tropical Eastern Pacific, the following account has special significance: on May 6, 1952, the Danish Reasearch Vessel *Galathea*, working off the coast of Costa Rica at 9°32'N and 89°32'W over the abyssal depth of 3,570 meters, lowered her huge otter trawl (32 meters across the mouth) and proceeded to make the most remarkable deep-sea haul ever taken (Wolff 1961). A single 55-minute run of the trawl on the bottom resulted in the capture of about 2,100 specimens representing 132 species of benthic metazoans! So far, 55 of the species have been satisfactorily identified, and 30 of them are apparently restricted to the Eastern Pacific. Furthermore, a number of the unidentified species appear to represent undescribed forms that may also prove to be endemic to the area. Twelve of the 55 were considered to be cosmopolitan since they were known to occur in all three oceans (8 echinoderms, 2 coelenterates, 1 ascidian, and 2 pycnogonid).

Regional Studies

Madsen (1961:212), in his discussion of the zoogeography and origin of the abyssal fauna, observed that the East Pacific Barrier, the region stretching from the western North Pacific southward to the east of Polynesia, is a boundary for the abyssal fauna since it is a poor feeding ground for benthic animals.

Summary   Although the abyssal plain has been very poorly explored, some animal groups are better represented in collections than others probably because specimens have been relatively easy to catch with deep-water dredges and trawls. Among the groups that are best represented and have also been worked on by experts, the data so far indicate some interesting distributional contrasts.

Preliminary indications about the fishes that are apparently restricted to the abyssal plain show a prevalence of species with limited distributions. It seems clear that abyssal mollusks except for the Bivalvia are highly restricted and that each of the major ocean basins probably has a complement of endemic species. The isopods may be the most restricted of all. This was emphasized by the demonstration that the isopod fauna of two adjacent basins in the South Atlantic was highly dissimilar. In a like manner, the amphipods, at least in the low latitude regions of the world, seem to be very provincial. The sponges and ascidians also give indications of being highly restricted.

In contrast to the above groups, an impressive number of echinoderm species have broad distributional patterns. This is especially true for the Porcellanasteridae, a primitive sea star family entirely confined to the deep sea. Six of the total of 14 best-known species were considered to be cosmopolitan. It was estimated that about 5 percent of the asteroids in general and 10 percent of the ophiuroids and holothurians were cosmopolitan. For abyssal holothurians, another author observed that a worldwide distribution may even prove to be the rule. A relatively large number of hydroids and sea pens have extensive horizontal distributions. The abyssal bivalve mollusks appear to have generally broad distributions.

Why should the fishes, most mollusks, isopods, amphipods, ascidians, and sponges appear to be so highly restricted, while the echinoderm groups, hydroids, sea pens, and bivalve mollusks are characterized by broadly distributed species? Have the latter groups been slower to adapt to and spread out on the abyssal plain? Is it logical to assume that one can detect an early stage of an invasion into a new environment by the presence of many widely distributed species? Conversely, should one expect the groups that have been in the abyss the longest to demonstrate adaptation by speciation in local basins that are separated by barriers? If the abyss had to be reinvaded by higher animal life after achieving its present cold temperature, the groups with species that are now restricted may have moved in first, while those with wide-ranging species may be comparatively recent arrivals.

In regard to regional distribution, it seems apparent that three major geographic areas are very distinct. These are the Arctic Basin (probably together with the Greenland and Norwegian Basins), the Antarctic Region (and its subareas), and the tropical Eastern Pacific. How about the other major oceanic areas and their basins? Judging from the work done so far on such groups as the mollusks, isopods, and amphipods, we may eventually find enough endemism to separate every one of the major basins—all 47 of them. In the meantime, the map of faunal areas and subareas published by Vinogradova (1959) will serve as a useful guideline with which to compare future discoveries.

At least, we can now say that a surprisingly diversified fauna, representing almost all the phyla and classes of marine animals, exists on the abyssal plain and that there is an interesting faunal contrast among the various oceanic basins. May the dearth of information about this enormous but populated area stimulate the curiosity of those biologists who have become sophisticated enough to enjoy studying whole animals and their ways of life. As Forbes (1859:11) once said, "he who venturously brings up from the abyss enough of their inhabitants to display the physiognomy of the country, will taste that cup of delight, the sweetness of whose draught only those who have made a discovery know."

Hadal Zone

Our knowledge about the hadal fauna of the trenches is almost entirely recent. The *Princesse-Alice*, royal yacht of Albert I Prince of Monaco, barely reached over the 6,000 meters mark in the Cape Verde Basin in 1901 (6,035 meters). It remained for the Swedish *Albatross* in 1948 to make the first successful catch from well within the trench environment when she took a haul from between 7,625 and 7,900 meters on the east side of the Puerto Rico Trench. Since then, a number of additional trenches have been sampled, but we still know very little about the animal life that exists at such depths.

In 1960, Wolff published the first comprehensive article on the hadal fauna. At that time, it appeared that the species comprising the fauna were not only distinct systematically but also exhibited other peculiarities such as an absence of body pigmentation, blindness, and gigantism (this in the Crustacea). Later, Menzies and George (1967) studied these and other characters in the abyssal and hadal isopods and were unable to find any consistant morphological differences between species from these two depth zones. In regard to size, however, Wolff (1970) published additional data showing that some hadal species of isopods and other crustaceans did demonstrate an increase in size with depth. He also noted the absence of decapod crustaceans.

In fishes, it is interesting to see that consistant, depth-related, morphological changes do occur among the benthic species. The large genus (54 species) *Careproctus* in the family Liparidae provides good study material since the species exist over an extreme depth range (Rass 1964). About 75 percent of the species are found on the slope, 23 percent on the shelf, and one (*C. amblystomopsis*) is a hadal species living between 6,100 and 7,579 meters. On the shelf, most of the species have bright body colors, an unpigmented peritoneum, and a normal eye size (17 to 29 percent of head length). On the upper slope, some of the species are a little darker, the peritoneum tends to become dusky, and the eye is markedly larger (up to 33 percent of head). On the lower slope, the body is very dark, the peritoneum black, and the eye reduced to normal size. In the hadal species, the body is colorless, the peritoneum also without pigment, and the eye is very small.

Apparently, no *Careproctus* species typically inhabit the abyssal plain. However, abyssal fishes in general appear to have variable amounts of body pigmentation, the overall extent of peritoneum pigmentation is unknown, and the eyes tend to be small to moderately large. As a group, they do appear to show some reduction of bone and muscle apparently to compensate for the reduced positive buoyancy of the swimbladder at great depths (Marshall 1960:113). So far, all the hadal fish species appear to lack body coloration and peritoneal pigmentation, and to have very small eyes (Nielsen 1964). Upon examining the eye structure of three deep benthic fishes, one from the lower abyss and two from the trenches, Munk (1964) found extensive retinal degeneration.

Most of the collecting in the trenches has been accomplished by the Russian vessel *Vitiaz*. These catches have been analyzed by Belyaev (1966a and b), and in 1966 at the International Oceanographic Congress in Moscow, he presented a review of the ultra-abyssal (hadal) fauna of the world ocean. His conclusions may be summarized as follows:

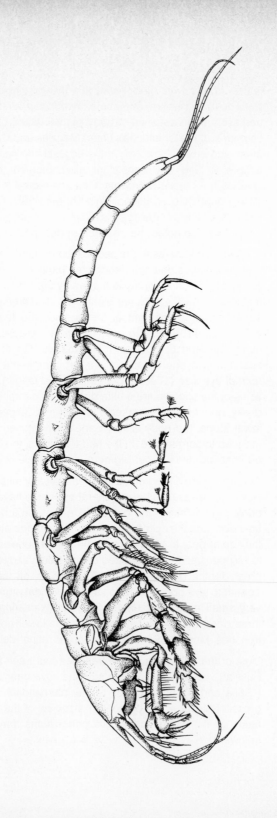

Figure 11-7 A blind tanoid crustacean (*Apseudes galatheae*). Taken by the *Galathea* Expedition from the Kermadec Trench. After Wolff (1956).

1   The bottom fauna is comprised of many Foraminifera and multicellular animals belonging to 30 classes, 140 families, and 200 genera. About 300 species have been identified, and after all the collected materials have been worked up, the total number of species will probably be about 700.

2   About 68 percent of the species, 10 percent of the genera, and one family appear to be endemic to the hadal zone. There is very little overlap between the abyssal and hadal faunas.

3   Within the boundaries of the zone, the bottom fauna becomes impoverished with depth. About 88 percent of the endemic species have a vertical range of less than 1,500 to 2,000 meters so that the composition of the fauna changes with increasing depth. This permits us to recognize three subzones: 6,000 to 7,000 meters, 7,000 to 8,500 meters, and below 8,500 meters.

4. Each deep-sea trench has its own more or less characteristic fauna, with the Banda Trench showing the least amount of endemism (33 percent); about 50 percent of its fauna consists of eurybathic species from the East Indies.

5   The tremendous hydrostatic pressure is a limiting factor, restricting the diversity of the hadal fauna, but the high rate of sedimentation creates favorable feeding conditions so that some species are present in large numbers. Holothurians are dominant in both numbers and biomass, followed by the bivalve mollusks and polychaetes.

An analysis of the bivalve mollusk fauna by Knudsen (1970) showed that all hadal zone specimens were apparently endemics but that all the genera to which they belonged were known from the abyssal plain.

Literature Cited

Alton, N. S. 1966. Bathymetric distribution of sea stars (Asteroidea) off the northern Oregon coast. *J. Fisheries Res. Board Can.*, 23(11):1673–1714.

Barnard, J. L. 1961. Gammaridean Amphipoda from depths of 400 to 6,000 meters. *Galathea Rep.*, 5:23–128, 83 figs.

———. 1966. Submarine canyons of southern California. Part V. Systematics: Amphipoda. *Allan Hancock Pacific Expedition*, 27(5):1–166, 46 figs.

Belyaev, G. M. 1966a. Bottom fauna of the ultraabyssal of the world ocean. *Abstr. 2nd Int. Oceanog. Congr., Moscow*, pp. 32–33.

———. 1966b. Bottom fauna of the ultraabyssal depths of the world ocean (in Russian). *Akad. Nauk. SSSR, Inst. Okeanol.* pp. 1–247.

Bigelow, H. B., and W. C. Schroeder. 1957. A study of the sharks of the suborder Squaloidea. *Bull. Museum Comp. Zool. Harvard*, 117(1):1–150, 16 figs., 4 plates.

———, and ———. 1962. New and little known batoid fishes from the Western Atlantic. *Bull. Museum Comp. Zool. Harvard*, 128(4):159–244, 23 figs., 1 plate.

———, and ———. 1965a. A further account of batoid fishes from the Western Atlantic. *Bull. Museum Comp. Zool., Harvard*, 132(5):443–477, 9 figs., 2 plates.

————, and ————. 1965b. Notes on a small collection of rajids from the Sub-Antarctic region. *Limnol. Oceanog.*, 10(suppl.): R38–R49, 5 figs.

————, and ————. 1968. Additional notes on batoid fishes from the Western Atlantic. *Breviora*, (281):1–23.

Blacker, R. W. 1962. Rare fishes from the Atlantic slope fishing grounds. *Ann. Mag. Nat. Hist., Ser. 13*, 5(53):261–271.

Bolin, R. L. 1957. Deep-water biological provinces of the Indo-Pacific. *Proc. 8th Pacific Sci. Congr.*, 3:373–376, 1 map.

Broch, H. 1957. Pennatularians. *Rept. Swedish Deep-Sea Expedition*, 2(21):349–364, 7 figs., 1 plate.

Brunn, A. F. 1956. The abyssal fauna: its ecology, distribution and origin. *Nature*, 177:1105–1108, 2 figs.

————. 1957. Deep-sea and abyssal depths. *in* J. W. Hedgpeth (editor), *Treatise on marine ecology and paleoecology*, vol. 1. Mem. Geol. Soc. Am., (67):641–672, 9 figs., 3 plates.

Bullis, H. R. 1967. Depth segregations and distribution of sex-maturity groups in the marbled catshark, *Galeus arae. in* P. W. Gilbert, R. F. Mathewson, D. P. Rall (editors), *Sharks, skates, and rays.* Johns Hopkins Press, Baltimore, pp. 141–148, 6 figs.

————, and P. J. Struhsaker. 1970. Fish fauna of the western Caribbean upper slope. *Quart. Florida Acad. Sci.*, 33(1):43–76, 5 figs.

Castle, P. H. J. 1960. Two eels of the genus *Synaphobranchus* from the Gulf of Mexico. *Fieldiana, Zool.*, 39(35):387–398, 2 figs.

————. 1961. Deep-water eels from Cook Strait, New Zealand. *Zool. Pub. Victoria Univ., Wellington*, (27):1–30, 6 figs.

————. 1964. Deep-sea eels: family Synaphobranchidae. *Galathea Rep.*, 7:29–42, 2 figs.

————. 1968. Synaphobranch eels from the Southern Ocean. *Deep-Sea Res.*, 15:393–396, 1 fig.

Clarke, A. H., Jr. 1962a. On the composition, zoogeography, origin and age of the deep-sea mollusk fauna. *Deep-Sea Res.*, 9:291–306.

————. 1962b. Annotated list and bibliography of the abyssal marine molluscs of the world. *Bull. Nat. Museum Can.*, (181):vi + 1–114.

————. 1963. On the origin and relationships of the Arctic Ocean abyssal mollusk fauna. *Proc. Int. Congr. Zool. Wash., D. C.*, 1:202.

Day, D. S., and W. G. Pearcy. 1968. Species associations of benthic fishes on the continental shelf and slope off Oregon. *J. Fisheries Res. Board Can.*, 25(12):2665–2675, 3 figs.

Dell, R. K. 1956. The archibenthal Mollusca of New Zealand. *Dominion Museum Bull.*, (18):1–235, 6 figs., 27 plates.

Ekman, S. 1953. *Zoogeography of the sea.* Sidgwick & Jackson, London, xiv + 417 pp., 121 figs.

Eliason, A. 1957. Polychaeta. *Rep. Swedish Deep-Sea Expedition*, 2(11):131–148, 5 figs., 2 plates.

Fell, H. B. 1958. Deep-sea echinoderms of New Zealand. *Zool. Pub. Victoria Univ., Wellington*, (24):1–40, 5 plates.

Filatova, Z. A. 1957. General review of the bivalve mollusks of the northern seas of the U.S.S.R. *in* B. N. Nikitin (editor), *Marine biology. Trans. Inst. Oceanology*, 20:1–44, 7 figs. (English trans. published by *Amer. Inst. Biol. Sci.*, Washington, 1959).

Forbes, E. 1844. Report on the Mollusca and Radiata of the Aegean sea, and on their distribution, considered as bearing on geology. *Rep. 13th Meet. Brit. Ass.* (not seen).

———. 1859. *The natural history of European seas* (edited and continued by Robert Godwin-Austen). John Van Voorst, London, viii + 306 pp., 1 map.

Garman, S. 1899. Reports on an exploration off the west coasts of Mexico, Central and South America, and off the Galapagos Islands, XXVI. The fishes *Mem., Mus. Comp. Zool.*, 24:1–431, 97 plates.

Garrick, J. A. F. 1960. Studies on New Zealand Elasmobranchii. Part XII. The species of *Squalus* from New Zealand and Australia; and a general account and key to the New Zealand Squaloidea. *Trans. Roy. Soc. N. Z.*, 88(3):519–557, 6 figs.

Gilbert, C. H. 1905. The deep-sea fishes. *in* D. S. Jordan and B. W. Evermann, *The aquatic resources of the Hawaiian Islands.* Bull. U. S. Fish Comm. 1903, part 2:575–713, 46 figs., 35 plates.

Grey, Marion. 1956. The distribution of fishes found below a depth of 2000 meters. *Fieldiana, Zool.*, 36(2):75–337.

Hansen, B. 1967. The taxonomy and zoogeography of the deep-sea holothurians in their evolutionary aspects. *Studies Trop. Oceanog. Miami*, 5:480–501, 13 figs.

Hartman, O. 1965. Deep-water benthic polychaetous annelids off New England to Bermuda and other North Atlantic areas. *Allan Hancock Found., Occas. Papers*, (28):1–378, 52 plates.

———, and J. L. Barnard. 1958. The benthic fauna of the deep basins off southern California. *Allan Hancock Pacific Expedition*, 22(1):1–297, 19 plates.

Hedgpeth, J. W. 1957. Classification of marine environments. *in Treatise on marine ecology and paleoecology.* vol. 1. Mem. Geol. Soc. Am. (67):17–28, 5 figs.

Hessler, R. R., and H. L. Saunders. 1967. Faunal diversity in the deep-sea. *Deep-Sea Res.*, 14(1):65–78, 3 figs.

Kamohara, T. 1964. Revised catalogue of fishes of Kochi Prefecture, Japan. *Rept. Usa Marine Biol. Sta.*, 11(1):1–99, 63 figs.

Knudsen, J. 1961. The bathyal and abyssal *Xylophaga* (Pholadidae, Bivalvia). *Galathea Rep.*, 5:163–209, 41 figs.

———. 1967. The deep-sea Bivalvia. *Sci. Rep. John Murray Expedition*, 1933–34, 11(3):237–343, 38 figs., 3 plates.

———. 1970. The systematics and biology of abyssal and hadal Bivalvia. *Galathea Dept.*, 11:7–236, 132 figs., 20 plates.

Leim, A. H., and W. B. Scott. 1966. Fishes of the Atlantic coast of Canada. *Bull. Fisheries Res. Board Can.*, (155):1–485, illus.

Levi, C. 1964. Spongiaries des zones bathyale, abyssale et hadale. *Galathea Rep.*, 7:63–112, 63 figs., 10 plates.

Madsen, F. J. 1961. On the zoogeography and origin of the abyssal fauna in view of the knowledge of the Porcellanasteridae. *Galathea Rep.*, 4:177–218, 2 figs.

Marshall, N. B. 1960. Swimbladder structure of deep-sea fishes in relation to their systematics and biology. *Discovery Rep.*, 31:1–122, 47 figs., 3 plates.

———. 1965. Systematic and biological studies of the macrourid fishes (Anacanthini-Teleostii). *Deep-Sea Res.*, 12(3):299–322, 9 figs.

———. 1967. The organization of deep-sea fishes. *Studies Trop. Oceanog. Miami*, 5:473–479, 2 figs.

Matsubara, K., and A. Ochiai. 1950. Studies on Hoplichthyidae, a family of mail-checked fishes, found in Japan and its adjacent waters I, II, III. *Japan. J. Ichthyol*, 1(2):73–88; (3):145–156, 3 figs.

———, ———, K. Amaoka, and I. Nakamura. 1964. Revisional study of the trachinoid fishes of the family Champsodontidae from the waters around Japan, and Tonking Bay. *Bull. Misaki Marine Biol. Inst.*, (6):1–20, 6 figs., 3 plates.

Mead, G. W. 1966. Family Bathysauridae. *in* G. W. Mead, (editor), *Fishes of the western North Atlantic*. Sears Found. Mar. Res., Mem. 1, part 5:103–112, 2 figs.

Menzies, R. J. 1965. Conditions for the existence of life on the abyssal sea floor. *Oceanog. Marine Biol. Ann. Rev.*, 3:195–210, 8 figs.

———, and R. Y. George. 1967. A re-evaluation of the concept of hadal or ultra-abyssal fauna. *Deep-Sea Res.*, 14:703–723, 9 figs.

Millar, R. H. 1970. Ascidians, including specimens from the deep sea, collected by the R. V. Vema and now in the American Museum of Natural History. *Zool. J. Linn. Soc.*, 49(2):99–159, 39 figs.

Mills, E. L. 1966. The distribution of benthic Amphipoda (Crustacea) in the deep-sea between Massachusetts and Bermuda. *Abstr., 2nd Int. Oceanog. Congr., Moscow*, pp. 254–255.

Mortensen, T. 1933. Papers from Dr. Th. Mortensen's Pacific Expedition 1914–16. LXV. Echinoderms of South Africa. *Vidensk. Medd. fra Dansk naturb. Foren.*, 93:215–400, 91 figs., 12 plates.

Munk, O. 1964. The eyes of three benthic deep-sea fishes caught at great depths. *Galathea Rep.* 7:137–149, 1 fig., 3 plates.

Nielsen, J. G. 1965. On the genera *Acanthonus* and *Typhlonus* (Pisces, Brotulidae). *Galathea Rep.*, 8:33–47, 10 figs.

———. 1966. Synopsis of the Ipnopidae (Pisces, Iniomi). *Galathea Rep.*, 8:49–75, 15 figs., 3 plates.

Nilsson-Cantell, C. A. 1957. Cirripedia. *Rep. Swedish Deep-Sea Expedition*, 2(17):215–220, 1 fig.

Norman, J. R. 1937. Coast fishes. Part II. The Patagonian region. *Discovery Rep.*,, 16:1–150, 76 figs., 5 plates.

———. 1939. Fishes. *Sci. Rep. John Murray Expedition*, 7(1):1–116, 41 figs.

Nybelin, O. 1957. Deep-sea bottom-fishes. *Rep. Swedish Deep-sea Expedition*, 2(20):247–345, 7 plates, 50 figs.

Parker, R. H. 1964. Zoogeography and ecology of some macro-invertebrates, particularly mollusks, in the Gulf of California and the continental slope off Mexico. *Dansk Naturh. Foren., Vidensk. Medd.*, 126:1–178, 29 figs., 15 plates.

Pawson, D. L. 1965. The bathyal holothurians of the New Zealand region. *Zool. Pub. Victoria Univ., Wellington*, (39):1–33, 7 figs.

Phleger, F. B. 1960. *Ecology and distribution of recent Foraminifera.* Johns Hopkins, Baltimore, viii + 1–297 pp., 83 figs., 11 plates.

Poll, M. 1951. *Expédition Océanographique Belge dans les Eaux Côtières Africaines de l'Atlantique Sud (1948–1949).* Résultats Scientifiques, vol. 4, fasc. 1, Poissons. Institut Royal des Sciences Naturelles de Belgique, Bruxelles, 154 pp., 67 figs., 13 plates.

———. 1953. *Expédition Océanographique Belge dans les Eaux Côtières Africaines de L'Atlantique Sud (1948–1949).* Résultats Scientifiques, vol. 4, fasc. 2, Poissons. Institut Royal des Sciences Naturelles de Belgique, Bruxelles, 258 pp., 104 figs., 8 plates.

———. 1954. *Expédition Océanographique Belge dans les Eaux Côtières Africaines de L'Atlantique Sud (1948–1949).* Résultats Scientifiques, vol. 4,fasc. 3A, Poissons. Institut Royal des Sciences Naturelles de Belgique, Bruxelles, 390 pp., 107 figs., 9 plates.

———. 1959. *Expédition Océanographique Belge dans les Eaux Côtières Africaines de l'Atlantique Sud (1948–1949).* Résultats Scientifiques, vol. 4, fasc. 3B, Poissons. Institut Royal des Sciences Naturelles de Belgique, Bruxelles, 417 pp., 127 figs., 7 plates.

Rass, T. S. 1964. Changes in eye size and body coloration in secondary deep-sea fishes. *in* T. S. Rass (editor), *Fishes of the Pacific and Indian Oceans, biology and distribution.* Trans. Inst. Oceanology, Acad. Sci. U.S.S.R., 73:1–9 (English trans. Jerusalem, 1966).

Rofen, R., and H. H. DeWitt. 1960. Antarctic fishes. *in Sciences in Antarctica.* Part I: *The life sciences in Antarctica.* Pub. 839, National Academy of Science, National Research Council, pp. 94–112.

Sanders, H. L., R. R. Hessler, and G. R. Hampson. 1965. An introduction to the study of deep-sea benthic faunal assemblages along the Gay Head-Bermuda transect. *Deep-Sea Res.*, 12:845–867.

Schroeder, W. C. 1955. Report on the results of exploratory otter-trawling along the continental shelf and slope between Nova Scotia and Virginia during the summers of 1952 and 1953. *Papers Mar. Biol. Oceanog. Deep-Sea Res.*, 3(suppl.):358–372.

Schultz, G. A. 1966. Submarine canyons of southern California. Part IV. Systematics: Isopoda. *Allan Hancock Pacific Expedition*, 27(4):1–56, 15 plates.

Shmidt, P. Y. 1950. *Fishes of the Okhotsk Sea.* Academy of Science, U.S.S.R., Moscow, pp. 1–370, 20 plates, 51 figs. (in Russian).

Silén, L. 1957. Bryozoa. *Rep. Swedish Deep-Sea Expedition*, 2(5):63–69, 1 fig.

Smith, J. L. B. 1949. *The sea fishes of southern Africa.* Central News Agency, Capetown, xvi + 550 pp., 1,245 figs.

Southward, E. C., and Southward, A. J. 1967. The distribution of Pogonophora in the Atlantic Ocean. Aspects of Marine Zoology. *Symp. Zool. Soc. London*, (19):145–158.

Springer, S. 1966. A review of Western Atlantic cat sharks, Scyliorhinidae, with descriptions of a new genus and five new species. *U.S.F.W.S. Fish. Bull.*, 65(3):581–624, 27 figs.

Thompson, J. R. 1963. The bathyalbenthic caridean shrimps of the southwestern North Atlantic. Ph. D. dissertation, Duke University. xii + 504 pp., 50 figs.

Vaughan, T. W., and J. W. Wells. 1943. Revision of the suborders, families, and genera of the Scleractinia. *Geol. Soc. Amer., Spec. Papers*, (44):xv + 1–363, 39 figs., 51 plates.

Vervoort, W. 1966. Bathyal and abyssal hydroids. *Galathea Rep.*, 8:97–174, 66 figs.

Vinogradova, N. G. 1959. The zoogeographical distribution of the deep-water bottom fauna in the abyssal zone of the ocean. *Deep-Sea Res.*, 5:205–208, 3 figs.

———. 1962. Vertical zonation in the distribution of deep-sea benthic fauna in the ocean. *Deep-Sea Res.*, 8:245–250, 5 figs.

———. 1964. Geographical distribution of deep-water bottom fauna of the antarctic. *Soviet Antarctic Expedition*, 1:121–122. Elsevier, New York.

———. 1966. Characteristic features in the geographical distribution of bottom fauna in the abyssal of the Pacific Ocean. *Abstr., 2nd Int. Oceanog. Congr., Moscow*, pp. 383–384.

Wesenberg-Lund, E. 1957. Sipunculidae. *Rep. Swedish Deep-Sea Expedition*, 2(15):199–201, 1 fig.

Wolff, T. 1960. The hadal community, an introduction. *Deep-Sea Res.*, 6:95–124.

———. 1961. Animal life from a single abyssal trawling. *Galathea Rep.*, 5:129–162, 26 figs., 4 plates.

———. 1962. The systematics and biology of bathyal and abyssal Isopoda Asellota. *Galathea Rep.*, 6:7–320, 184 figs., 19 plates.

———. 1970. The concept of the hadal or ultra-abyssal fauna. *Deep-Sea Res.*, 17(6):983–1003, 3 figs.

Wyville, Thomson, C. 1873. *The depths of the sea*. An account of the general results of the dredging cruises of H.M.S.S. "Porcupine" and "Lightning" during the summers of 1868, 1869, and 1870. Macmillan and Co., London, xxi + 527 pp., 84 figs.

Zenkevitch, L. 1963. *Biology of the seas of the U.S.S.R.* George Allen and Unwin Ltd., London, pp. 1–955, 427 figs.

Origin
and Dispersal
of Life

Zoogeography and Evolution

Part I - A History of Marine Life

*In the beginning and for a
billion years, Earth was
stark, sterile, and forlorn
And then God touched the
moon, so it came close and
did enormous tidal stresses
spin*

*Generating heat and melting and
expelling of gases so atmosphere
and hydrosphere were born
Was the moon the original
priceless gift causing the stage
to be set and the evolutionary
play to begin?*

J. C. B.

The total age of our earth, and indeed that of the entire solar system, has been determined through comparisons of lead isotope composition with meteoric lead to be about 4.5 to 4.8 billion years (Eicher 1968:138). The oldest known metamorphic and granitic rocks together with the isotope composition of various minerals indicate that crustal melting probably took place about 3.5 to 3.6 billion years ago. Cloud (1968a) pointed out that if this thermal event had been caused by the close approach of the moon, samples of moon rock representing its original crust should give radiometric ages of about 3.5 to 3.6 billion years.

On July 20, 1969 astronauts Armstrong and Aldren picked up rocks from the surface of the Sea of Tranquility and brought them back to earth. Age studies of the samples indicated that the Sea of Tranquility was formed 3.65 ± 0.05 billion years ago (Albee et al. 1970) and that the total age of the moon was about 4.66 billion years (Tatsumoto and Rosholt 1970). Moon material from succeeding trips of the Apollo series is currently being studied. As a result, we will, in the near future, know considerably more about the early history of both the moon and the earth.

At this time, it appears that the earth possibly began as a relatively cool agglomerate (Engel 1969:471) which was then subjected to considerable internal heating from the radioactive decay of long-lived nuclides. It seems likely that, perhaps a billion years later, the crusts of both the moon and earth were profoundly affected by the same thermal event. As an aftermath, the earth's atmosphere and hydrosphere were probably formed from plutonic gases escaping from volcanoes, fumaroles, and hot springs. This outgassing still takes place and provides a continuous supply of vital materials such as H_2O and CO_2 (Rubey 1964).

It was J. B. S. Haldane in 1929 and 1932 (Haldane 1954) who first stated that ultraviolet radiation of an evolving hydrosphere would produce a vast variety of organic substances and that the initial life forms must have been anaerobic and heterotrophic. Direct evidence of the oldest known organisms was published by Engel et al. (1968), who found alga-like forms in the Onverwacht Series of South Africa that appeared to be as old as 3.4 billion years, and by Barghoorn and Schopf (1966), who discovered bacterium-like organisms and filamentous structures in the Fig Tree Series of South Africa with an age determination of about 3.1 billion years. So it seems that biogenesis took place within a few hundred million years after the provision of an atmosphere and a hydrosphere.

In his review of premetazoan evolution and the origins of the Metazoa,

Cloud (1968b) suggested that the first living organisms were probably aquatic and confined to relatively deep water since water was the most convenient shield against excess ultraviolet radiation. Weyl (1968) suggested that aggregations of organic molecules produced by abiotic processes probably accumulated along the steepest density gradients. In the low latitude seas, density increases rapidly with depth between about 50 and 100 meters. Weyl supposed that the earliest organisms were probably planktonic and restricted to this density-gradient layer.

Following biogenesis, the next great step was the evolution of autotrophs, organisms capable of manufacturing their own food. Cloud (1968a) expressed the opinion that the first autotrophs were associated with ionized iron. If ferrous iron acted as the oxygen acceptor, then one could account for the extensive sedimentary deposits comprising the Banded Iron Formation. The living organism responsible could have been very much like our modern blue-green algae.

The Banded Iron Formation is found among sediments deposited from about 3 to 1.8 billion years ago (Cloud 1968a). As green plants became more efficient at photosynthesis and as the ferrous ions in the hydrosphere became used up, it seems reasonable to suppose that oxygen began to accumulate and to escape into the atmosphere. The presence of free oxygen was presumably followed by the evolution of the eucaryotic cell with a nucleus, chromosomes, and the capacity for sexual reproduction. Also, as oxygen (O_2) built up in the atmosphere, some of it was converted to ozone (O_3) forming an effective ultraviolet shield. This probably opened up the surface waters of the oceans to occupation by phytoplankton and vastly increased the level of O_2 in the atmosphere.

With protection against ultraviolet radiation and a plentiful supply of O_2, the evolution of multicellular or metazoan animals could take place (Nursall 1959). It is interesting to note that, as the Precambrian drew to a close, atmospheric CO_2 apparently underwent a significant reduction (as indicated by an increase in the deposition of limestone and dolomite). The sudden decrease in CO_2 may have led to a temperature drop, setting off a widespread glaciation (Harland 1964, Cloud 1968a). So, early in their history, metazoan animals may have been exposed to a temperature fluctuation of their environment.

The Paleozoic Era

In his review article, Cloud (1968a) noted that, on various grounds, different geologists have arrived at the notion that the Precambrian-

Paleozoic boundary lies somewhere near 600 million years. Although no undoubted metazoan fossils have been reported from the Precambrian, a number of different phyla of the Metazoa appeared over a relatively short time interval in the early Paleozoic. It has been suggested (Rudwick 1964) that the end of a lengthy and worldwide Precambrian glaciation, marked by a warm climate and the flooding of continental areas, may have stimulated a rapid metazoan evolution.

The *Atlas of Palaeobiogeography* (edited by A. Hallem) published in the spring of 1973 has proved to be an indispensable aid in its provision of a series of up-to-date review articles. In this atlas, the emphasis is on Paleozoic and Mesozoic marine animal groups.

In his article on the distribution of Cambrian trilobites, Palmer (1973) observed that a major unresolved problem is the lack of a suitable geographic base. He noted that the major faunal contrasts were between those areas that had unrestricted access to the open ocean and those that were restricted to broad expanses of shallow sea over carbonate or terrigenous substrates. Palmer also found striking similarities between the faunas of Antarctica and those of Siberia and other areas bordering the western Pacific. A comparable resemblance was found between the faunas of Argentina and North America. He suggested that such similarities may be explained most easily by the idea of common paleolatitudes for regions that share similar faunas.

The Ordovician period began about 500 million years ago. Bergström (1973) noted that the conodont faunas provided a striking illustration of provincialism. He was able to recognize two main provinces (North Atlantic and Midcontinent) that existed throughout the period. For corals, Kaljo and Klaamann (1973) identified a Middle Ordovician developmental center situated in the seas of North America. These authors were able to identify two main coral provinces (Americo-Siberian and Euroasiatic). For graptolites, Skevington (1973) found that, in the early Ordovician, two major provinces (Pacific and Atlantic or European) developed. But, in the late Ordovician, the graptolite faunas took on a more cosmopolitan aspect.

Also during the Ordovician, the articulate brachiopods underwent an explosive evolution (Jaanusson 1973). For this group, northern and southern faunal regions were recognized and, during the Middle and Upper Ordovician, two northern subprovinces were distinguishable. In regard to trilobites, Whittington (1973) determined that, during the Lower Ordovician, independent evolution of faunas apparently took place in the shelf waters around three land masses. Eventually, four distinct trilobite faunas developed, and in the early Upper Ordovician, genera of families previously confined to one faunal region appeared in others.

It is interesting to note that, by the Upper Ordovician, both graptolites and trilobites appeared to become more widely distributed. As Skevington (1973) pointed out, Wilson (1966) proposed that, in Lower Paleozoic time, a proto-Atlantic Ocean existed so as to form a boundary between two realms and that, during Middle and Upper Paleozoic time, the ocean closed by stages, bringing dissimilar facies together. Such closing action may account for the appearance of broader distributions in the Upper Ordovician.

The Silurian Period began about 450 million years ago. By this time, a large group of corals had become extinct (Kaljo and Klaamann 1973), and this loss was accompanied by an increasing importance of widely distributed coral general. By the Late Silurian, many of the corals typical of the Period had disappeared, and only two provinces (European and Asiatic) could be distinguished. For brachiopods, the Silurian opened with almost complete cosmopolitanism (Boucot and Johnson 1973) which gave way to a low degree of provincialism at the end of the period.

Although the graptolites almost became extinct prior to the Silurian, they recovered to develop into a unique new fauna (Berry 1973). These planktonic, offshore animals demonstrated an essentially cosmopolitan distribution until their extinction in the early Devonian. The ostracoderms, the most primitive known vertebrates, first made their appearance in the Middle Ordovician but did not become numerous until the Silurian. The most diverse and best-known group of ostracoderms is that called the Heterostraci (Halstead and Turner 1973). The heterostracans, found in the western United States in the Ordovician, apparently spread (in the Silurian) to the Canadian Arctic, eastern United States, the Baltic, and Anglo-Wales. By the Devonian, it was possible to identify an "Old Red Continent Realm" (northern North America to northeastern Europe) with three evolutionary centers and a "Tungussian Realm" (Siberia) with one evolutionary center.

The beginning of the Devonian, about 400 million years ago, apparently was a time when the epicontinental seas were becoming reduced in size. This process supposedly continued until the Emsian Interval of the Middle Devonian when marine seaways were highly restricted. From that time onward, the epicontinental seas enlarged, reaching their maximum in the Upper Devonian. In their article on Devonian brachiopods, Johnson and Boucot (1973) observed that provinciality increased when seaways became restricted and decreased as inundation proceeded. They recognized three distinct brachiopod provinces in the Middle Devonian but only one worldwide province in the Upper Devonian. Apparently, a major faunal change took place in the benthic shelf habitat at the close of the period (Bretsky 1968).

The Carboniferous period began about 340 million years ago and lasted approximately 60 million years. The distribution of the foraminiferids, those associated particularly with the shelf and nearshore areas, was studied by Ross (1973). He was able to delineate for the early Carboniferous a Eurasian-Arctic Province and a North American Midcontinent Province (south of a line between southwestern Colorado and the Great Lakes). For the middle and late Carboniferous, the latter province was modified to include parts of South America, and its name was changed to the Midcontinent-Andean Province.

With regard to Lower Carboniferous corals, Hill (1973) recognized three distinctive zoogeographic regions—those of North America, Eurasia, and Australia. These regions were also subdivided into smaller units. In Hill's view, the coral distribution patterns of the Lower Carboniferous fit well with a distribution of continents about the Polar Sea very similar to that of today. He also observed that, if the abundance of corals then, as now, indicated shallow warm waters, then the temperature of the Polar Sea and the waters leading into it must have been very much warmer than now.

The Permian, the last period of the Paleozoic, extending from about 280 to 230 million years ago, is considerably better known than the preceding periods. Stehli (1964) was able to plot diversity gradients in brachiopods, bivalves, and the fauna of large terrestrial reptiles. These diversity gradients demonstrated a close relationship to latitude and indicated that the continents must have been in approximately the same latitudinal relationship to one another as they are now. Furthermore, the gradients indicated that the north rotational pole was in about its present position (Helsey and Stehli 1964).

More recently, Stehli (1973) was able to concentrate on certain temperature-sensitive brachiopod families and to correlate their distributions with Vebeekinid fusulines, some coral families, and the algal genus *Mizzia*. This made it possible for him to determine that the warm-water (Tethyan) region of the Permian occupied a fairly well-defined belt, asymmetrical about the present geographic equator, yet having borders roughly parallel to the equator. This belt was found to extend latitudinally between about 60°N and 30°S. To the north and south of the belt, a cool-water (boreal) assemblage was recognized. Gobbett (1973) noted that the fusuline Foraminifera underwent important changes in their distribution, going from a nearly worldwide pattern in the early Permian to a strictly Tethyan one in the late Permian.

One of the most striking crises in the history of life took place at the end of the Permian when some major animal groups became extinct (trilobites, eurypterids) and many others were decimated (bryozoans, corals, brachiopods, ammonoids, crinoids, etc.). Since the Permian seas were largely withdrawn from the continental shelves, it had been assumed by some paleontologists that the reduction in habitat for the shallow-water animals had drastic effects. However, as Fischer (1964) pointed out, the Pleistocene brought about the same kind of sea-level change, yet a corresponding faunal crisis did not occur.

Beurlen (1956) was the first to propose the theory that the late Permian crisis for marine life was caused by a marked reduction in oceanic salinity. This theory was further elaborated by Fischer (1964) and may be summarized as follows: As the Permian seas withdrew from the continents, large shallow basins were left which, in some areas, remained connected to the world ocean. Evaporation in the basins produced large amounts of brine which, having a greater density than sea water, drained into the oceanic abyss and stagnated. This led to a marked reduction in salinity in the circulating part of the oceanic water mass, resulting in an almost worldwide extinction of stenohaline organisms.

Oxygen isotope measurements of Permian sea surface temperatures, based on brachiopod material, were reported for northwestern Australia by Lowenstam (1964). The earliest data indicated a temperature of about 8° C. But, within 5 million years or less the temperature climbed to 24 to 26°C. This change possibly reflects the transition of continental Australia from a glaciated to a deglaciated state.

The Mesozoic Era

The Lower Triassic marine communities were noted for cosmopolitan distribution, sparseness, and for the peculiar absence of many groups that had been diverse and abundant (Newell 1962). For example, foraminiferans, sponges, echinoderms, articulate brachiopods, gastropods, crustaceans, corals, and bryozoans were extremely rare. In fact, the only marine fossils that are abundant in the Lower Triassic are bivalve mollusks, ammonoids, inarticulate brachiopods, and chondrostean fishes.

Fischer (1964) suggested that the early Triassic marine faunas were derived largely from the brackish-water, estuarine faunas of the Permian, the stenohaline species having become eliminated except for perhaps a few favored places in the world where higher salinities were maintained. The surviving remnants of the Permian stenohaline faunas then repopulated the seas during the later Triassic when normal salinities were restored. Worldwide warm temperatures apparently prevailed during this period. Kummel (1973), in his study of Lower Triassic mollusks, observed that, most probably, there was only a slight temperature differential from the high to the low latitudes.

In his definitive monograph on the Jurassic geology of the world, Arkell (1956:610) concluded that "boreal" marine faunas first appeared during the Callovian stage of the Middle Jurassic. Imlay (1965), in his work on North American ammonites and pelecypods, found that a differentiation, suggesting a cooling of the northern seas, developed after the Middle Bajocian of the Jurassic. More recent work on belemnites (Stevens 1973a) and ammonites (Carion 1973, Enay 1973) has substantiated the fact that distinct "boreal" and tethyan (tropical) faunas existed from the Middle Jurassic onward.

Although it is often assumed that a worldwide tropical fauna existed in the Lower Jurassic, Howarth (1973) was able to separate the ammonites of the midpoint of that time into boreal and tethyan faunas. Both Stevens (1973a) and Enay (1973) were able to distinguish three provinces within the tropics, thus showing that barriers to longitudinal dispersal must have existed. Beauvais (1973) plotted the distribution of Jurassic hermatypic corals and found that they occupied a belt from about the present equator to 50° N. It appears likely that the Jurassic boreal area was equivalent, in temperature, to our present warm-temperate regions.

Oxygen isotope analyses determined largely from belemnites indicated a Cretaceous period fluctuation of polar temperatures between about 10°C and 16 to 17°C (Emiliani 1961:144, Lowenstam 1964:244, Bowen 1966). The 16 to 17°C peak apparently took place near the close of the Cretaceous about 80 million years ago. In New Zealand the Upper Cretaceous climate was probably warm-temperate, since the temperature was not warm enough to support the growth of the tropical rudistids and corals (Fleming 1962:66). These data appear to correlate well with the Northern Hemisphere paleodistributional studies of Bandy (1960) on the Foraminifera and of Durham (1959a;13) on the Mollusca. Reid (1967:178), who worked on the Porifera, called attention to the fact that the northern area commonly referred to as the "boreal" province of the Cretaceous was actually a subtropical or warm-temperate region. The latter three works suggest that, during the Upper Cretaceous, the northern boundary of the tropics was in the vicinity of 50 to 60°N.

Kauffman (1973) published a detailed study of the Cretaceous Bivalvia based on the data provided by Moore (1969) in the *Treatise on Invertebrate Paleontology.* Kauffman, who was able to follow the history of zoogeographic development throughout the period, used as primary divisions, Tethyan, North Temperate, and South Temperate Realms. These realms were in turn subdivided into regions, provinces, sub-provinces, and endemic centers. Of particular interest in his analysis, is the appearance, in the Tethyan Realm, of a distinct Caribbean (New World) Province in the Aptian stage of the Middle Cretaceous. Also, the appearance, in the South Temperate Realm, of a South Atlantic Sub-province by the next stage (the Albian).

The foregoing Middle Cretaceous events appear to be related to the opening of the present Atlantic Ocean. Along this same line, Stevens (1973b) felt that the belemnite faunas provided information indicating that the North Atlantic began to open up in the uppermost Jurassic or the lowermost Cretaceous and that the South Atlantic opened in the Lower Cretaceous. For the Cretaceous belemnites, Stevens also recognized three main zoogeographic realms (Boreal, Tethyan, Austral).

The Cenozoic Era By the early Tertiary (Paleocene-Eocene), it is possible that polar temperatures may have dropped to the point where the development of a cold-temperate fauna could have been initiated. Durham (1952:map) noted the presence of a fossil fauna at Spitzbergen that indicated winter surface temperatures of 5 to 8°C. This evidence agrees with that of Chaney (1940, 1964) and Dorf (1955, 1964) who described the remains of a terrestrial, cold-temperate Eocene flora from several places on the periphery of the Arctic Ocean. From the work of Davis and Elliot (1957) and Woodring (1960) on the paleogeography of the London Clay Sea of the Paleocene–Lower Eocene, there are convincing indications that the northern border of the tropics was about 52°N. The composition of the fish fauna of London Clay Sea (Casier 1966) lends support to this idea.

In the mid-Oligocene, the polar sea water temperature may have been about 10.4°C, according to Emiliani (1961), but the tropical boundary on the Pacific coast of North America had moved south to about 47°N (Durham 1959a:8), so the polar basin was probably somewhat colder. Durham (1959b:579) observed that, by the beginning of the Oligocene, the North Pacific shallow-water molluscan faunas had become almost entirely provincial. MacNeil (1965) reported that the earliest known *Mya* came from the late Eocene or early Oligocene beds of Japan. The initial southward dispersal of the terrestrial, Arcto-Tertiary flora apparently took place in the Oligocene (MacGinitie 1958, Barghoorn 1964).

By the Lower to Middle Miocene, the polar sea surface temperature had apparently dropped to about 7°C (Emiliani 1961), and the tropical boundary on the Pacific coast had moved down to about 37°N (Durham 1959a). Sorgenfrei (1958:415), who examined the Middle Miocene molluscan fauna of South Jutland, concluded that the temperature of the North Sea Basin was in general about 5°C higher than today with a similar annual variation. This would place the Miocene tropical boundary in the Eastern Atlantic at just about 37°N (the Strait of Gibraltar). Fossil fish evidence (Arambourg 1965) showed that the Mediterranean began to be affected by cooler temperatures in the Oligocene and, by Miocene times, had lost much of its tropical character. Davies (1958:487) was of the opinion that both seals and walruses were confined to the Arctic until the Miocene when they moved southward.

By the Upper Pliocene, the northern boundary of the tropics on the Pacific coast of North America had dropped to about 27°N—only about 4° north of its present position (Durham 1959a)—and the temperature of the Arctic waters was apparently about 2.2°C (Emiliani 1961). Soot-Ryan (1932) concluded that a large part of the European pelecypod fauna had originated in the North Pacific and that the most extensive migrations had taken place during the Pliocene. Djakonov (1945), on the basis of echinoderm studies, emphasized the importance of Pliocene faunal movements in the same direction and Nesis (1962) considered that several of the northwestern Atlantic deeper water benthic invertebrates had migrated through the Arctic Basin during the Upper Pliocene.

In looking over the earlier literature on paleotemperature determinations, one can easily get the impression that a gradual climatic deterioration took place throughout the Tertiary (which then culminated in the ice ages of the Pleistocene). However, there are now good indications that considerable climatic fluctuation occurred. In New Zealand waters, there were apparently low temperatures during the Paleocene (13°C) rising to 22°C in the late Eocene, a drop back to 13°C in the Oligocene, another warm peak in the middle Miocene, and finally a decrease to about present temperatures in the late Pliocene (Stonehouse 1969, Jenkins 1968). In the northeastern Pacific, a mid-Miocene temperature reversal, or warm peak, was detected by means of an analysis of molluscan distributional patterns by Addicott (1969).

In general, people tend to equate the onset of the most recent series of glacial stages with the beginning of the Pleistocene. It was only a few years ago that Kulp (1961), in his competent review of information on the geologic time scale, observed that it appeared best to assume a date of 1.0 million years for the Pliocene-Pleistocene boundary. Since then, this time boundary has been pushed back farther and farther. Bandy and Wade (1967) provided a foraminiferal definition of the boundary giving a date of about 3 million years, and Eicher (1968) utilized a date of 2.5 million years.

Major glaciations apparently began considerably earlier than the Pliocene-Pleistocene boundary. Curry (1966) reported evidence of glaciation in the Sierra Nevada dated at about 3 million years, and Goodell et al. (1968) concluded that Antarctic glaciation was initiated prior to 5 million years ago. Now, we have evidence that high altitude, continental glaciation may have begun more than 6 million years ago (Herman 1970).

The North Atlantic and the North Pacific

Compared to the earlier periods already discussed, a relatively huge amount of fossil and geological data is available for the Pleistocene. In general, the northern oceans are better known than those of the Southern Hemisphere, and the North Atlantic is better known than the North Pacific. Unfortunately, many of those who have studied the North Atlantic have tended to apply their findings to the world ocean. But it is becoming clear that the North Atlantic is a singular ocean with a distinct history and that conclusions reached about it may not be applicable to other oceans (Briggs 1970).

It is instructive to compare the faunas of the North Atlantic and North Pacific Oceans. Both cover large geographic areas, have similar overall physical and chemical characteristics, and are connected via the Arctic Ocean. However, the faunas of these two oceanic areas, while demonstrating a basic relationship, differ so markedly from one another in diversity and geographic distribution that one is compelled to search for an explanation.

Upon examining the distribution of the species that comprise the relatively poor shore or shelf fauna of the boreal (cold-temperate) North Atlantic, one can recognize the presence of two zoogeographic provinces, one on each side of the ocean (p. 248). In contrast, it can be said the boreal North Pacific is divisible into five such provinces, each identifiable by a significant amount of endemism (pp. 266–267).

In the North Atlantic, a comparatively broad latitudinal distribution of individual species appears to be the rule. On the eastern side, many boreal shore species extend from about 50 to about 72°N, and many warm-temperate species range from about 15 to 50°N. In the eastern North Pacific, on the other hand, latitudinal ranges seem to be much more restricted; Valentine (1967:155) found that along the shelf the average latitudinal range of shelled molluscan species was only about 9°. Although no comparable data have been published, it seems apparent that the average latitudinal range of the Atlantic mollusks is much greater.

The marked contrast in faunal diversity between the North Atlantic and North Pacific is apparent in most of the phyla represented on the continental shelf. For example, about 60 species of shore fishes (aside from occasional summer visitors) reside in the Gulf of Maine between Cape Cod and Nova Scotia (Bigelow and Schroeder 1953), but along the coast of British Columbia from Vancouver to the Queen Charlotte Islands there are more than three times as many species (Clemens and Wilby 1961). Furthermore, several fish families and a host of genera are confined to the North Pacific, but there are no endemic families and few such genera in the North Atlantic. For the marine invertebrate fauna in Canadian waters, it has been found that, at comparable latitudes, the number of Pacific species is more than three times that of the Atlantic coast (Powell and Bousfield 1969:14).

In the western North Atlantic, the Florida peninsula extends far enough south so that its distal end lies in tropical waters (p. 63). This has the effect of dividing the Carolina Warm-Temperate Region into an Atlantic coast portion, extending from about Cape Kennedy to Cape Hatteras, and a Gulf coast portion that occupies the northern Gulf of Mexico. Despite their geographic separation, the faunas of the two parts are very similar. Almost all of the warm-temperate species found in the Atlantic coast segment also occur in the northern Gulf of Mexico.

In the eastern North Pacific, the Baja California peninsula also extends south so that its tip lies in tropical waters (p. 46). This divides the California Warm-Temperate Region into an outer coast portion, extend-

The Shore Faunas

ing from about Magdalena Bay to Point Conception, and an inner portion that is confined to the Gulf of California. Here, the geographic separation has had a profound effect on the species that comprise the two faunas. Each area is characterized by a high degree of endemism, and only a few of the warm-temperate species of the outer coast are also found in the Gulf of California. So, we find that on the Atlantic coast of North America, the tip of the Florida peninsula does not function as a significant zoogeographic barrier while, on the Pacific coast, the end of the Baja California peninsula is a highly effective barrier.

The Pelagic Fauna

The contrast between the pelagic faunas of the two oceans is even more striking. As was pointed out in Chapter 10 (The Pelagic Realm), the North Atlantic does not even possess an endemic boreal epipelagic fauna. The broad expanse of open ocean between the arctic and warm-temperate waters is almost entirely occupied by animals with three basic patterns of distribution: arctic-boreal, eurythermic temperate, and broad eurythermic tropical. Furthermore, the boreal waters are comparatively depauperate in numbers of species. The adjoining warm-temperate belt exhibits a much greater species diversity mainly because of the presence of many eurythermic tropical forms. There are a few warm-temperate endemics but not enough to recognize this belt as a distinct region. Consequently, only a single, large temperate region has been delineated for this part of the upper pelagic world.

In the North Pacific, the boreal waters possess the same kind of distributional groups discussed above, but in addition, one can also identify in this richer fauna a general boreal group made up of a large number of species confined to such cold-temperate waters, and even further, discrete assemblages that are restricted to the northern or southern portions of these waters. This situation makes it possible to recognize both North and South Boreal Zoogeographic Regions. For the epipelagic fishes, a division of this kind has been pointed out by Rass (1967:134) and Parin (1968:120). In addition, one can also recognize a distinct North Pacific Warm-Temperate Region, since a relatively large number of endemic species have been found (p. 319).

Oceanic Islands

We know that evolutionary change tends to occur very rapidly in the small populations of marine shore animals that become isolated around oceanic islands. As the result, those islands that are relatively old should possess faunas that show a high degree of evolutionary divergence. The best indication of the extent of such divergence is probably the level of endemism. If the level of endemism at such islands is very low, one may suspect that a major alteration of the environment has occurred, resulting in an extinction followed by repopulation by migrants from other areas.

When one examines the shore fish faunas of the old and well-isolated (300 miles or more from the nearest land) oceanic islands of the North Atlantic and the North Pacific, a very interesting pattern of endemism is revealed (Briggs 1966). In the North Atlantic, the Azores, a group of nine islands of probable Miocene age at 36 to 39°N, have only a 1 percent endemism; Madeira, two islands of probable Miocene age at 32 to 33°N, shows about 3 percent endemism; Bermuda, consisting of about 360 small islands of Eocene or Oligocene origin at 32°N, has about 5 percent endemism; and the Cape Verde Islands, with 10 main islands of Lower Cretaceous age at 14 to 17°N, have an endemic level of about 4 percent. In the North Pacific, the only old and well-isolated island group that has a well-known fish fauna is Hawaii. This group of 20 islands, of a late Miocene or earlier origin at 18 to 23°N, has about a 34 percent endemism.

Here again, we find a remarkable contrast between the northern parts of the two oceans. Both the Cape Verde Islands and Bermuda occupy positions comparable to that of Hawaii in that each is within the tropical zone but lies close to its northern boundary, but these islands show endemic rates of only 4 and 5 percent while that of Hawaii is 34 percent.

In the North Atlantic, a variety of studies have shown that the fauna has undergone, during the late Tertiary and the Pleistocene, a series of replacements, latitudinal shifts, and a general depauperization. Extinctions and climate-influenced changes of coiling direction in the Foraminifera have been reported by Ericson and Wollin (1954); Ericson (1959); Ericson, Ewing, and Wollin (1964); Ericson and Wollin (1964); and Bé (1966). Woodring (1959, 1966) called particular attention to the degradation that has taken place in the shallow-water molluscan fauna. In the Middle Miocene, when the Isthmus of Panama was inundated, practically identical molluscan faunas were found in the tropical waters of the western Atlantic and eastern Pacific. Now, about 43 of the genera and subgenera are extinct on the Atlantic side but survive in the Pacific. In contrast, only 4 genera are extinct on the Pacific side but survive in the Atlantic.

Fossil
Distributions

Perhaps most revealing of all is the detailed study of coccolith distribution in the North Atlantic that was carried out by McIntyre (1967). He compared the distribution of selected species from both recent and mid-Wisconsin glacial sediments and concluded that the maximum cooling of the glacial period resulted in an approximate 15° southward shift in latitude for the planktonic populations, with the greatest shift occurring in the eastern Atlantic. Furthermore, using the population boundaries of species with known temperature ranges, he was able to construct a paleoisotherm map of the mid-Wisconsin North Atlantic. By comparing this map with modern surface temperature charts, one can obtain a clear picture of the effect of the most recent glacial period on the sea surface temperature. From this evidence, it appears that the mid-Wisconsin surface temperatures of the open North Atlantic were about 3°C colder than the average February (coldest month) and about 6°C colder than the average annual temperatures today.

In the North Pacific, there is a limited amount of fossil evidence for major faunal disturbances during the late Tertiary and Pleistocene. Some work on the planktonic foraminiferal faunas off the California coast has shown indications of temperature-related changes. This evidence has been summarized by Bandy (1968) who described marked southward shifts in cold-temperate populations of *Globigerina pachyderma*. He noted that 50 to 90 percent sinistral populations, presently not existing south of northern Oregon, had invaded southern California in the Upper Miocene, Middle Pliocene, and four times during the Pleistocene.

In contrast, the fossil evidence for long-term stability in the North Pacific is convincing. Emerson (1956) found, in his investigation of Pleistocene invertebrate deposits of southern California and Baja California, that his material did not clearly indicate a general drop in sea surface temperature during the glacial periods. Woodring (1957), in his report on the marine Pleistocene of California, concluded that the recognition of glacial and interglacial faunas was doubtful. Valentine and Meade (1960) investigated isotopic and zoogeographic paleotemperatures of California Pleistocene mollusks but could find no good evidence of vastly cooler ocean temperatures. Ericson and Wollin (1964:255), who studied the foraminiferans from deep-sea cores, noted that the well-defined climatic zones they had been able to spot in the Atlantic cores were absent from those taken in the Pacific. Finally, Wiles (1967) who also studied foraminiferan species concluded that the Pacific Ocean, or at least the eastern equatorial portion of it, did not experience any marked lowering of temperature during the Pleistocene glacial stages.

The oxygen isotope (O^{18}/O^{16}) method of paleotemperature measurement, developed by H. C. Urey in 1947, at first showed great promise but later, conflicting results by different investigators raised some doubts. The initial findings of Emiliani (1955) led the authors of major works to jump to conclusions such as "surface seawater temperatures in the tropics have fluctuated within a range of 6°C throughout much of Pleistocene time" (Flint 1957:439), "the approach of glacial conditions was widely felt in the world's seas with cold water animals displacing the warm water species" (Charlesworth 1957:696), and "the surface temperatures of the oceans fluctuated by some 6°C" (Schwarzbach 1963:199).

In a later paper, Emiliani (1964) suggested that Pleistocene fluctuations for the equatorial eastern Pacific were about 4 to 6°C and that those of the tropical Atlantic and Caribbean amounted to about 9°C. More recent oxygen isotope determinations were carried out by Olausson (1965, 1967), who concluded that the North Atlantic cores suggested an amplitude of only about 3°C for the tropical and subtropical regions. He also expressed the opinion that if there was any change in the equatorial eastern Pacific, it was likely too small ($\pm$ 1°C or less) to influence the distribution of the foraminiferans. It is encouraging to see that these recent determinations appear to correlate very well with temperatures inferred from the distribution of many fossil species.

The obvious impoverishment of the North Atlantic fauna, compared to that of the North Pacific, has puzzled biologists for a long time. A favorite and persistent explanation has been that a land bridge across the North Atlantic must have blocked off connection with the boreal North Pacific via the Arctic Ocean. This theory was apparently first published by Forbes (1859), who, in turn, gave credit to Sir John Richardson for suggesting this explanation. Forbes described the bridge as probably extending from 70 to 75°N and completing in its northern coastline the symmetrical form of the Arctic Basin.

More recently, Model (1943), on the basis of freshwater bivalve distribution, suggested an Oligocene connection. Ekman (1953:164) maintained that it was almost certain that a land bridge existed at the end of the Tertiary and the beginning of the Quaternary. The comprehensive survey of Lindroth (1957) presented evidence to show that a complete bridge could not have existed during the Pleistocene, but it was suggested that Greenland and Iceland were connected with Europe some time before the Wisconsin glaciation. Löve (1958), on botanical

Paleotemperature
Measurements

Land Bridge
Effects

evidence, said that the main discontinuity was between Greenland and Iceland and that a land connection between Iceland and Western Europe existed during the penultimate glaciation. Lindberg (1963a, 1963b), who analyzed the distribution of seven freshwater fish families, concluded that North America and northwestern Europe were joined in late Miocene–early Pliocene times.

On the other hand, Simpson (1947:219) pointed out that, after the early Eocene, evidence from mammalian paleontology clearly favors the existence of a single migratory route (which probably was the Bering Land Bridge) between Eurasia and North America. Darlington (1957:591) stated definitely that the record of mammals does not indicate any connection across the Atlantic between Europe and North America. He further observed that, up to the early Eocene, the evidence is indecisive but that it does not require an Atlantic bridge. From the standpoint of the marine shore fauna, it is important to note that the late Miocene and Pliocene were periods of frequent movement of molluscan species from the Arctic Ocean to the North Atlantic (MacNeil 1965).

There is now available an impressive amount of biogeographic and stratigraphic evidence about the existence and the periods of operation of the Bering Land Bridge (Hopkins 1967). Durham and MacNeil (1967:342) were unable to find any evidence of the migration of marine invertebrates through the Bering Strait during the early Tertiary. Also, during this time, the marine mammal faunas on each side of the land bridge were strikingly different; the desmostylians, sea lions, and ancestral walruses were confined to the North Pacific, while the true seals of the family Phocidae were found only in the Arctic-Atlantic (Hopkins 1967:453). But, in the late Miocene, about 12 to 10 million years ago, a very few marine organisms of Atlantic ancestry reached the North Pacific and a substantially larger number of Pacific organisms reached the North Atlantic.

The late Miocene seaway must have been short-lived, since repeated dispersals of land mammals took place across the land bridge during most of the Pliocene from 10 to 4 million years ago (Hopkins, op. cit.). Near the close of the Pliocene, about 3.5 to 4.0 million years ago, the seaway opened up again. As the result, the North Atlantic molluscan fauna was transformed by the arrival of boreal mollusks from the Pacific, but the molluscan fauna of the North Pacific remained little affected (Durham and MacNeil 1967). This spectacular dispersal may have been made possible partly by a somewhat warmer climate and, as Hopkins (1967:460) suggested, partly by passages through the Queen Elizabeth Islands providing continuous channels of marine communication as low as 72°N. The Bering Strait seaway was, of course, in operation during several of the interglacial stages of the Pleistocene as it is now.

A fundamental relationship between the boreal faunas of the North Atlantic and North Pacific has been recognized for many years. In *On the Origin of Species*, Darwin (1859:372) made the following observation: "As on the land, so in the waters of the sea, a slow southern migration of a marine fauna, which during the Pliocene or even a somewhat earlier period, was nearly uniform along the continuous shores of the Polar Circle, will account, on the theory of modification, for many closely allied forms now living in areas completely sundered."

Relationship of the Faunas

Probably most of the relationship that is apparent between the boreal faunas of the North Atlantic and North Pacific is attributable to the migrations of Pacific species that took place through the Bering Strait in the late Pliocene. This means that the two faunas have been effectively separated for about 3.5 million years. Fish species with amphiboreal distributions (not existing in Arctic waters) are either very rare or completely absent. Some forms previously considered to be amphiboreal, such as the Greenland halibut *Reinhardtius hippoglossoides* reported by Hubbs and Wilimovsky (1964), are actually arctic-boreal species with interrupted distributions. Zenkevitch (1963:749) quoted Andriashev as giving 50 cases of amphiboreal distributions among the fishes, but an examination of Andriashev's (1954) most recent work shows no true amphiboreal species.

Among some of the invertebrate groups, amphiboreal distributions seem to be quite common. The classic contribution on the subject is that of Berg (1934). More recently, Ushakov (1955) found 46 species of amphiboreal polychaetes in the northwestern Pacific. Nesis (1962) examined the distribution of 13 species of well-known benthic invertebrates that now live only in the northwestern Atlantic and the North Pacific. He concluded that all of them originated in the Pacific and migrated to the Atlantic via the Canadian archipelago. Zenkevitch (1963:748) summarized the Russian literature on the question and said that all groups of the fauna and flora of the northern part of the Pacific are characterized by their marked amphiboreal distribution. However, one needs to be aware that in some of the Russian publications the term amphiboreal is used in a broader sense so that related as well as identical species are included.

Attention has been drawn to the importance of the interglacial periods, especially to the postglacial climatic optimum (the "Littorina" era of

4,000 to 6,000 years ago), as likely times for the trans-Arctic migration of boreal species (Nesis 1962:94). But, the general evidence for such recent migrations is not satisfactory and, indeed, the distributional patterns of the boreal fishes seem to argue against it. One may suspect that, in groups where there are many truly amphiboreal species, evolutionary divergence has simply been very slow.

Continental Drift   In 1885 Edward Suess, the famous German geologist, made the suggestion that the southern continents were at one time parts of a great common land mass—Gondwana Land (Woodford 1965). Essentially, the same proposal was made by the American F. B. Taylor in 1910 and the German Alfred Wegener in 1912. Wegener's publications, especially the English translation of the third edition of his book *The Origin of Continents and Oceans* printed in 1924, were particularly influential. In 1937, Alex. L. du Toit, South African geologist and enthusiastic admirer of Wegener's work, published a book entitled *Our Wandering Continents.*

Despite the appearance of du Toit's work, the concept of continental drift lost popularity from about 1930 to the early 1950s. Paleontologic investigations, particularly the work on mammals (Matthew 1915, Simpson 1947:219, Darlington 1957:591), had shown that, at least during the Cenozoic, it was not necessary to move the continents in order to account for distributional patterns. The discovery of palaeomagnetism in the early 1950s, with its implication that a major shifting of continental blocks may have occurred, gave the continental drift theory a new respectability. This was followed by significant advances in marine geology and geophysics, especially concerning the nature of the midoceanic ridges, which resulted in the theory of ocean basin evolution by sea-floor spreading (Dietz 1961).

The decade of the 1960s saw a veritable flood of literature on continental drift, and the level of this output has continued to date. Sea-floor spreading was soon incorporated into the more comprehensive theory of plate tectonics. Symposia were held, books were published, and hundreds of articles appeared in scientific and popular journals. Most accounts begin with a single supercontinent called Pangaea which then breaks up into a northern part called Laurasia and a southern part called Gondwana. Laurasia is depicted as including Eurasia and North America, while Gondwana is supposed to comprise the southern continents plus India.

For drift advocates, the most widely accepted account, at this time, appears to be that of Dietz and Holden (1970). Their reconstruction visualized Pangaea as existing 200 million years ago. By the end of the Triassic, some 180 million years ago, Laurasia and Gondwana are supposed to have separated from one another, and Gondwana supposedly then began to break up into its smaller components. India and then Australia are considered to have broken their connections with Antarctica and to eventually drift far to the north. The separation of North America from Eurasia and of South America from Africa is shown to involve mainly longitudinal rather than latitudinal movements.

Some biologists have embraced continental drift with unquestioning enthusiasm, while others have remained skeptical. As things now stand, both the stratigraphic and the paleontologic evidence for a previous joining or at least a close approximation of South America and Africa is most impressive. The data up to 1964 were analyzed by Darlington (1965) and additional information was brought to bear by Romer (1968). Myers' (1967) work on the relationship of the freshwater fish faunas provided more affirmative evidence. Cretaceous bivalve and belemnite data has already been referred to (p. 409). The South Atlantic Ocean, or at least the major portion of it, must have been formed as the result of continental drift that probably took place sometime in the middle to late Mesozoic.

Biological Evidence

The discovery of a nearly complete early Triassic *Lystrosaurus* fauna (a complex of archaic amphibian and reptilian species) in Antarctica has shed important light on the early relationship of that continent with Africa (Colbert 1971, Kitching et al. 1972). The faunas of the two areas resemble each other so closely that they are considered to have been parts of a single, continuous complex. Since a *Lystrosaurus* fauna has also been found in India, it has been used as evidence to promote the Gondwana concept. However, the same fauna has also been found in the Lower Triassic sediments of Sinkiang and Shansi, China.

It seems evident that continental drift has taken place between Africa and Antarctica in the early Mesozoic and between Africa and South America in the later Mesozoic. North America and Europe may have separated somewhat earlier (p. 409). What about India and Australia? Did the future Indian peninsula drift far north to ground on the southern margins of Asia, and did Australia, like a great Noah's Ark, float to the northeast, across the Indian Ocean, to its present position near the East Indies (phraseology from Romer 1968)? As Woodring (1954:719) so aptly observed, "Paleogeography starts as a concoction of essential ingredients, generally too meager, and winds up as a heady essence distilled through the imagination of the perpetrator."

For a long time, it has been known that species diversity gradients show an inverse relationship to latitude with the greatest diversity occurring in the low latitudes of the tropics (p. 12). In our discussion of the Permian period (p. 406), reference was made to Stehli's (1964) research on fossil diversity gradients which indicated that the southern continents (South America, Africa, Australia) must have occupied about the same latitudinal position then that they do today. In a more recent contribution, Stehli, Douglas, and Newell (1969) gave evidence to show that such latitudinal gradients have apparently retained their equilibrium for the last 270 million years.

Some of the most dependable evidence for continental relationships is provided by the freshwater fishes. Many families belong to a "primary" freshwater category (Myers 1949) and are, and have been, strictly intolerant of salt water. At the 17th International Zoological Congress held at Monaco, Patterson (1972) presented a review of the distribution of Mesozoic freshwater fishes. He concluded that, although the good correlation between the early Cretaceous freshwater fish faunas of Brazil and West Africa could be attributed to a pre-drift union, the Gondwana-Laurasia model was not otherwise a great improvement over the fixed continent arrangement.

For the freshwater and terrestrial faunas in general, the theory of a southeast Asian (Oriental Region) center of evolutionary radiation (Darlington 1957) seems as good as ever. Because Africa, South America, and Australia have always been hard to reach, they serve as refuges for phylogenetic relics. Of the three, Africa has been the easiest to invade and thus has the more modern fauna, South America has been more difficult (except during the Pleistocene) and has many ancient groups, and Australia, the hardest to reach, has the most archaic fauna. The presence of marsupials and other old animal groups in both Australia and South America but not in Africa (Keast 1971) can be explained by the fact that such groups were once very widespread, and in the more accessible parts of the world, including Africa, they were replaced by modern groups.

For the marine environment, it is tempting to speculate that the relatively poor fauna of the North Atlantic, compared to that of the North Pacific, may be due to the more recent establishment of the former. However, it appears that this difference is explainable on the basis of differences in climatic history (p. 415). In general, the present distribution patterns of the marine faunas do not require past continental movements. Furthermore, the distributions of some marine fossils appear to give evidence for a long-term, latitudinal stability of the continents (except Antarctica).

Part I
Summary

1   The total age of the earth and our solar system is about 4.5 to 4.8 billion years. About 3.6 billion years ago, the earth and moon were apparently both affected by a thermal event of enormous magnitude. This could have been caused by the moon on its initial entry into earth's orbit. As an aftermath, the earth's atmosphere and hydrosphere were formed from plutonic gases.

2   Since the oldest fossils of algae and bacterium-like organisms are about 3.1 to 3.4 billion years old, it appears that biogenesis took place within a few hundred million years after the provision of a hydrosphere. The first living organisms were probably aquatic and confined to relatively deep water, since water was the most convenient shield against excess ultraviolet radiation. The first autotrophs may have been associated with ionized iron resulting in the deposition of the Banded Iron Formation.

3   As the ferrous ions in the hydrosphere were used up, oxygen began to accumulate and escape into the atmosphere. Some of the oxygen was then converted to ozone, forming a new, effective ultraviolet shield. This opened up the surface waters of the oceans to occupation by phytoplankton. With protection against ultraviolet radiation and a plentiful supply of oxygen from phytoplankton photosythesis, the evolution of multicellular animals took place.

4   Several different metozoan phyla appeared over a relatively short time interval in the early Paleozoic. This may have been the result of a rapid evolution during a period of warm climate at the end of a lengthy and worldwide Precambrian glaciation.

5   The distributional patterns of marine animals during most of the Paleozoic are very poorly known. However, the Permian, last period of the Paleozoic extending from about 280 to 230 million years ago, is much better known. Diversity gradients in brachiopods, bivalves, and large terrestrial reptiles show the same relationship to latitude that modern faunas do. Permian fossils indicate that a broad tropical belt extended around the earth between about 60°N and 30°S. A crisis in the history of life took place at the end of the Permian when some major marine groups became extinct and many others were decimated. A reduction in the salinity of the circulating part of the oceanic water mass may have been the primary cause.

6   Worldwide warm temperatures with very broad distributions of tropical

animals apparently prevailed during the early Mesozoic. A cooling of the polar seas took place in about the Middle Jurassic and continued, with some fluctuation, through the Cretaceous. Both the distribution of fossils and oxygen isotope temperature determinations indicate that Cretaceous polar temperatures were probably in the warm-temperate range.

7 There is faunal and floral evidence of cold-temperate (5 to 8°C) conditions in the Eocene of the Arctic Basin, and such lower temperatures may have actually become established during the Paleocene. From this evidence, we can surmise that our modern boreal fauna began its development in the Arctic Basin, and probably also in the North Pacific, in the early Tertiary (Paleocene-Eocene). Climatic deterioration continued enabling the boreal faunas to spread southward and occupy large areas in the North Atlantic and North Pacific. Possibly by the Lower Pliocene and certainly by the Upper Pliocene, the temperature of the Arctic Ocean in general became too cold for the boreal fauna and a new arctic fauna began to develop.

8 Contrary to opinions expressed only a short time ago, it now seems likely that the Pleistocene began as early as 2.5 million or even possibly 3 million years ago. Compared to the seas of the Southern Hemisphere, the North Atlantic and North Pacific Oceans are relatively well known. It is instructive to compare the faunas of these two northern areas since they demonstrate the effects of contrasting climatic histories during the Cenozoic and especially the Pleistocene.

9 The marine shore fauna of the boreal Atlantic is depauperate compared to that of the Pacific. Only two zoogeographic provinces may be identified, one on each side of the ocean. Five such provinces are found in the boreal Pacific. In the North Pacific, the latitudinal ranges of the individual species are relatively restricted and the diversity of species is much greater. Also, many genera and families are found that are not present in the Atlantic.

10 Although the tip of the Florida peninsula divides the Carolina Warm-Temperate Region, the shore faunas of the two parts are so similar that the Florida peninsula cannot be called a significant zoogeographic barrier. On the other hand, the tip of the Baja California peninsula, which divides the California Warm-Temperate Region, separates two vastly different faunas so that it must be recognized as a highly significant barrier.

11 The North Atlantic does not possess a distinct boreal pelagic fauna, and the animals found in these waters apparently all have distributional patterns that extend well outside the boreal area. In contrast, the North Pacific has a rich fauna that is divisible into North and South Boreal Regions. Also, in the latter area, there is a large group of species with general boreal distributions occurring in both the North and South Regions.

12 The shore fish faunas of the old well-isolated oceanic islands of the North

Atlantic demonstrate very little evolutionary divergence as determined by the level of endemism. The Azores show only 1 percent endemism, and it is 5 percent or less at Madeira, Bermuda, and Cape Verdes. At Hawaii, about 34 percent of the shore fish species are considered to be endemic.

13   A variety of studies in the North Atlantic have shown that, during the late Tertiary and the Pleistocene, a series of faunal replacements, latitudinal shifts, and extinctions have taken place. These have occurred in both the pelagic and shore faunas. Most revealing was a study of coccolith distribution which led to the construction of a paleoisotherm map for the mid-Wisconsin glacial period. This showed that the sea surface temperatures for the open North Atlantic were about 3°C colder than the average winter temperatures and about 6°C colder than the average annual temperatures today. In the North Pacific, the only fossil evidence for major faunal disturbances appears to be confined to a single species of planktonic foraminiferan. It may be that this species is unusually sensitive to minor temperature changes.

14   Although the early works dealing with temperature determinations based on oxygen isotope measurements were interpreted as indicating general, widespread, high-amplitude fluctuations of sea surface temperature during the Pleistocene, later work has revealed that such temperature changes were considerably less and that they may have been most pronounced in the North Atlantic. These more recent determinations correlate very well with temperature conditions indicated by the distribution of many fossil organisms.

15   Evidence that the movement of boreal species from the Arctic to the North Atlantic may have been impeded by a land bridge appears to be weak. However, the Bering Land Bridge was in existence until the late Miocene. This means that separate boreal faunas could have developed in the Arctic–North Atlantic and in the North Pacific for as long as 40 to 50 million years. The late Miocene dispersal through the Bering Strait was mainly northward, then eastward.

16   The second Tertiary opening of the Bering seaway took place near the close of the Pliocene. This resulted in so many successful dispersals of Pacific boreal mollusks to the North Atlantic that the character of the fauna in the latter area was transformed. Such migrations were probably facilitated by slightly warmer climatic conditions than we have today and possibly by the presence of a lower latitude passage through the Canadian Arctic. Successful migrations in the opposite direction were few and the boreal fauna of the North Pacific remained little affected. The Bering seaway was also open during several of the interglacial stages of the Pleistocene as it is now.

17   Despite the fundamental relationship between the boreal faunas of the North

Atlantic and North Pacific, they are well separated at the species level. In the fishes, species with true amphiboreal distributions (not ranging into Arctic waters) are rare or completely absent. In some of the benthic invertebrate groups, a good many amphiboreal species have been identified, but these probably represent cases where evolutionary change has been very slow. The interglacial stages of the Pleistocene, including the postglacial climatic optimum of 6,000 to 4,000 years ago, were probably not warm enough to permit the trans-Arctic migration of boreal species.

18 The theory of continental drift has once again become very popular. It seems evident that drift took place between Africa and Antarctica in the early Mesozoic and between South America and Africa in the later Mesozoic. North America and Europe may have separated somewhat earlier. Neither the terrestrial and freshwater faunas nor the marine faunas require additional continental movement. The distributions of some marine fossils appear to give evidence for long-term (270 million years) latitudinal stability of the southern continents (except Antarctica).

Conclusions   In order to place our current knowledge of zoogeography in the proper perspective, we need to keep in mind that biogenesis took place a very long time ago (over 3 billion years in the past) and that, while life was undergoing its marvelous evolutionary diversification, the earth was subjected to a series of long-term climatic cycles. Such cycles have apparently occurred at least as long as there have been multicellular organisms, perhaps longer.

There are probably several factors that contribute to the operation of the major climatic cycles, but so far, their interrelationships are not at all clear. Major glaciations have not occurred except during periods of high land elevation, so there must be some orogenic effect. There are small, regular variations in the earth's tilt and in its orbit that, in turn, cause gradual changes in the amount of insolation received by a given area on the earth's surface (Broecker et al. 1968). Also, there may be slow changes in the rate at which the sun radiates its energy (Opik 1958). The latter two factors have been combined into an astronomical theory of glaciation (first proposed by Milankovitch) that has received considerable support (Veeh and Chappel 1970).

The zoogeographic effects of the ancient glaciations and of the climatic deteriorations that must have preceded them are difficult to evaluate because of the lack of paleontologic information. But, the fossil records for the Mesozoic and Cenozoic are comparatively good, so that we can attempt to piece together the events that led to the establishment of our present world zoogeographic patterns.

It now seems clear that our modern boreal faunas, despite their present obvious relationship, had a dual origin. In Paleocene-Eocene times, one cold-temperate evolutionary center became established in the Arctic Basin; then, as the climate grew colder, another center south of the Bering Land Bridge began its development. Possibly the ancestral lines of these early boreal faunas became established in the warm-temperate polar waters of the Upper Jurassic and Cretaceous. Some of the early warm-temperate species were probably able to adapt to boreal conditions, while others were forced to migrate southward.

During the Oligocene, boreal species were able to move out of the Arctic Basin at least as far as the Greenland and Norwegian Seas. In a like manner, a greater area of the North Pacific (probably including the Gulf of Alaska and Okhotsk Sea) became available to its embryonic boreal fauna. It is interesting to find that by the late Miocene, when a seaway first became established through the Bering Strait, the movement of boreal species was predominately northward into the Arctic Basin and thence to the North Atlantic. This indicates a competitive superiority for the Pacific species and suggests that they came from an area that already possessed a greater species diversity and ecosystem stability. The second opening of the seaway in the late Pliocene had an even more striking result, transforming the character of the North Atlantic fauna yet scarcely affecting that of the North Pacific.

The boreal fauna of the Arctic Basin apparently got the earliest start and, at first, had a greater geographic area for its development. Why did it not flourish to the same or a greater extent than that of the North Pacific? We know that the general climatic deterioration of the Tertiary was a worldwide phenomenon so that the North Pacific was affected as well as other areas. However, general climatic trends are themselves made up of minor cycles. Were the early tertiary cycles more severe in the Arctic and the North Atlantic? Faunal diversity is related not only to geographic area but to climatic stability; that is, the more stable the climate, the greater the diversity (Sanders 1968). If present conditions in these parts of the world offer a good insight into the past, the effect of short-term temperature oscillations may have prevented a rich fauna from developing.

It may be concluded that, despite a dual origin, the present close relationship of the North Atlantic and North Pacific boreal faunas is primarily due to the late Miocene and especially the late Pliocene invasions from the Pacific to the Atlantic. During the Pleistocene and following the Wisconsin ice age, the Arctic Basin temperatures have apparently been too cold to permit such invasions to recur so that the two faunas have been effectively separated for about 3.5 million years. Now, although many genera are shared, there are relatively few truly amphiboreal species.

It has been suggested that historic fluctuations in sea level have been responsible for mass extinctions in the shallow-water marine life (Newell 1962). Would this account for some of the faunal differences between the North Atlantic and North Pacific? Probably not, for both Newell (1963:16) and Fischer (1964) pointed out that it is well known that the sea level oscillations of the Pleistocene did not produce numerous extinctions in the shallow-water communities. Also, since the area occupied by the continental shelf in the North Atlantic appears to be roughly equivalent to that of the North Pacific shelf and since glacio-eustatic changes in sea level should have affected each ocean about equally, it does not seem reasonable that sea level change could have been an important factor.

In the contemporary North Atlantic, such characteristic features as a depauperate shore fauna; species with broad latitudinal ranges; the complete lack of an endemic, boreal pelagic group; and the very low level of endemism at the oceanic islands present a decided contrast to conditions in the North Pacific. Also, it seems clear that the Florida peninsula has not been able to function as an effective zoogeographic barrier because the lowering of sea surface temperature during the glacial stages permitted circumvention of the peninsula by warm-temperate species. The Baja California peninsula has been a highly efficient barrier apparently because similar temperature drops did not occur in the North Pacific. These facts together with good evidence of Pleistocene faunal replacements, latitudinal shifts, and extinctions reveal that the North Atlantic has provided for its marine fauna a more rigorous environment than the North Pacific.

Since the above phenomena could have been caused by a history of temperature fluctuation and since, during the Pleistocene such temperature changes did in fact occur in the Atlantic but apparently not in the Pacific, it seems clear that alteration in temperature, the one evident variable, was the primary cause.

Why has the North Atlantic been subjected to more severe temperature cycles? Lamb (1964:337) attributed a greater ice age cooling in the Atlantic to the fact that the opening between Greenland and Europe was the only effective outlet for sea ice from the Arctic Ocean. Broecker (1965:742) noted the greater proximity of the continental ice sheets to the North Atlantic. It has also been pointed out that the North Atlantic is a relatively small ocean with a correspondingly smaller heat budget, and during the most recent glaciation the land was covered with ice as far south as the southern British Isles on the east and New Jersey on the west (Briggs 1966:160).

It may be concluded that the geographic setting of the North Atlantic with its open exposure to the Arctic Ocean and its relatively small size are the primary factors that are responsible for its history of varied surface temperature. The most severe drops in temperature probably took place during the ice ages of the Pleistocene and these appear to have averaged about 3°C below the present winter minimum and about 6°C below the present annual mean temperatures. Moreover, such temperature oscillations were probably not confined to the Pleistocene but may have extended well back into the Tertiary. The shelf and pelagic surface fauna that becomes established under such conditions is vastly different, in terms of diversity and local geographic distribution, than a fauna occupying a more stable environment.

The foregoing account, which emphasizes the origin of our present boreal faunas and involves a close comparison between the North Atlantic and North Pacific Oceans, illustrates the profound effect of rather small (to us) climatic changes on the diversity and local distributions of the marine faunas. Of even greater importance to the history of marine life is that climatically unstable areas such as the North Atlantic do not appear to function as evolutionary centers so that species which are produced there do not give rise to continuing phyletic lines. Such evolutionary consequences are considered next.

Part II—Worldwide Patterns

Who can explain why one species ranges widely and is very numerous, and why another allied species has a narrow range and is rare? Yet these relations are of the highest importance, for they determine the present welfare, and, as I believe, the future success and modification of every inhabitant of this world.

Charles Darwin *in* On the Origin of Species, *1859*

The discovery of crinoids and other primitive animals in relatively deep water excited the early naturalists and indicated to them that the deep sea might prove to be a treasure trove of ancient phylogenetic relicts. However, the extensive collections made by the *Challenger* Expedition did not disclose an abyssal fauna that was primarily archaic. In 1953, a colloquium on the distribution and origin of the deep sea bottom fauna was held at the 14th International Congress of Zoology in Copenhagen.

In his summary of the 1953 colloquium, Spärck (1954) pointed out that the bottom fauna of the ocean floor and especially of the deep trenches consists to a great extent of species belonging to the same genera and families that inhabit the shallow water. This information, which has been reinforced by more recent investigations, raises some important questions: Why is the abyssal and hadal fauna essentially modern instead of archaic? Where are the most ancient animals to be found? Can we recognize in the world ocean a fundamental relationship between evolution and distribution?

The Great Depths

In order to explain the modern nature of the fauna of the deepest parts of the ocean, we need to examine briefly the history of the environment. Since there are indications of cold-temperate (5 to 8°C) conditions in the Eocene of the Arctic Basin (Durham 1952) which may have actually become established during the Paleocene, and since the early Paleocene sea surface temperatures in New Zealand were very low—comparable to present-day temperatures (Stonehouse 1969)—there is some evidence to indicate that the Mesozoic era may have closed with a dramatic climatic change that was reflected by a sudden drop in the surface temperatures of the high latitude seas.

We know that new water is supplied to the abyssal depths of the ocean by means of a slow thermohaline circulation. Cold, dense surface water in the Weddell Sea area of the Antarctic and in the North Atlantic south of Greenland sinks to supply eventually the deep basins of the world (Pickard 1964). Thus, the low temperatures of the depths are maintained by a continuous flow of cold water from the surface. Since the polar temperatures near the close of the Cretaceous were about 16 to 17°C (Emiliani 1961, Lowenstam 1964, Bowen 1966), similar temperatures probably prevailed at the great depths. If, at the end of the Mesozoic, a dramatic temperature drop did take place in the polar seas, it must have also had its effect in the deep sea.

It is possible that a warm-water, abyssal fauna may have accumulated during the whole of the Mesozoic only to have been exterminated at the end of this era by a sudden change to cold-temperate conditions. Furthermore, the warm peaks that evidently took place in the late Eocene and Middle Miocene and the intervening cold period of the Oligocene (p. 410) were probably all reflected by similar temperature changes in the great depths. So, although one is tempted to think of the deep sea below 3,000 meters as a very stable environment, since it does not undergo perceptible seasonal changes, its Tertiary temperature history has probably been quite unstable. Such temperature changes could account for the scarcity of an old fauna and at least partly for the present relatively low level of species diversity.

As Madsen (1961) pointed out, Murray long ago (1895) suggested it is possible that the deep seas were anaerobic during most or all of the Mesozoic. Vertical circulation may have been prevented by the presence of relatively warm surface waters everywhere so that multicellular abyssal life could not exist until the poles became cold and the dense, oxygenated water could sink into the depths. Regardless of whether a Mesozoic deep-sea fauna was decimated by falling temperature at the end of that era or whether such a fauna existed at all, the present fauna of the great depths seems to be a relatively new one. The present deep, cold-temperature regime (1 to 2.5°C) may not have become established until the late Pliocene so it is possible that most of the invasions into the abyssal and trench habitats have taken place since that time.

The Ancient Ones

Where are the phylogenetic relics, those ancient animals that can reveal so much about the evolutionary story? A variety of modern works help to provide an interesting answer to this question.

Invertebrates

In his work on the molluscan fauna of the New Zealand slope, Dell (1956) noted that a number of Tertiary genera which had previously been considered extinct in the New Zealand area were found to be still living on the continental slope. Knudsen (1961:199) reported that more species of the pelecypod genus *Xylophaga* were found on the slope than in any other environment; he also noted that it would be feasible to assume that the genus originated in the shallow water of the tropics but was eliminated from that environment by competition from other wood-borers. Clarke (1961, 1962) found that in the deeper waters of the world the primitive pelecypod order Protobranchiata and the primitive gastropod order Archaeogastropoda were relatively well represented. In the latter order, 102 of the 126 deep-water species were listed as occurring on the continental slopes, and 56 of the species appeared to be confined or almost confined to the lower slope between about 1,000 and 3,000 meters.

In his survey of the bathymetric distribution of the deep-sea crinoids, Gislén (1957) observed: (1) among the comatulids, the families with the primitive traits did not reach the great depths and (2) among the stalked crinoids, the archaic types also stopped at moderate depths. In the opposite direction, none of these primitive forms generally occurred above 200 meters. Madsen (1961:192) found that the majority of recent stalked crinoids, including all the larger forms, were primarily bathyal (slope) in their distribution.

Madsen (1961), who revised the primitive sea star family Porcellanasteridae, stated that in all probability this group evolved from astropectinid-like ancestors, which, in Jurassic and Cretaceous times, migrated from the sublittoral region of the Tethys Sea to the slope habitat where the most primitive genera still occur. An invasion into the abyssal region, where the most advanced genera are now found, began in the early Tertiary. This invasion into the greatest depths appears to be still going on.

In regard to the decapod crustaceans, Wolff (1961:28) found that the primitive, still-symmetrical Pylochelidae (hermit crabs) have survived primarily at slope depths where the possibilities of a less competitive existence are apparently best. Madsen (1961:192) noted that the Eryonidea, an ancient decapod order that dates back to the Jurassic, is represented by two recent genera and both are distributed primarily on the slope. In their study of evolution in the spiny lobsters of the family Palinuridae, George and Main (1967) found the older and more primitive genera to exist at slope depths.

In discussing the distribution of the gammaridean amphipods, Barnard (1961:124) pointed out that the slope fauna is richer in phylogenetic relicts than is that of the abyssal depths. He further observed that this pattern would fit the abyssal cooling theory that called for a newly evolved fauna at the greatest depths. Voss (1967:531), in his work on the deep-sea cephalopods, emphasized that the Nautiloidea is an ancient assemblage with only five or six species surviving to modern times. He suggested that these few species have probably survived by adapting to the slope environment where the physical conditions have remained stable for a long period of time.

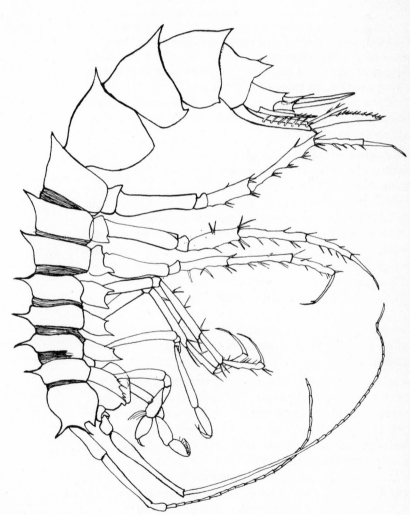

Figure 12-1 A gammaridean amphipod (*Lepechinella monocuspidata*). A phylogenetic relict taken by the *Galathea* Expedition from the lower slope off the eastern coast of Africa near Kenya. After Barnard (1961).

Finally, a notable paper was published by Parker (1962), who called attention to the importance of the lower continental slopes that descend directly to the outer sea floor. He noted the occurrence in such places of many ancient groups such as the segmented mollusk *Neopilina*, stalked crinoids, paleoconch and protobranch pelecypods, archaeo-gastropods, certain primitive enchinoderms, an archaic suborder of isopods, and specimens of Pogonophora. He suggested that many of these species came originally from shelf and epicontinental bottom faunas and had migrated down the slope in Paleozoic and Mesozoic times in response to competition from newly evolved forms.

The ancient subteleostean bony fishes, represented by only a few living species are, with one notable exception, found primarily in freshwater habitats. The exception is the coelacanth *Latimeria chalumnae* which lives in marine waters at the lower edge of the East African continental shelf. In comparison, the teleostean bony fishes comprise a huge group of 413 families and about 20,000 species—by far the largest assemblage of vertebrate animals.

Fishes

Utilizing the most recent classification of teleostean fishes (Greenwood, Rosen, Weitzman, and Myers 1966), one sees that the group is divided into 30 orders. The species belonging to the 27 most primitive orders (containing 256 families) are found primarily in fresh water, pelagic waters, the cold waters toward the poles, and the deep sea. The shelf waters of the tropics, where the diversity of fish species is the greatest, are dominated by species belonging to the three most advanced orders. This general pattern was first noted by Jordan (1901:566) who said, "The fresh waters, the arctic waters, the deep sea, and the open sea, represent forms of ichthyic back woods, regions where change goes on more slowly, and in them we find survivals of archaic or generalized types." Mead (1970) called attention to the fact that in the Pacific Ocean, the more primitive groups of shore fishes were concentrated toward the poles.

There are some exceptions to the basic, worldwide pattern of fish distribution. We do find represented in tropical shelf waters some primitive teleostean groups such as certain eel families and their relatives the elopids, the herrings (Clupeidae), anchovies (Engraulidae), toadfishes (Batrachoididae), clingfishes (Gobiesocidae), and frogfishes (Antennariidae). Also, there are the pipefishes (Syngnathidae), scorpionfishes (Scorpaenidae), triglids (Triglidae), and flying gurnards (Dactylopteridae). However, these represent but a handful compared to the more than 200 primitive families that are primarily relegated to the peripheral areas. It is the species belonging to the 157 families of the three most advanced orders (Perciformes, Pleuronectiformes, Tetraodontiformes) that are by far the most numerous in the shallow, tropical waters of the world. Some of these advanced species have invaded fresh waters and the open sea, but they are almost entirely absent from the deep waters of the lower slope or the abyssal plain or the trenches.

The fact that our modern concepts of fish systematics still bear out Jordan's original observation has important evolutionary connotations. It is the tropical shelf regions in general and the Indo-West Pacific in particular that possess the most advanced groups of fishes and, at the same time, support the greatest diversity of species. In regard to the idea of the continental slope as a refuge for primitive relicts, it can be said that the fishes of that habitat do belong almost entirely to families of the more primitive orders. One family of the advanced order Tetraodontiformes, the spike fishes (Triacanthodidae) does primarily inhabit the slopes; but it is considered to be the most primitive family in the order (Tyler 1968).

Conclusion There was at one time considerable controversy among marine biologists about the relative antiquity of the abyssal fauna. These arguments were summarized by Madsen (1961) in his thorough review article on the zoogeography and origin of the abyssal fauna. Madsen also concluded that the modern nature of the abyssal fauna compared to that of the slope must be an indication that colonization of the great depths took place markedly later than that of the slope environment.

The foregoing data indicate that many ancient, phylogenetic relicts have accumulated in the slope habitat. Such relicts are also found in other places, but on the slope, and particularly on the lower slope, they appear to comprise an impressive proportion of the fauna. The slope seems to have offered a good refuge, since competition from advanced species tends to be less, and the temperature has probably remained fairly stable. Some ancient fishes are found in fresh waters, but possibly due to osmotic difficulties, this habitat has not been penetrated by many of the primitive invertebrates.

Regional Relationships In characterizing regions, it is necessary to recognize the presence of boundaries or barriers that separate them. Often barriers have simply been looked upon as areas that are relatively difficult for animals to cross. Because migration is impeded, gene flow is cut down, and the separated regions tend, over a period of time, to develop their own individual faunas. In the terrestrial environment it has been known for some time that certain regions act as centers of dispersal, producing advanced, dominant species that, if they succeed in surmounting barriers, can establish themselves in other regions (Matthew 1915, Darlington 1957). Furthermore, it appears that such centers of dispersal possess the most diverse faunas and therefore have relatively stable ecosystems.

Animals that transgress barriers and attempt to invade new regions, particularly those that try to penetrate a more stable ecosystem, are apt to get into the situation so graphically described by Elton (1958:116): "they will find themselves entering a highly complex community of different populations, they will search for breeding sites and find them occupied, for food that other species are already eating, for cover that other animals are sheltering in, and they will bump into them and be bumped into—and often bumped off."

The relationship among the four tropical shelf regions of the world was discussed in Chapter 5. It was concluded that the Indo-West Pacific Region operated as an important marine evolutionary center. Some of the species produced in this center are and were capable of migrating across its boundaries and establishing themselves in adjacent regions, but species arising in the peripheral regions apparently have been unable to penetrate and successfully colonize the Indo-West Pacific. To judge from the general indications of relationship among the tropical shelf faunas, the Indo-West Pacific has been donating species to the other regions for many millions of years. In the shore fishes, for example, virtually all the tropical families and most of the genera are probably of Indo-West Pacific origin.

The warm-temperate shelf regions of the world, which border the tropics to the north and south, demonstrate in general a strong tropical relationship. Although many of the species which inhabit such regions are restricted to them, they belong mainly to tropical families and genera. Another strong faunal element in the warm-temperate regions is the eurythermic tropical group composed of species that range through both tropical and warm-temperate regions. The latter group, plus those endemics of tropical derivation, usually comprise the great majority of species found in warm-temperate waters.

The cold-temperate shelf regions, called the Boreal region in the

Northern Hemisphere, usually possess faunas that are sharply distinct from those of the warm-temperate and tropical regions. As was related earlier (p. 409), boreal faunas probably began their development in the North Pacific and in the Arctic–North Atlantic in the early Tertiary (Paleocene-Eocene). Of the two early Boreal areas, the North Pacific developed the most diverse fauna, so when the Bering Land Bridge was inundated in the late Miocene and again in the late Pliocene, dominant Pacific species entered the Arctic and the North Atlantic in great numbers. Also, North Pacific species have bypassed the equatorial region by means of isothermic submergence to invade the cold waters of the Southern Hemisphere.

The cold-temperate and Antarctic shelf waters of the Southern Hemisphere are inhabited principally by animals that demonstrate by their relationships an extended evolutionary history in that part of the world. Many families and genera are endemic and probably date back at least to the Paleocene when cold-temperate conditions were probably first established. Although these southern waters have also been invaded by species belonging to genera and families that were originally confined to the North Pacific, it is significant that no reciprocal migrations have taken place.

In considering the relationships and influence of all the marine zoogeographic regions, one can see a fundamental difference between the warm-water and the cold-water faunas of the world. In the warm waters, the Indo-West Pacific is and has been for a long time the dominant center of evolution and dispersal. After completion of the Old World Land Barrier in the Lower Miocene and the New World Land Barrier at the end of the Tertiary, the direct influence of the Indo-West Pacific diminished. The Western Atlantic tropics then began to play a role in the Atlantic and lately have supplied many species to the eastern side.

In the cold waters, the North Pacific has been the dominant center. It has supplied species to and to a large extent has controlled the faunal complexion of the North Atlantic and the Arctic. This means that we can identify for the warm waters of the world one primary distributional center, the Indo-West Pacific. The Western Atlantic is a lesser center and has exerted some influence in the Atlantic Ocean during the Pleistocene. For the cold waters of the world we can also recognize one primary distributional center, the North Pacific, and one secondary center, the general Antarctic area—both approximately the same age, but the former is more influential than the latter.

Data gathered about the operation of various, major zoogeographic barriers, in both marine and terrestrial environments, appear to indicate that such barriers affect the distribution of animals in a consistent manner. This has made it possible to propose a general theory of barrier function (Briggs 1974):

A zoogeographic barrier will almost always separate regions of different geographic area and/or different climatic history. Consequently, the degree of species diversity and thus of ecosystem stability will also differ. The region that develops the greatest ecosystem stability will function as the more important evolutionary center and will supply species to the lesser area but will accept few or no species in return. Therefore, the direction and extent of successful migration across a zoogeographic barrier is an expression of a basic faunal relationship between adjacent regions.

Evolutionary Connotations

Another effective approach to the geography of evolution, a method that becomes more feasible as we learn more about the broadly distributed groups of marine animals, is to examine the patterns exhibited by individual genera, families, and orders. There are some revealing examples among the marine shore fishes. In the clingfishes (family Gobiesocidae), it was found that a clear-cut relationship existed between the distributional and evolutionary patterns (Briggs 1955:157). The tropical Indo-West Pacific was found to contain the most advanced genera and species with the more primitive forms generally being the most remote, either in terms of geographic distance or of inaccessibility to invasion.

In his work on the anchovies (family Engraulidae) of the Indian Ocean, Whitehead (1967:17) noted that the species *Stolephorus buccaneeri* had a peculiar distribution (Hawaii, Formosa, Hong Kong, Red Sea, Comora Islands, Durban) and observed that it may be a relatively primitive form that has been forced to the fringes of its region by later and more specialized species. Springer (1967), who worked on the genus *Entomacrodus* (family Blenniidae), identified relict distributions for several Indo-West Pacific species. To explain such patterns, he used the analogy of the fairy-ring mushroom, which starts from a central area and spreads as an ever-enlarging ring around a central area that no longer harbors the species. If the suitable habitat is limited, the mushroom (or fish) will have its last stand along the periphery of the habitat.

Similar patterns showing peripheral relicts may be found among the invertebrates. Abbot (1960), in his work on the molluscan genus *Strombus* in the Indo-West Pacific, depicted such distributions for two of the species (*S. vomer* and *S. haemastoma*). At the generic level, in at least two groups, there is a positive relationship between latitude and age. In both the hermatypic corals (Stehli and Wells 1971:121) and the benthic foraminifera (Durazzi and Stehli 1973:386), it appears that the higher the latitude the more ancient are the genera.

Among the marine mammals, the worldwide cetacean pattern is interesting (Davies 1963). Three strange, monotypic genera—*Balaena*, the Greenland whale; *Monodon*, the narwhale; and *Delphinapterus*, the white whale—are restricted to the Arctic. Other cetaceans that are confined to either the northern or southern temperate zone are generally considered to be archaic types, while the advanced genera and species are found in the tropics.

It is true that, occasionally, very primitive species are found to occur within the areas of the major evolutionary centers. It should be recognized, however, that there are probably two effective ways for a species to escape competition from its more advanced relatives. The usual method is a move out to the geographic periphery, but in most instances, this probably only postpones rather briefly the inevitable fate of extinction. The second and less common method is a shift into a different niche. This method may insure an immunity from competition that would permit a primitive form to persist for millions of years. Such species have been referred to as "hanging relicts" (Schmidt 1943:248).

The Two
Evolutions
The zoogeographic patterns described above can exist only if the evolutionary process is confined to certain favorable centers or if there are two kinds of evolution taking place. We know, of course, that evolution does take place in peripheral areas. In fact, it has been stated that the greatest amount of active speciation takes place in locations that are richest in geographic barriers (Mayr 1963:565). Such barriers seem to be most numerous in insular and other similar areas that are peripheral to the main centers of dispersal, both terrestrial and marine.

Since the main zoogeographic patterns tell us that dominant, advanced species apparently come only from certain favorable centers and since we also know that speciation is very active in areas peripheral to such centers, we should recognize that two kinds of evolutionary change are possibly taking place—one that may be successful in terms of a phyletic future and another that is unsuccessful.

It has not been generally realized that these two obvious kinds of evolution are geographically separated. Darlington (1948:109, 1957:565, 1959:488) proceeding from a different viewpoint, distinguished three kinds of evolution, "differentiation of species," "adaptation to special environments," and "general adaptation." Brown (1957) suggested a "centrifugal speciation" process whereby populations of a central species that had become isolated in peripheral pockets would, upon speciation, occasionally be able to compete and share a niche with the central species. Brown (1958:161) also discussed "special adaptations" and "general adaptations." But none of these categories are useful for the concept of geographically distinct *successful* and *unsuccessful* evolutions.

Aside from the evidence provided by major zoogeographic patterns, are there other reasons why peripheral evolution should be considered unsuccessful? One kind of peripheral evolution has already been recognized as being mainly or almost entirely unsuccessful. This is the process that takes place in situations that have been described as evolutionary traps, or blind alleys (Simpson 1953:306). The usual examples are islands, lakes, or mountain peaks where small populations are likely to become highly isolated. The term *founder principle* was introduced by Mayr (1942:237) to designate the establishment of such populations by a few original migrants.

Unsuccessful Evolution

There seems to be little doubt that, when a population has existed in an evolutionary trap long enough to speciate, it becomes particularly liable to extinction. The basic cause of this vulnerable state is apparently the loss of genetic variability. The reasons for such a loss have been summarized by May (1963:538): (1) the founders represent only a fraction of the variability of the species; (2) owing to inbreeding, more recessives will become homozygous and thus will be exposed to selection; (3) owing to the reduced population size, there will be changes in the selective value of alleles and certain alleles will be eliminated; (4) during the reconstitution of the epigenotypes, many genes will lose the advantage of being a part of a balanced system and will be selected against; and (5) as long as the new population is small it may lose additional genes through errors of sampling.

The problem has also been approached experimentally using *Drosophila* (Dobzhansky and Pavlovsky 1957). The results demonstrated that populations which have passed through the bottleneck of small size show, in succeeding generations, far greater morphological variance than the continuously large populations. Such experiments indicate that the smaller the number of founders, the more variable and less predictable the genetic processes (Dobzhansky 1961). The high degree of morphological variance is presumed to be caused by gene loss and the accompanying breakdown of genetic homeostasis.

The loss of genetic variability by peripheral populations has been observed cytologically in the genus *Drosophila*. In a study of the populations of *D. willistoni*, it was found that the number of heterozygous inversions per individual was highest in northern South America, less in the peripheral mainland areas of Central America and Florida, and least on the oceanic islands of the Antilles (Dobzhansky 1957). The number of such inversions was considered to be positively correlated with the degree of genetic variability.

Ayala (1968:1455) stated that central populations of *Drosophila* have been observed to possess greater genetic variability than marginal populations. Furthermore, he concluded that the observed superior performance in the laboratory of central populations is most likely the result of natural selection being more efficient in the populations where more genotypes are available for selection.

A total of 38 possible models of speciation were discussed by Grant (1963:501). However, only two of them, both involving allopatric differentiation, would seem to be of importance to the great majority of animal populations. Of the two, one would apparently be characteristic of the major evolutionary centers, since it requires large populations and takes place in stable or closed biotic communities; the other is considered to occur where small, peripheral populations are isolated. But Grant, together with Simpson (1953:335) and others, has emphasized that the second process is the one that is probably responsible for quantum evolution—a very rapid change that is said to result in the relatively sudden appearance of new major groups (genera, families, and orders).

The difficulty in utilizing the second model to account for quantum evolution is that the evidence now available indicates that the effects of isolation upon small populations are debilitating so that their chances of success must be greatly reduced. In this case, it may be reasonable for the biologist to consider the dictum of a famous geologist (James Hutton) who maintained that a knowledge of present events is the best key to an understanding of the past. Since there is no recent or current information to show that small, isolated populations might enjoy continued evolutionary success, in terms of their ability to compete elsewhere, why should it be assumed that this process has taken place in the past?

The important point to bear in mind at this time is this: if we are to accept the indications provided by the study of zoogeographic regions, the distributional patterns of individual animal groups, and the work on *Drosophila* populations, unsuccessful evolution cannot be the rule *only* in obvious evolutionary traps, but it also is probably characteristic of most kinds of peripheral areas.

In recent years, an impressive amount of work has been accomplished in an attempt to document evolutionary rates. Most of it was apparently done in order to find out if rates differed among various groups of animals or, in some cases, if marine groups differed from the terrestrial.

Evolutionary Rates

Several modern summaries of evolutionary rates, based on evidence from fossil materials, have appeared (Simpson 1953, Zeuner 1958, Rensch 1959, Mayr 1963). Although the rates vary widely, all can be called relatively slow. The age of most living species of mammals has been estimated at 1 million years and upward and the majority of living mollusks can be traced back for about 25 million years (Simpson, op. cit., 36–37). For insects, the rate appears to be about the same as that given for mollusks (Zeuner quoted by Sylvester-Bradley 1963:126). It has been emphasized that the fastest known species development took place in the elephant (500,000 years, Zeuner, op. cit., 387) and the most sedate species has probably been the fairy shrimp *Triops cancriformis* (170 million years, Simpson, loc. cit.). More recently (Day 1963:47), an average estimated age of 5 million years was given for each species of the heart-urchin *Micraster*. Simpson (op. cit., 333) also noted that slow evolution is less likely to occur in difficult or highly variable environments.

Those who have concentrated on the history of living populations have found that speciation often can proceed with almost unbelievable rapidity. For example, in Lake Lanao on Mindanao Lsland in the Philippines, a body of water with an estimated age of 10,000 years, no less than 18 species and 4 genera of fishes have evolved from a single ancestral species (Myers 1960). Lake Waccamaw in North Carolina is almost certainly of late Pleistocene origin, yet it contains 4 distinct endemic fish species (Hubbs and Raney 1946). The postpluvial disruption of streams, lakes, and springs in the Great Basin area took place about 10,000 to 12,000 years ago, and this period of time has been sufficient to allow the formation of some full species in the fishes (Miller 1961). In Lake Nabugabo, Uganda, five endemic species of cichlid fishes evolved in about 4,000 years (Greenwood, 1965). Bananas were introduced into the Hawaiian Islands about 1,000 years ago and, since that time, five species of *Hedylepta* moths apparently developed on this food source (Zimmerman 1960:138). In the Faeroe Islands, the house mouse (*Mus musculus*) introduced as recently as 300 years ago, has become so different that some authors consider it to be a full species (Mayr 1963:579). It is even possible that, in the Salton Sea, a distinct species of copepod (*Cyclops dimorphys*) has evolved in less than 30 years (Johnson 1953).

Why are there such vast differences in the time that it takes for a species to develop? It has been suggested that terrestrial groups are liable to evolve faster than marine groups, since the latter live under more equable conditions and are less affected by climatic fluctuations (Zeuner 1958:389, Day 1963:47, Mayr 1963:583). However, it can be shown that, in some situations, evolution among marine animals has taken place very rapidly. Cocos Island in the eastern tropical Pacific is apparently of Pleistocene origin (Vinton 1951:373), yet there are now at least six endemic shore fishes, five endemic mollusks, and five endemic crustaceans (Hertlein 1963). Bouvet Island in the sub-Antarctic area was also probably formed in the Pleistocene (Wilson 1963:536), but now has endemic species belonging to three different marine groups (Ekman 1953:221, Nybelin 1947, Powell 1951).

From a distributional standpoint, it may be said that there appears to be a consistent correlation between evolutionary rate and geographic location. In this context, it is important to note that *every example of rapid speciation* has been described from a situation that is obviously an evolutionary trap, or at least an isolated peripheral location. This leads one to suspect that unsuccessful evolution may be generally identified by the rapid rate at which it occurs. This idea is apparently compatable with the paleontologic evidence. That is, fossil materials well represented in the geological record are most likely to pertain to species that were widespread, numerous, and successful (at least for a reasonable period of time). It is suggested, therefore, that the paleontologists have given, primarily, estimates of the rates of successful evolution, while students of contemporary populations have shown how fast unsuccessful evolution can occur.

There is a persistent belief among some biologists that, since the tropics possess the greater diversity of species, including usually the more advanced forms, evolution must have proceeded more rapidly there than in the temperate or cold regions of the world. Furthermore, it has been suggested that the cause of this rapid tropical evolution was ultimately the greater capture of solar energy in the lower latitudes (Stehli, Douglas, and Newell 1969). However, this theory is incompatable with that of successful-unsuccessful evolution which states that the rate in the tropical centers tends to be relatively slow but successful, while that in the peripheral areas (tropical or nontropical) is faster but unsuccessful.

It can, with some justification, be said that the foregoing account is an oversimplification. Certainly, the speciation rate in a given population is apt to depend also on such factors as the size of the isolated area, the ability of the population to shift into a new niche, and the selection pressure (Mayr 1963:585). However, it does appear that geographic location (including the effectiveness of the isolation) is the most important of all. Instead of comparing all evolutionary rates on the same basis, one should distinguish between that which occurs in the main center of dispersal and that which takes place in traps and other peripheral areas—one is slow but often successful, while the other is rapid and seems to be unsuccessful.

So far, it has been noted that dominant species seem to be produced in certain important evolutionary centers and that they have the ability to cross zoogeographic barriers and establish themselves in distant areas. In so doing, they are likely to dispossess peripheral species that happen to be occupying the desirable niches. This is probably the primary reason for the high rate of extinction in peripheral species. It has been found that most of the species of birds that have become extinct within the last 200 years have been island birds (Mayr 1963:74).

Dominant Species

It is interesting to find that dominant species cannot only succeed in colonizing when they manage to migrate across barriers but often show spectacular success as the result of man-made introductions, either purposeful or accidental. Elton (1958) gives a useful summary of the history of man-caused invasions. Zoogeographically, the most important aspect of this work and a review that appeared soon after (Pearsall 1959) is the attention drawn to the question of ecological stability versus instability.

Hutchinson (1959) noted that a complex trophic organization is more stable than a simple one, and Bates (1960) observed that a general principle was beginning to emerge from a variety of different sorts of ecological investigations to the effect that the more diverse the composition of the community the more stable it is. Margalef (1963), in pointing out the difference between mature and immature ecosystems, stated that succession can build history only when the environment is stable. In the case of a changing environment, the selected ecosystem will be composed of fewer species with high reproductive rates and lower special requirements.

Varying degrees of dominance need to be recognized, since dominant species may arise in either major or lesser evolutionary centers. For example, a tropical shelf echinoderm from the western Pacific would be expected to do well in the western Indian Ocean where the competition is less; but a western Indian Ocean species would probably dominate in the western Atlantic, and a western Atlantic species would be likely to succeed in the eastern Pacific. If two areas are occupied by about the same number of species, and thus have ecosystems of approximately equal stability, successful invasions could occur in both directions. This has evidently been the case in regard to the recent terrestrial introductions that have taken place between Europe and North America (Elton 1958).

How are dominant species produced? There is no good reason to suspect that such species arise by any fundamentally different process than that which results in unsuccessful species. The stable ecosystem with its large number of species and high level of competition provides the proper environment. Centers of successful evolution tend to include large areas that apparently allow enough room so that sufficient geographic isolation can take place. Further, if the process is to be successful it must involve relatively large populations with adequate genetic resources.

For the biologist who is interested in speciation, I believe it is worthwhile to point out that the geographic location of the population that is selected for study may be exceedingly important. The researcher should take the opportunity to weigh the advantages of concentrating on a peripheral species that is probably on its way out of the evolutionary picture compared to a dominant species that has a possible phyletic future. In the first instance, one would be studying a population that may have accumulated many deleterious genetic changes; in the second case, the population would comprise a better genetic reservoir and the individuals would be apt to be more viable.

1   Contrary to the expectations of the early naturalists, the contemporary fauna of the great depths seems to be a relatively new one. The present extremely cold temperatures may not have become established until the late Pliocene (although previous cold cycles probably did occur in the Tertiary) so that it is possible that most of the invasions into the abyssal and trench habitats have taken place since that time.

2   Many ancient phylogenetic relicts have accumulated in the slope habitat. Such relicts can be found in other places, too, but on the slope, and particularly on the lower slope, they appear to comprise an impressive proportion of the fauna. The slope seems to offer a good refuge for relicts, since competition from advanced species tends to be less and the temperature has probably remained relatively stable.

3   In the surface layers, one can see a fundamental difference between the warm-water and cold-water faunas of the world. For the warm waters, the Indo-West Pacific is and has been for a long time the dominant center of evolution and dispersal. In the Quaternary, the Western Atlantic tropical region began to play a role and has supplied many species to the Eastern Atlantic. For the cold waters, the North Pacific has been the dominant center. It has supplied species to, and to a large extent has controlled, the faunal complexion of the North Atlantic and the Arctic. The Antarctic center, important to the Southern Hemisphere, has been of lesser general influence.

4   A regional zoogeographic barrier will almost always separate two areas of contrasting geographic size and/or climatic history. Consequently, the degree of species diversity and thus of ecosystem stability will also differ. The region that develops the greatest ecosystem stability will function as the more important evolutionary center and will supply species to the lesser area but will accept few or no species in return. Therefore, the direction and extent of successful migration across a zoogeographic barrier is an expression of a basic faunal relationship between adjacent regions.

5   Examination of distributional patterns in some of the better-known groups of marine animals, including fishes, mollusks, and cetaceans, indicate that the most advanced genera and species inhabit the center of dispersal, while the more primitive forms are generally the most remote, in terms of either geographic distance or inaccessibility to invasion.

6   Since it is known that speciation is very active in areas peripheral to the major dispersal centers, we should consider that two geographically distinct kinds of evolution may be taking place—one that may be successful in terms of a phyletic future and one that is unsuccessful.

7   It is known that unsuccessful evolution is the rule in situations that have been described as evolutionary traps, or blind alleys. However, the accumulation of phylogenetic relicts on the slopes, the horizontal relationships among the

Part II
Summary

major zoogeographic regions, the distributional patterns of individual animal groups, and some of the work in *Drosophila* genetics, all appear to provide evidence that unsuccessful evolution may be a basic characteristic of any peripheral situation.

8 The rate at which speciation takes place in the evolutionary centers is apparently very slow. In contrast, the rate in evolutionary traps or other isolated peripheral locations can be exceedingly rapid.

9 Dominant species are produced in certain important evolutionary centers. If they manage to get across zoogeographic barriers, they are often able to colonize other areas. Such species appear to evolve from relatively large populations that exist in stable ecosystems under conditions of maximum biological competition.

Conclusions I hope that the material presented in this final section will help clarify the relationship between zoogeography and evolution. It has led to the formation of a hypothesis which may be stated as follows: Two contrasting kinds of evolutionary change are taking place simultaneously. Successful evolution, which occurs in the major centers of dispersal, develops slowly and is likely to produce dominant species that can become widespread and have a potential phyletic future. Unsuccessful evolution, typical of areas peripheral to the major centers, takes place rapidly and produces species that are limited in their ability to penetrate other areas and in their phyletic future (Briggs 1966b:288).

Let us imagine that a new species has just arisen, and fortunately for its future prospects, it has done so in the richest area (in terms of species diversity) of a large region that acts as an evolutionary center. It can be expected to expand its range as rapidly as its powers of dispersal and its competitive advantage will permit. In so doing, it can be expected to eliminate, from the habitat which is being exploited, an older species, most likely its closest relative. As the older species disappears from the range it once occupied, its distributional pattern is converted from a continuous type to a fringe or relict type. If such an older species has made the most of its prime (lasting possibly a few million years), it may have succeeded in populating some localities that are extremely hard for its competition to follow. We know that in some places relict populations persist for a very long time.

Suppose that a species under competitive displacement has given up the mainland continental shelf but manages to hang on as a relict around certain oceanic islands. Long-continued survival in such evolutionary traps is not very probable. In reaching very isolated places a population usually has to pass through the bottleneck of small size. As was related earlier (p. 441), the resultant loss of genetic variability would lead to further competitive weakness, and the population would then be highly vulnerable to extinction. It appears that a more successful way for a relatively primitive species to avoid extinction is to make the necessary physiological and/or morphological adjustments in order to enter a new environment. A notably successful strategy has been to abandon the shelf for the deeper waters of the continental slope. This means becoming adapted to colder temperatures, less light, and greater pressure and, in addition, the establishment of new biological relationships.

Another successful ruse, for species under competitive harassment, is to move across one of the horizontal temperature boundaries (tropical to warm-temperate, warm-temperate to cold-temperate, cold-temperate to arctic). For benthic shelf species, an alternative that does not require a temperature adjustment is a switch to the pelagic habitat. This usually requires an adaptation to different food organisms plus a series of rather drastic morphological changes, but it has been done repeatedly. This kind of change appears to be easiest for fishes and crustaceans but has also taken place in such diverse groups as polychaetes, gastropods, and cephalopods. These strategies have been going on for a long time. We know this because some of the relict species are very old. On the slopes there are many species that have changed only slightly since the Paleozoic era.

Some groups have become so adept at avoiding competition that they have made several successive moves into contrasting environments. One rather frequent route has been shelf →epipelagic →mesopelagic, and another, shelf →upper slope →lower slope →abyssal plain →trench. In the bathypelagic zone, some of the species apparently came from the mesopelagic zone above and some from the benthic habitat of the lower slope. Such multiple habitat changes probably took many millions of years and often involved whole lineages requiring changes in genera as well as species. Animals belonging to at least two North Pacific groups (fishes and gastropods) apparently accomplished the following: in the North Pacific, shelf →upper slope →lower slope; migration to the Southern Hemisphere via the west coast of the America: then, lower slope →upper slope →shelf; and finally a further horizontal distribution in the Southern Hemisphere!

In the introduction to this section, a question was posed. Can we recognize in the world ocean a fundamental relationship between

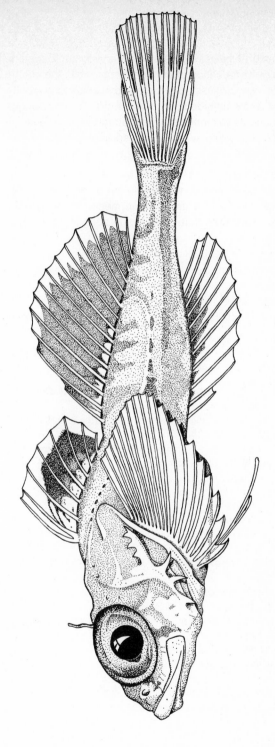

Figure 12-2 The cottid (*Antipodocottus galatheae*). A species found on the upper slope in the Tasman Sea. Its closest relatives, and the family, are found in the North Pacific. After Bolin (1952).

evolution and distribution? Indeed we can. There is a dynamic system requiring the continuous production of dominant species in the evolutionary centers by the slow process of successful evolution. Evolutionary change also takes place outside such centers, where it may occur very rapidly, but the species so formed are the products of unsuccessful evolution and will eventually give way to those spreading out from the main centers.

Distributional and phylogenetic patterns form an inextricable mosaic. One cannot be understood without reference to the other. The investigator, interested in the history of a given group, needs to develop an appreciation for the natural laws that govern both zoogeography and evolution.

Competition is the universal stimulus for progress. It is the vital component of biogeography, organic evolution, and almost all forms of human endeavor.

Epilogue

Literature Cited

Abbot, R. T. 1960. The genus *Strombus* in the Indo-Pacific. *Indo-Pacific Mollusca*, 1(2):33–146, 117 plates.

Addicott, W. O. 1969. Tertiary climatic change in the marginal northeastern Pacific Ocean. *Science*, 165(3893):583–586, 3 figs.

Albee, A. L., et al. 1970. Ages, irradiation history, and chemical composition of lunar rocks from the Sea of Tranquility. *Science*, 167(3918):463–466, 4 figs.

Andriashev, A. P. 1954. Fishes of the northern seas of the U.S.S.R. Keys to the fauna of the U.S.S.R. *Zool. Inst. Acad. Sci. U.S.S.R.*, (53):1–617 (English translation, Jerusalem, 1964).

Arambourg, C. 1965. Considerations nouvelles au sujet de la faune ichtyologique paleomediterraneenne. *Senckenbergiana, Lethaea, Frankfurt am Main*, (46a):13–17.

Arkell, W. J. 1956. *Jurassic geology of the world*. Hafner, New York, pp. 1–806, 46 plates.

Ayala, F. J. 1968. Genotype, environment, and population numbers. *Science*, 162(3861):1453–1459, 2 figs.

Bandy, O. L. 1960. Planktonic foraminiferal criteria for paleoclimate zonation. *Sci. Rep. Tohoku Univ., 2nd Ser. (Geol.), Spec. vol.*, 4:1–8 (not seen).

———. 1968. Cycles in neogene paleoceanography and eustatic changes. *Palaeogeogr., Palaeoclimatol., Palaeoecol.*, (5):63–75.

———, and M. E. Wade. 1967. Miocene-Pliocene-Pleistocene boundaries in deep-water environments. *Progr. Oceanog.*, 4:51–66, 6 figs.

Barghoorn, E. S. 1964. Quantitation of sequential change in North American

Cenozoic floras as a clue to palaeoclimates. *in* A. E. M. Nairn (editor), *Problems in palaeoclimatology.* Interscience, New York, pp. 31–39, 2 figs.

——, and J. W. Schopf. 1966. Microorganisms three billion years old from the Precambrian of South Africa. *Science*, 152(3723):758–763, 12 figs.

Barnard, J. L. 1961. Gammaridean Amphipoda from depths of 400 to 6,000 meters. *Galathea Rep.*, 5:23–128, 83 figs.

Bates, M. 1960. Ecology and evolution. *in* Sol Tax (editor), *Evolution after Darwin.* University of Chicago Press, Chicago, 1:547–568.

Bé, A. W. H. 1966. Distribution of planktonic foraminifera in the world oceans. *Abstr. 2nd Int. Oceanog. Congr., Moscow*, p. 26.

Beauvais, L. 1973. Upper Jurassic hermatypic corals. *in* A. Hallam (editor), *Atlas of palaeobiogeography.* Elsevier, Amsterdam, pp. 317–328, 4 figs., 2 plates.

Berg, L. S. 1934. Über die Amphiboreale (diskontinuierliche) Verbreitung der Meeresfauna in der nördlichen Hemisphäre. *Zoogeographica*, 2(3):393–409.

Bergström, S. M. 1973. Ordovician conodonts. *in* A. Hallam (editor), *Atlas of palaeobiogeography.* Elsevier, Amsterdam, pp. 47–58, 5 figs.

Berry, W. B. N. 1973. Silurian-Early Devonian graptolites. *in* A. Hallam (editor), *Atlas of palaeobiogeography.* Elsevier, Amsterdam, pp. 81–87, 3 figs.

Beurlen, K. 1956. Der Faunenschnitt an der Perm-Triasgrenze. *Z. Deut. Geol. Ges.*, 108:88 (not seen).

Bigelow, H. B., and W. C. Schroeder. 1953. Fishes of the Gulf of Maine. *U. S. Fish Wildlife Service, Fishery Bull.*, (74):viii + 577 pp., 288 figs.

Boucot, A. J., and J. G. Johnson. 1973. Silurian brachiopods. *in* A. Hallam (editor), *Atlas of Palaeobiogeography.* Elsevier, Amsterdam, pp. 59–65, 3 figs.

Bowen, R. 1966. *Paleotemperature analysis.* Elsevier Co., New York, pp. x + 1–265, 35 figs.

Bretsky, P . W. 1968. Evolution of Paleozoic marine invertebrate communities. *Science*, 159(3820):1231–1233, 1 fig.

Briggs, J. C. 1955. A monograph of the clingfishes (Order Xenopterygii). *Stanford Ichthyol. Bull.*, 6:1–224, 114 figs.

——. 1966a. Oceanic islands, endemism, and marine paleotemperatures. *Syst. Zool.*, 15(2):153–163, 4 figs.

——. 1966b. Zoogeography and evolution. *Evolution*, 20(3):282–289.

——. 1970. A faunal history of the North Atlantic Ocean. *Syst. Zool.*, 19(1):19–34, 3 figs.

——. 1974. The operation of zoogeographic barriers. (In manuscript.)

Broecker, W. S. 1965. Isotope geochemistry and the Pleistocene climatic record. *in* H. E. Wright, Jr., and David G. Frey (editors), *The Quarternary of the United States.* Princeton University Press, Princeton, N. J. pp. 737–753, 7 figs.

—— et al. 1968. Milankovitch hypothesis supported by precise dating of coral reefs and deep-sea sediments. *Science*, 159(3812):297–300, 2 figs.

Brown, W. L., Jr. 1957. Centrifugal speciation. *Quart. Rev. Biol.*, 32(3):247–277, 10 figs.

——. 1958. General adaptation and evolution. *Syst. Zool.*, 7(4):157–168, 3 figs.

Cariou, E. 1973. Ammonites of the Callovian and Oxfordian *in* A. Hallam (editor),

Atlas of palaeobiogeography. Elsevier, Amsterdam, pp. 287–295, 3 figs. 2 plates.

Casier, E. 1966. *Faune ichthyologique du London Clay*. British Museum, London, xiv + 496 pp., 82 figs., 68 plates.

Chaney, R. W. 1940. Tertiary forests and continental history. *Bull. Geol. Soc. Amer.*, 51:469–488, 2 plates, 3 figs.

——. 1964. Some observations on climatic relations of tertiary floras bordering the north Pacific Basin. *in* A. E. M. Nairn (editor), *Problems in palaeoclimatology*. Interscience, New York, pp. 40–43, 2 figs.

Charlesworth, J. K. 1957. *The Quarternary era*. vol. 2. Edward Arnold, London, pp. 595–1700, figs. 113–326, plates 25–32.

Clarke, A. H., Jr. 1961. Structure, zoogeography and evolution of the abyssal mollusk fauna. *Ann. Rep. Bull., Amer. Malacol. Union* (27):(reprint, no pagination).

——. 1962. Annotated list and bibliography of the abyssal marine molluscs of the world. *Bull. Nat. Museum Can.*, (181):vi + 1–114.

Clemens, W. A., and G. V. Wilby. 1961. Fishes of the Pacific coast of Canada. *Bull. Fisheries Res. Board Can.*, 2d edition, 68:1–443, 281 figs.

Cloud, P. E., Jr. 1968a. Atmospheric and hydrospheric evolution on the primitive earth. *Science*, 160(3829):729–736.

——. 1968b. Pre-metazoan evolution and the origins of the Metazoa. *in* E. T. Drake (editor), *Evolution and environment*. Yale University Press, New Haven, Conn., pp. 1–72, 10 figs.

Colbert, E. H. 1971. Tetrapods and continents. *Quart. Rev. Biol.*, 46(3):250–269, 8 figs.

Curry, R. R. 1966. Glaciation about 3,000,000 years ago in the Sierra Nevada. *Science*, 154(3750):770–771, 1 fig.

Darlington, P. J., Jr. 1948. The geographical distribution of cold-blooded vertebrates. *Quart. Rev. Biol.*, 23(2):105–123, 5 figs.

——. 1957. *Zoogeography: the geographical distribution of animals*. John Wiley, New York, xi + 1–675 pp., 80 figs.

——. 1959. Area, climate, and evolution. *Evolution*, 13(4):488–510, 8 figs.

——. 1965. *Biogeography of the southern end of the world*. Harvard University Press, Cambridge, x + 236 pp., 38 figs.

Darwin, C. 1859. *On the origin of species*. John Murray, London, ix + 1–490 pp., 1 chart.

Davies, J. L. 1958. The Pinnipedia: an essay in zoogeography. *Geograph. Rev.*, 48(4):474–493, 10 figs.

——. 1963. The antitropical factor in cetacean speciation. *Evolution*, 17(1):107–116, 2 figs.

Davis, G., and G. E. Elliott. 1957. The paleogeography of the London clay sea. *Proc. Geologist's Ass. (England)*, 68(4):255–277, 2 figs.

Day, J. H. 1963. The complexity of the biotic environment. J. P. Harding and N. Trebble (editors), *in* Symposium: *Speciation in the sea. Syst. Ass. Publ.*, (5):31–49.

Dell, R. K. 1956. The archibenthal Mollusca of New Zealand. *Dominion Museum Bull.*, (18):1–235, 6 figs., 27 plates.

Dietz, R. S. 1961. Continents and ocean basins, evolution by spreading of the sea floor. *Nature*, 190:854–857.

———, and J. C. Holden. 1970. The breakup of Pangaea. *Sci. Amer.*, 223(4):30–41, illus.

Djakanov, A. M. 1945. On the relationship between the Arctic and the North Pacific marine faunas based on the zoogeographical analysis of the Echinodermata. *J. Gen. Biol.*, 6:125–155. (English summary).

Dobzhansky, T. 1957. Genetics of natural populations. XXVI. Chromosomal variability in island and continental populations of *Drosophila willistoni* from Central America and the West Indies. *Evolution*, 11(3):280–293, 3 figs.

———. 1961. Biological evolution on islands. Abstracts of symposium papers, *10th Pacific Sci. Congr., Honolulu, Hawaii*, pp. 472–473.

———, and O. Pavlovsky. 1957. An experimental study of interaction between genetic drift and natural selection. *Evolution*, 11(3):311–319, 2 figs.

Dorf, E. 1955. Plants and the geologic time scale. Crust of the Earth (a symposium). *Geol. Soc. Amer., Spec. Papers*, (62):575–592, 2 figs.

———. 1964. The use of fossil plants in palaeoclimatic interpretations. *in* A. E. M. Narin (editor), *Problems in palaeoclimatology*. Interscience, New York, pp. 13–31, 8 figs.

Durazzi, J. T., and F. G. Stehli. 1973. Average generic age, the planetary temperature gradient, and pole location. *Syst. Zool.*, 21(4):384–389, 6 figs.

Durham, J. W. 1952. Early tertiary marine faunas and continental drift. Amer. Jour. Sci., 250:321–343, 1 map.

———. Palaeoclimates *in* Physics and chemistry of the earth, vol. 3:1–16, 8 figs. Pergamon Press, New York.

———. 1959b. Tertiary land masses and the shallow water North Pacific molluscan and echinoid faunas. Preprints, *Int. Oceanog. Cong., Amer. Ass. Adv. Sci.*, pp. 578–580.

———, and F. S. MacNeil. 1967. Cenozoic migrations of marine invertebrates through the Bering Strait region. *in* D. M. Hopkins (editor). *The Bering land bridge*. Stanford University Press, Stanford, pp. 326–349.

Du Toit, A. L. 1937. *Our wandering continents*. Oliver and Boyd, London, xiii + 366 pp., 48 figs.

Eicher, D. L. 1968. *Geologic time*. Prentice-Hall, Englewood Cliffs, N. J., pp. 1–149, illus.

Ekman, S. 1953. *Zoogeography of the sea*. Sidgwick & Jackson, London, xiv + 417 pp., 121 figs.

Elton, C. S. 1958. *The ecology of invasions by animals and plants*. Methuen, London, pp. 1–181, 51 figs., photos.

Emerson, W. K. 1956. Pleistocene invertebrates from Punta China, Baja California, Mexico. With remarks on the composition of the Pacific Coast Quarternary faunas. *Bull. Amer. Museum Nat. Hist.*, 111(4):313–342, 1 fig., 2 plates.

Emiliani, C. 1955. Pleistocene temperatures. *J. Geol.*, 63:538–578.

———. 1961. The temperature decrease of surface seawater in high latitudes and of abyssal-hadal water in open oceanic basins during the past 75 million years. *Deep-Sea Res.*, 8:144–147.

———. 1964. Paleotemperature analysis of the Caribbean cores A254–BR–C and CP–28. *Bull. Geol. Soc. Amer.*, 75:129–144, 7 figs.

Enay, R. 1973. Upper Jurassic (Tithonian) Ammonites. *in* A. Hallam (editor), *Atlas of palaeobiogeography.* Elsevier, Amsterdam, pp. 297–307, 3 figs.

Engel, A. E. J. 1969. Time and the earth. *Amer. Sci.*, 57(4):458–483, 12 figs.

———, et al. 1968. Alga-like forms in Onverwacht Series, South Africa: oldest recognized life-like forms on earth. *Science*, 161(3845):1005–1008, 4 figs.

Ericson, D. B. 1959. Coiling direction of *Globigerina pachyderma* as a climatic index. *Science*, 130(3369):219–220.

———, and G. Wollin. 1954. Coiling direction of *Globorotalia truncatulinoides* in deep-sea cores. *Deep-Sea Res.*, 2:152–158.

———, and ———. 1964. *The deep and the past.* Alfred Knopf, New York, pp. xiv + 292 + ix, 29 figs.

———, M. Ewing, and G. Wollin. 1964. The Pleistocene Epoch in deep-sea sediments. *Science*, 146(3645):723–732, 5 figs.

Fischer, A. G. 1964. Brackish oceans as the cause of the Permo-Triassic marine faunal crisis. *in* A. E. M. Nairn (editor), *Problems in palaeoclimatology.* John Wiley, New York, pp. 566–579, 2 figs.

Fleming, C. A. 1962. New Zealand biogeography. A paleontologist's approach. *Tuatara*, 10:53–108, 15 figs.

Flint, R. F. 1957. *Glacial and Pleistocene geology.* John Wiley, New York, xiii + 553 pp., 5 plates.

Forbes, E. 1859. The natural history of European seas (edited and continued by Robert Godwin-Austen). John Van Voorst, London, viii + 306 pp., 1 map.

George, R. W., and A. R. Main. 1967. The evolution of spiny lobsters (Palinuridae): a study of evolution in the marine environment. *Evolution*, 21(4):803–820, 3 figs.

Gislén, T. 1957. Crinoidea, with a survey of the bathymetric distribution of the deep-sea crinoids. *Rep. Swedish Deep-Sea Expedition*, 2(4):51–59, 1 fig., 1 plate.

Gobbett, D. J. 1973. Permian Fusulinacea. *in* A. Hallam (editor), *Atlas of palaeobiogeography.* Elsevier, Amsterdam, pp. 151–158, 4 figs.

Goodell, H. G., N. D. Watkins, T. T. Mather, and S. Koster. 1968. The antarctic glacial history recorded in sediments of the Southern Ocean. *Palaeogeography, Palaeoclimatol., Palaeoecol.*, 5:41–62.

Grant, V. 1963. *The origin of adaptations.* Columbia University Press, New York, x + 1–606 pp., 102 figs.

Greenwood, P. H. 1965. The cichlid fishes of Lake Nabugabo, Uganda. *Bull. Brit. Museum (Nat. Hist.), Zool.*, 12(9):313–357, 12 figs.

———, D. E. Rosen, S. H. Weitzman, and G. S. Myers. 1966. Phyletic studies of teleostean fishes, with a provisional classification of living forms. *Bull. Amer. Museum Nat. Hist.*, 131(4):341–455, illus.

Haldane, J. B. S. 1954. The origins of life. *New Biol.*, 16:12–27.

Hallam, A. (editor) 1973. *Atlas of palaeobiogeography.* Elsevier, Amsterdam, pp. 1–531, illus.

Halstead, L. B., and S. Turner. 1973. Silurian and Devonian ostracoderms. *in* A.

Hallam (editor), *Atlas of palaeobiogeography.* Elsevier, Amsterdam, pp. 67–79, 9 figs.

Harland, W. B. 1964. Evidence of late Precambrian glaciation and its significance. *in* A. E. M. Nairn (editor), *Problems in palaeoclimatology.* Interscience, New York, pp. 119–149, 4 figs.

Helsley, C. E., and F. G. Stehli. 1964. Comparison of Permian magnetic and zoogeographic poles. *in* A. E. M. Nairn (editor), *Problems in palaeoclimatology.* John Wiley, New York, pp. 558–565, 3 figs.

Herman, Y. 1970. Arctic paleo-oceanography in late Cenozoic time. *Science,* 169(3944):474–477, 3 figs.

Hertlein, L. G. 1963. Contribution to the biogeography of Cocos Island, including a bibliography. *Proc. Calif. Acad. Sci.,* 32:219–289, 4 figs.

Hill, D. 1973. Lower Carboniferous corals. *in* A. Hallam (editor), *Atlas of palaeobiogeography.* Elsevier, Amsterdam, pp. 133–142, 1 fig.

Hopkins D. M. 1967. The Cenozoic history of Beringia—a synthesis. *in* D. M. Hopkins (editor), *The Bering Land Bridge.* Stanford University Press, Stanford, pp. 451–484, 4 figs.

Hubbs, C. L., and E. C. Raney. 1946. Endemic fish fauna of Lake Waccamaw, North Carolina. *Misc. Publ. Museum Zool., Univ. Mich.,* (65):1–30, 1 plate.

———, and N. J. Wilimovsky. 1964. Distribution and synonymy in the Pacific Ocean, and variation, of the Greenland halibut, *Reinhardtius hippoglossoides* (Walbaum). *Fisheries Res. Board Can.,* 21(5):1129–1154, 5 figs.

Hutchinson, G. E. 1959. Homage to Santa Rosalia or why are there so many kinds of animals? *Amer. Nat.,* 93:145–159.

Imlay, R. W. 1965. Jurassic marine faunal differentiation in North America. *J. Paleont.,* 39(5):1023–1038, 6 figs.

Jaanusson, V. 1973. Ordovician articulate brachiopods. *in* A. Hallam (editor), *Atlas of palaeobiogeography,* Elsevier, Amsterdam, pp. 19–25, 3 figs.

Jenkins, D. G. 1968. Variations in the numbers of species and subspecies of planktonic Foraminiferida as an indicator of New Zealand Cenozoic paleotemperatures. *Palaeogeography, Palaeoclimatol, Palaeoecol.,* (5):309–313, 1 fig.

Johnson, J. G., and A. J. Boucot. 1973. Devonian brachiopods. *in* A. Hallam (editor), *Atlas of palaeobiogeography.* Elsevier, Amsterdam, pp. 89–96, 6 figs.

Johnson, M. W. 1953. The copepod *Cyclops dimorphys* Kiefer from the Salton Sea. *Amer. Midl. Nat.,* 49:188–192.

Jordan, D. S. 1901. The fish fauna of Japan, with observations on the geographical distribution of fishes. *Science,* new series 14(354):545–567.

Kaljo, D., and E. Klaamann. 1973. Ordovician and Silurian corals. *in* A. Hallam (editor), *Atlas of palaeobiogeography,* Elsevier, Amsterdam, pp. 37–45, 4 figs.

Kauffman, E. G. 1973. Cretaceous Bivalvia. *in* A. Hallam (editor), *Atlas of Palaeobiogeography.* Elsevier, Amsterdam, pp. 353–383, 10 figs.

Keast, A. 1971. Continental drift and the evolution of the biota on southern continents. *Quart. Rev. Biol.,* 46(4):335–378, 4 figs.

Kitching, J. W., J. W. Collinson, D. H. Elliot, and E. H. Colbert. 1972. *Lystrosaurus* Zone (Triassic) fauna from Antarctica. *Science,* 175(4021):524–527, 3 figs.

Knudsen, J. 1961. The bathyal and abyssal *Xylophaga* (Pholadidae, Bivalvia). *Galathea Rep.*, 5:163–209, 41 figs.

Kulp, J. L. 1961. Geological time scale. *Science*, 133(3459):1105–1114, 1 fig.

Kummel, B. 1973. Lower Triassic (Scythian) molluscs. *in* A. Hallam (editor), *Atlas of palaeobiogeography*. Elsevier, Amsterdam, pp. 225–233, 4 figs., 3 plates.

Lamb, H. H. 1964. The role of atmosphere and oceans in relation to climatic changes and the growth of ice-sheets on land. *in* A. E. M. Nairn (editor), *Problems in palaeoclimatology*. Interscience, New York, pp. 332–348, 10 figs.

Lindberg, G. V. 1963a. Fishes of our time reveal earth's past; hypotheses, suggestions, conjectures. *Nauka i Zhizn, Moscow*, (11):46–49.

———. 1963b. On the connection between the continents of Europe and America, *in Soviet fisheries investigations in the northwest Atlantic.* Israel Program for Scientific Translations, Jerusalem. pp. 68–81, 3 figs.

Lindroth, C. H. 1957. *The faunal connections between Europe and North America*. John Wiley and Sons, New York; Almquist and Wiksells, Uppsala. pp. 1–344, 61 figs.

Love, A. 1958. Transatlantic connections and long-distance dispersal. *Evolution*, 12(3):421–423.

Lowenstam, H. A. 1964. Palaeotemperatures of the Permian and Cretaceous Periods *in Problems in palaeoclimatology* (A. E. M. Nairn, Editor). Interscience, New York. pp. 227–248, 9 figs.

MacGinitie, H. D. 1958. Climate since the Late Cretaceous *in Zoogeography* (C. L. Hubbs, Editor). Amer. Assoc. Adv. Sci., 51:61–79.

MacNeil, F. S. 1965. Evolution and distribution of the genus *Mya*, and Tertiary migrations of Mollusca. *U.S. Geol. Survey Prof. Paper 483-G.* pp. 1–51, 11 plates.

Madsen, F. J. 1961. On the zoogeography and origin of the abyssal fauna in view of the knowledge of the Porcellanasteridae. *Galathea Report*, 4:177–218, 2 figs.

Margalef, R. 1963. On certain unifying principles in ecology. *American Nat.*, 97:357–374.

Matthew, W. D. 1915. Climate and evolution. *Ann. New York. Acad. Sci.*, 24:171–318, 33 figs.

Mayr, E. 1942. *Systematics and the origin of species.* Columbia University Press, New York. xiv + 334 pp., 28 figs.

———. 1963. *Animal species and evolution.* Harvard University Press, Cambridge, Mass., xiv + 1–797 pp., 201 figs.

McIntyre, A. 1967. Coccoliths as paleoclimatic indicators of Pleistocene glaciation. *Science*, 158(3806):1314–1317, 3 figs.

Mead, G. W. 1970. A history of South Pacific fishes. *in Scientific exploration of the South Pacific*. National Academy of Sciences, Washington, D. C. pp. 236–251.

Miller, R. R. 1961. Speciation rates in some fresh-water fishes of western North America. *in Symposium on vertebrate speciation, University of Texas*. University of Texas Press, Austin, pp. 537–560, 10 figs.

Modell, H. 1943. Tertiäre Najaden. III. Nordamerikanische Najaden im bayrischen Oliogozän. *Arch. für Molluskenk. Frankfurt am Main*, 75:107–117 (not seen).

Moore, R. C. (editor) 1969. *Treatise on invertebrate paleontology*. Part N. Mollusca 6. *Bivalvia* (1, 2, 3). Geological Society of America, University of Kansas Press, pp. 1–952.

Murray, J. 1895. A summary of the scientific results. I-II. *Challenger Report Summary*, pp. xix + 1608, 22 plates, 2 vols.

Myers, G. S. 1949. Salt-tolerance of fresh-water fish groups in relation to zoogeographical problems. *Bijdragen tot de Dierkunde*, 28:315–322.

———. 1960. The endemic fish fauna of Lake Lanao, and the evolution of higher taxonomic categories. *Evolution*, 14(3):323–333.

———. 1967. Zoogeographical evidence of the age of the South Atlantic Ocean. *Studies Trop. Oceanog. Miami*, 5:614–621.

Nesis, K. N. 1962. Pacific elements in northwest Atlantic benthos. *in Soviet fisheries investigations in the northwest Atlantic*. Rybnoe Knozyaistvo, Moscow (English translation, Jerusalem, 1963), pp. 82–99, 3 figs.

Newell, N. D. 1962. Paleontological gaps and geochronology. *J. Paleontol.*, 36:592–610, 13 figs.

———. 1963. Crises in the history of life. *Sci. Amer.* February, pp. 1–16, 13 figs (reprint from W. H. Freeman, San Francisco).

Nursall, J. R. 1959. Oxygen as a prerequisite to the origin of the Metazoa. *Nature*, 183:1170–1172.

Nybelin, O. 1945. Antarctic fishes. Scientific Results of the Norwegian Antarctic Expeditions 1927–1928 et Sqq. No. 26, Det Norske Videnskaps—Akademi i Oslo, 76 pp., 6 plates.

Olausson, E. 1965. Evidence of climatic changes in North Atlantic deep-sea cores, with remarks on isotopic and paleotemperature analysis. *Progr. Oceanog.*, 3:221–252, 10 figs.

———. 1967. Climatological, geoeconomical and paleooceanographical aspects on carbonate deposition. *Progr. Oceanogr.*, 4:245–265, 3 figs.

Opik, E. J. 1958. Climate and the changing sun. Reprint from *Sci. Amer.*, June:1–8, 7 figs.

Palmer, A. R. 1973. Cambrian trilobites. *in* A. Hallam (editor), *Atlas of palaeobiogeography*. Elsevier, Amsterdam, pp. 3–11, 3 figs.

Parin, N. V. 1968. Epipelagic ichthyofauna of the oceans. *Acad. Sci. U.S.S.R.*, *Moscow*, pp. 1–186, 56 figs. (in Russian).

Parker, R. H. 1962. Speculations on the origin of the invertebrate faunas of the lower continental shelf. *Deep-Sea Res.*, 8(3–4):286–293, 3 figs.

Patterson, C. 1972. The distribution of Mesozoic freshwater fishes. *17th Int. Zool. Congr.*, *Monaco*, Theme No. 1:1–22, 7 figs.

Pearsall, W. H. 1959. The ecology of invasion: ecological stability and instability. *New Biol. (Penguin Books)*, 29:95–101.

Pickard, G. L. 1964. *Descriptive physical oceanography*. Pergamon Press, Oxford viii + 199 pp., 31 figs.

Powell, A. W. B. 1951. Antarctic and subantarctic Mollusca: Pelecypoda and Gastropoda. *Discovery Rep.*, 26:47–196, 14 figs., 6 plates.

Powell, N. A., and E. L. Bousfield. 1969. Canadian marine invertebrate life. *in Animal life in Canada today.* National Museum of Natural Sciences, Ottawa, pp. 14–15.

Rass, T. S. (editor) 1967. Biology of the Pacific Ocean, Book III. Fishes of the open waters. *Inst. Oceanology, Acad. Sci. U.S.S.R., Moscow,* pp. 1–273, 42 figs.

Reid, R. E. H. 1967. Tethys and the zoogeography of some modern and Mesozoic Porifera. *Syst. Ass. Publ.,* (7):171–181.

Rensch, B. 1959. *Evolution above the species level.* Methuen, London, xvii + 1–419 pp., 113 figs.

Romer, A. R. 1968. Fossils and Gondwanaland, *Proc. Amer. Phil. Soc.,* 112(5):335–343, 6 figs.

Ross, C. A. 1973. Carboniferous Foraminiferida. *in* A. Hallam (editor), *Atlas of palaeobiogeography.* Elsevier, Amsterdam, pp. 127–132, 4 figs.

Rubey, W. W. 1964. Geologic history of sea water. *in* P. J. Brancazio and A. G. W. Cameron (editors), *The origin and evolution of atmospheres and oceans.* John Wiley, New York, pp. 1–63, 4 figs.

Rudwick, M. J. S. 1964. The infra-Cambrian glaciation and the origin of the Cambrian fauna. *in* A. E. M. Nairn (editor), *Problems in palaeoclimatology.* Interscience, New York, pp. 150–155, 1 fig.

Sanders, H. L. 1968. Marine benthic diversity: a comparative study. *Amer. Nat.,* 102(925)243–282, 18 figs.

Schmidt, K. P. 1943. Corollary and commentary for "climate and evolution." *Amer. Midl. Nat.,* 30(1):241–253.

Schwarzbach, M. 1963. *Climates of the past.* Van Nostrand, London, xii + 328 pp., 134 figs.

Simpson, G. G. 1947. Evolution, interchange, and resemblance of the North American and Eurasian Cenozoic mammalian faunas. *Evolution,* 1:218–220.

Skevington, D. 1973. Ordovician graptolites. *in* A. Hallam (editor), *Atlas of palaeobiogeography.* Elsevier, Amsterdam, pp. 27–35, 4 figs.

Soot-Ryen, T. 1932. Pelecypoda, with a discussion of possible migrations of Arctic pelecypods in Tertiary times. *Sci. Res. Norwegian North Polar Expedition, the "Maud,"* 1918–25, 5(12):1–35, 2 plates.

Sorgenfrei, T. 1958. *Molluscan assemblages from the marine Middle Miocene of South Jutland and their environments.* 2 vols. Reitzel, Copenhagen, 503 pp., 38 figs., 76 plates.

Sparck, R. 1954. Summary of the colloquium. On the distribution and origin of the deep sea bottom fauna. *Int. Union Biol. Sci., Ser. B,* No. 16:89–90.

Springer, V. G. 1967. Revision of the circumtropical shorefish genus *Entomacrodus* (Blenniidae: Salariinae). *Proc. U.S. Nat. Museum,* 122(3582):1–150, 11 figs., 30 plates.

Stehli, F. G. 1964. Permian zoogeography and its bearing on climate. *in* A. E. M. Nairn (editor), *Problems in palaeoclimatology.* John Wiley, New York, pp. 537–549, 12 figs.

———. 1973. Permian brachiopods. *in* A. Hallam (editor), *Atlas of palaeobiogeography.* Elsevier, Amsterdam, pp. 143–149, 3 figs.

————, and J. W. Wells. 1971. Diversity and age patterns in hermatypic corals. *Syst. Zool.*, 20(2):115–126, 13 figs.

————, R. G. Douglas, and N. D. Newell. 1969. Generation and maintenance of gradients in taxonomic diversity. *Science*, 164(3882):947–949, 6 figs.

Stevens, G. R. 1973a. Jurassic belemnites. *in* A. Hallam (editor), *Atlas of palaeobiogeography.* Elsevier, Amsterdam, 259–274, 4 figs., 1 plate.

————. 1973b. Cretaceous belemnites. *in* A. Hallam (editor), *Atlas of palaeobiogeography.* Elsevier, Amsterdam, pp. 385–401, 5 figs., 1 plate.

Stonehouse, B. 1969. Environmental temperatures of Tertiary penguins. *Science*, 163(3868):673–675, 2 figs.

Sylvester-Bradley, P. C. 1963. Post-tertiary speciation in Europe. *Nature*, 199(4889):126–130, 1 fig.

Tatsumoto, M., and J. N. Rosholt. 1970. Age of the moon: an isotopic study of uranium-thorium-lead systematics of lunar samples. *Science*, 167(3918):461–463, 2 figs.

Tyler, J. C. 1968. A monograph on plectognath fishes of the superfamily Triacanthoidea. *Acad. Nat. Sci. Phila., Monogr.*, 16:1–364, 209 figs.

Ushakov, P. V. 1955. Polychaeta of the far eastern seas of the U.S.S.R. *Zool. Inst., Akad. Nauk. SSSR* (English translation, Jerusalem, 1965), 419 pp., 164 figs.

Valentine, J. W. 1967. The influence of climatic fluctuations on species diversity within the Tethyan provincial system. *Syst. Ass. Pub.* (7):153–166, 3 figs.

Valentine, W., and R. F. Meade. 1960. Isotopic and zoogeographic paleotemperatures of Californian Pleistocene Mollusca. *Science*, 132 (3430):810–811.

Veeh, H. H., and J. Chappell. 1970. Astronomical theory of climatic change: support from New Guinea. *Science*, 167(3919):862–865, 3 figs.

Vinton, K. W. 1951. Origin of life on the Galapagos Islands. *Amer. J. Sci.*, 249:356–376, 2 figs., 2 plates.

Voss, G. L. 1967. The biology and bathymetric distribution of deep-sea cephalopods. *Studies Trop. Oceanog. Miami*, 5:511–535, 4 figs.

Wegener, A. 1924. *The origin of continents and oceans.* Methuen, London, xx + 212 pp., 44 figs.

Weyl, P. K. 1968. Precambrian marine environment and the development of life. *Science*, 161:158–160.

Whitehead, P. J. P. 1967. Indian Ocean anchovies collected by the *Anton Bruun* and the *Te Vega*, 1963–64. *J. Marine Biol. Ass. India*, 9(1):13–37, 4 figs.

Whittington, H. B. 1973. Ordovician trilobites. *in* A. Hallam (editor), *Atlas of palaeobiogeography.* Elsevier, Amsterdam, pp. 13–18, 3 figs.

Wiles, W. W. 1967. Pleistocene changes in the pore concentration of a planktonic foraminiferal species from the Pacific Ocean. *Progr. Oceanog.*, 4:153–160, 3 figs.

Wilson, J. T. 1963. Evidence from islands on the spreading of ocean floors. *Nature*, 197(4867):536–538, 4 figs.

————. 1966. Did the Atlantic close and then re-open? *Nature*, 211:676–681.

Wolff, T. 1961. Description of a remarkable deep-sea hermit crab, with notes on the evolution of the Paguridea. *Galathea Rep.*, 4:11–32, 11 figs.

Woodford, A. O. 1965. *Historical geology.* W. H. Freeman, San Francisco, pp. 1–512, illus.

Woodring, W. P. 1954. Caribbean land and sea through the ages. *Bull. Geol. Soc. Amer.*, 65:719–732.

————. 1957. Marine Pleistocene of California. *Geol. Soc. Amer. Mem.*, (67):589–598, 1 fig.

————. 1959. Tertiary Caribbean molluscan faunal Province. *Preprints Inst. Oceanog. Conr. Amer. Ass. Adv. Sci.*, pp.299–300.

————. 1960. Paleoecologic dissonance *Astarte* and *Nipa* in the early Eocene London Clay. "Bradley Vol." (special number of the journal) *Amer. Sci.*, 258-A:418–419.

————. 1966. The Panama land bridge as a sea barrier. *Proc. Amer. Phil. Soc.*, 110(6):425–433, 3 figs.

Zenkevitch, L. 1963. *Biology of the seas of the U.S.S.R.* George Allen and Unwin, Ltd., London, pp.1–955, 427 figs.

Zeuner, F. E. 1958. *Dating the past.* Methuen, London, xx + 1–516 pp., 105 figs., 27 plates.

Zimmerman, E.C. 1960. Possible evidence of rapid evolution in Hawaiian moths. *Evolution*, 14(1):137–138.

Index